Lecture Notes on Data Engineering and Communications Technologies

Volume 65

Series Editor

Fatos Xhafa, Technical University of Catalonia, Barcelona, Spain

The aim of the book series is to present cutting edge engineering approaches to data technologies and communications. It will publish latest advances on the engineering task of building and deploying distributed, scalable and reliable data infrastructures and communication systems.

The series will have a prominent applied focus on data technologies and communications with aim to promote the bridging from fundamental research on data science and networking to data engineering and communications that lead to industry products, business knowledge and standardisation.

Indexed by SCOPUS, INSPEC, EI Compendex.

All books published in the series are submitted for consideration in Web of Science.

More information about this series at http://www.springer.com/series/15362

Leonard Barolli · Juggapong Natwichai ·
Tomoya Enokido
Editors

Advances in Internet, Data and Web Technologies

The 9th International Conference on Emerging
Internet, Data & Web Technologies
(EIDWT-2021)

 Springer

Editors
Leonard Barolli
Department of Information
and Communication Engineering
Fukuoka Institute of Technology
Fukuoka, Japan

Juggapong Natwichai
Faculty of Engineering
Chiang Mai University
Chiang Mai, Thailand

Tomoya Enokido
Rissho University
Tokyo, Japan

ISSN 2367-4512 ISSN 2367-4520 (electronic)
Lecture Notes on Data Engineering and Communications Technologies
ISBN 978-3-030-70638-8 ISBN 978-3-030-70639-5 (eBook)
https://doi.org/10.1007/978-3-030-70639-5

This Springer imprint is published by the registered company Springer Nature Switzerland AG
The registered company address is: Gewerbestrasse 11, 6330 Cham, Switzerland

Welcome Message of EIDWT-2021 International Conference Organizers

Welcome to the 9th International Conference on Emerging Internet, Data and Web Technologies (EIDWT-2021), which will be held from February 25 to February 27, 2021, in Chiang Mai, Thailand.

The EIDWT is dedicated to the dissemination of original contributions that are related to the theories, practices and concepts of emerging Internet and data technologies, yet most importantly of their applicability in business and academia toward a collective intelligence approach.

In EIDWT-2021, topics related to information networking, data centers, data grids, clouds, social networks, security issues and other Web 2.0 implementations toward a collaborative and collective intelligence approach leading to advancements of virtual organizations and their user communities will be discussed. This is because, current and future Web and Web 2.0 implementations will store and continuously produce a vast amount of data, which if combined and analyzed through a collective intelligence manner will make a difference in the organizational settings and their user communities. Thus, the scope of EIDWT-2021 includes methods and practices which bring various emerging Internet and data technologies together to capture, integrate, analyze, mine, annotate and visualize data in a meaningful and collaborative manner. Finally, EIDWT-2021 aims to provide a forum for original discussion and prompt future directions in the area.

An international conference requires the support and help of many people. A lot of people have helped and worked hard for a successful EIDWT-2021 technical program and conference proceedings. First, we would like to thank all authors for submitting their papers. We are indebted to program area chairs, program committee members and reviewers who carried out the most difficult work of carefully evaluating the submitted papers. We would like to give our special thanks to the honorary chair of EIDWT-2021 Prof. Makoto Takizawa, Hosei University, Japan, for his guidance and support. We would like to express our appreciation to our keynote speakers for accepting our invitation and delivering very interesting keynotes at the conference.

EIDWT-2021 International Conference Organizers

EIDWT-2021 Steering Committee Chair

Leonard Barolli   Fukuoka Institute of Technology (FIT), Japan

EIDWT-2021 General Co-chairs

Juggapong Natwichai   Chiang Mai University, Thailand
David Taniar   Monash University, Australia

EIDWT-2021 Program Committee Co-chairs

Pruet Boonma   Chiang Mai University, Thailand
Tomoya Enokido   Rissho University, Japan

EIDWT-2021 Organizing Committee

Honorary Chair

Makoto Takizawa   Hosei University, Japan

General Co-chairs

Juggapong Natwichai   Chiang Mai University, Thailand
David Taniar   Monash University, Australia

Program Co-chairs

Pruet Boonma   Chiang Mai University, Thailand
Tomoya Enokido   Rissho University, Japan

International Advisory Committee

Publicity Co-chairs

Prompong Sugunnasil   Chiang Mai University, Thailand
Kin Fun Li   University of Victoria, Canada
Keita Matsuo   Fukuoka Institute of Technology, Japan
Omar Hussain   University of New South Wales, Australia
Flora Amato   University of Naples "Frederico II", Italy

International Liaison Co-chairs

Paskorn Champrasert   Chiang Mai University, Thailand
Admir Barolli   Alexander Moisiu University, Albania
Santi Caballé   Open University of Catalonia, Spain
Elis Kulla   Okayama University of Science, Japan
Farookh Hussain   University of Technology Sydney, Australia
Nadeem Javaid   COMSATS University Islamabad, Pakistan

Local Organizing Committee Co-chairs

Web Administrators

Phudit Ampririt   Fukuoka Institute of Technology, Japan
Kevin Bylykbashi   Fukuoka Institute of Technology, Japan
Ermioni Qafzezi   Fukuoka Institute of Technology, Japan

Finance Chair

Makoto Ikeda   Fukuoka Institute of Technology, Japan

Steering Committee Chair

Leonard Barolli   Fukuoka Institute of Technology, Japan

PC Members

Akimitsu Kanzaki	Shimane University, Japan
Akio Koyama	Yamagata University, Japan
Akira Uejima	Okayama University of Science, Japan
Akshay Uttama Nambi S. N.	Microsoft Research India, India
Alba Amato	National Research Council (CNR)—Institute for High-Performance Computing and Networking (ICAR), Italy
Alberto Scionti	LINKS Foundation, Italy
Albin Ahmeti	TU Wien, Austria
Alex Pongpech	National Institute of Development Administration, Thailand
Ali Rodan	Higher Colleges of Technology, United Arab Emirates
Alfred Miller	Higher Colleges of Technology, United Arab Emirates
Amelie Chi Zhou	ShenZhen University, China
Amin M. Khan	Pentaho, Hitachi Data Systems, Japan
Ana Azevedo	ISCAP, Porto, Portugal
Andrea Araldo	Massachusetts Institute of Technology, USA
Animesh Dutta	National Institute of Technology, Durgapur, India
Anirban Mondal	Shiv Nadar University, India
Anis Yazidi	Oslo and Akershus University College of Applied Sciences, Norway
Antonella Di Stefano	University of Catania, Italy
Arcangelo Castiglione	University of Salerno, Italy
Beniamino Di Martino	University of Campania "Luigi Vanvitelli", Italy
Benson Raj	Higher Colleges of Technology, United Arab Emirates
Bhed Bista	Iwate Prefectural University, Japan
Bowonsak Srisungsittisunti	University of Phayao, Thailand
Carmen de Maio	University of Salerno, Italy
Chang Yung-Chun	Teipei Medical University, Italy
Chonho Lee	Osaka University, Japan
Chotipat Pornavalai	King Mongkut's Institute of Technology Ladkrabang, Thailand
Christoph Hochreiner	TU Wien, Austria
Congduc Pham	University of Pau, France
Dalvan Griebler	Pontifícia Universidade Católica do Rio Grande do Sul, Brazil
Dana Petcu	West University of Timisoara, Romania

Danda B. Rawat	Howard University, USA
Debashis Nandi	National Institute of Technology, Durgapur, India
Diego Kreutz	Universidade Federal do Pampa, Brazil
Dimitri Pertin	Inria Nantes, France
Dipankar Das	Jadavpur University, Kolkata, India
Douglas D. J. de Macedo	Federal University of Santa Catarina, Brazil
Dumitru Burdescu	University of Craiova, Romania
Dusit Niyato	NTU, Singapore
Elis Kulla	Okayama University of Science, Japan
Eric Pardede	La Trobe University, Australia
Fabrizio Marozzo	University of Calabria, Italy
Fabrizio Messina	University of Catania, Italy
Feng Xia	Dalian University of Technology, China
Francesco Orciuoli	University of Salerno, Italy
Francesco Palmieri	University of Salerno, Italy
Fumiaki Sato	Toho University, Japan
Gabriele Mencagli	University of Pisa, Italy
Gen Kitagata	Tohoku University, Japan
Georgios Kontonatsios	Edge Hill University, UK
Ghazi Ben Ayed	Higher Colleges of Technology, United Arab Emirates
Giovanni Masala	Plymouth University, UK
Giovanni Morana	C3DNA, USA
Giuseppe Caragnano	LINKS Foundation, Italy
Giuseppe Fenza	University of Salerno, Italy
Gongjun Yan	University of Southern Indiana, USA
Guangquan Xu	Tianjing University, China
Gustavo Medeiros de Araujo	Universidade Federal de Santa Catarina, Brazil
Harold Castro	Universidad de Los Andes, Bogotá, Colombia
Hiroaki Yamamoto	Shinshu University, Japan
Hiroshi Shigeno	Keio University, Japan
Houssem Chihoub	Grenoble Institute of Technology, France
Ilias Savvas	TEI of Larissa, Greece
Inna Skarga-Bandurova	East Ukrainian National University, Ukraine
Isaac Woungang	Ryerson University, Canada
Ja'far Alqatawna	Higher Colleges of Technology, United Arab Emirates
Jamal Alsakran	Higher Colleges of Technology, United Arab Emirates
Jamshaid Ashraf	Data Processing Services, Kuwait
Jaydeep Howlader	National Institute of Technology, Durgapur, India
Jeffrey Ray	Edge Hill University, UK
Ji Zhang	University of Southern Queensland, Australia
Jiahong Wang	Iwate Prefectural University, Japan
Jugappong Natwichai	Chiang Mai University, Thailand

Kazuyoshi Kojima	Saitama University, Japan
Ken Newman	Higher Colleges of Technology, United Arab Emirates
Kenzi Watanabe	Hiroshima University, Japan
Kiyoshi Ueda	Nihon University, Japan
Klodiana Goga	LINKS Foundation, Italy
Leila Sharifi	Urmia University, Iran
Leyou Zhang	Xidian University, China
Lidia Fotia	Università Mediterranea di Reggio Calabria (DIIES), Italy
Long Cheng	Eindhoven University of Technology, The Netherlands
Lucian Prodan	Polytechnic University Timisoara, Romania
Luiz Fernando Bittencourt	UNICAMP - Universidade Estadual de Campinas, Brazil
Makoto Fujimura	Nagasaki University, Japan
Makoto Nakashima	Oita University, Japan
Mansaf Alam	Jamia Millia Islamia, New Delhi, India
Marcello Trovati	Edge Hill University, UK
Mariacristina Gal	University of Salerno, Italy
Marina Ribaudo	University of Genoa, Italy
Markos Kyritsis	Higher Colleges of Technology, United Arab Emirates
Marwan Hassani	TU Eindhoven, The Netherlands
Massimo Torquati	University of Pisa, Italy
Matthias Steinbauer	Johannes Kepler University Linz, Austria
Matthieu Dorier	Argonne National Laboratory, USA
Mauro Marcelo Mattos	FURB Universidade Regional de Blumenau, Brazil
Mazin Abuharaz	Higher Colleges of Technology, United Arab Emirates
Minghu Wu	Hubei University of Technology, China
Mingwu Zhang	Hubei University of Technology, China
Minoru Uehara	Toyo University, Japan
Mirang Park	Kanagawa Institute of Technology, Japan
Monther Tarawneh	Higher Colleges of Technology, United Arab Emirates
Morteza Saberi	UNSW Canberra, Australia
Mohsen Farid	University of Derby, UK
Motoi Yamagiwa	University of Yamanashi, Japan
Mouza Alshemaili	Higher Colleges of Technology, United Arab Emirates
Muawya Aldalaien	Higher Colleges of Technology, United Arab Emirates
Mukesh Prasad	University of Technology, Australia

Muhammad Iqbal   Higher Colleges of Technology,
                                        United Arab Emirates
Naeem Janjua   Edith Cowan University, Australia
Naohiro Hayashibara   Kyoto Sangyo University, Japan
Naonobu Okazaki   University of Miyazaki, Japan
Neha Warikoo   Academia Sinica, Taiwan
Nobukazu Iguchi   Kindai University, Japan
Nobuo Funabiki   Okayama University, Japan
Olivier Terzo   LINKS Foundation, Italy
Omar Al Amiri   Higher Colleges of Technology,
                                        United Arab Emirates
Omar Hussain   UNSW Canberra, Australia
Osama Alfarraj   King Saud University, Saudi Arabia
Osama Rahmeh   Higher Colleges of Technology,
                                        United Arab Emirates
P. Sakthivel   Anna University, Chennai, India
Panachit Kittipanya-Ngam   Electronic Government Agency, Thailand
Paolo Bellavista   University of Bologna, Italy
Pavel Smrž   Brno University of Technology, Czech Republic
Peer Shah   Higher Colleges of Technology,
                                        United Arab Emirates
Pelle Jakovits   University of Tartu, Estonia
Per Ola Kristensson   University of Cambridge, UK
Philip Moore   Lanzhou University, China
Pietro Ruiu   LINKS Foundation, Italy
Pornthep Rojanavasu   University of Phayao, Thailand
Pruet Boonma   Chiang Mai University, Thailand
Raffaele Pizzolante   University of Salerno, Italy
Ragib, Hasan   The University of Alabama at Birmingham, USA
Rao Mikkilineni   C3dna, USA
Richard Conniss   University of Derby, UK
Ruben Mayer   University of Stuttgart, Germany
Sachin Shetty   Old Dominion University, USA
Sajal Mukhopadhyay   National Institute of Technology, Durgapur, India
Salem Alkhalaf   Qassim University, Saudi Arabia
Salvatore Ventiqincue   University of Campania "Luigi Vanvitelli", Italy
Samia Kouki   Higher Colleges of Technology,
                                        United Arab Emirates
Saqib Ali   Sultan Qaboos University, Australia
Sayyed Maisikeli   Higher Colleges of Technology,
                                        United Arab Emirates
Sazia Parvin   Deakin University, Australia
Sergio Ricciardi   Barcelonatech, Spain
Shadi Ibrahim   Inria Rennes, France
Shigetomo Kimura   University of Tsukuba, Japan

Shinji Sugawara	Chiba Institute of Technology, Japan
Sivadon Chaisiri	University of Waikato, New Zealand
Sofian Maabout	Bordeaux University, France
Sotirios Kontogiannis	University of Ioannina, Greece
Stefan Bosse	University of Bremen, Germany
Stefania Boffa	University of Insubria, Italy
Stefania Tomasiello	University of Salerno, Italy
Stefano Forti	University of Pisa, Italy
Stefano Secci	LIP6, University Paris 6, France
Suayb Arslan	MEF University, Istanbul, Turkey
Subhrabrata Choudhury	National Institute of Technology, Durgapur, India
Suleiman Al Masri	Higher Colleges of Technology, United Arab Emirates
Tatiana A. Gavrilova	St. Petersburg University, Russia
Teodor Florin Fortis	West University of Timisoara, Romania
Thamer AlHussain	Saudi Electronic University, Saudi Arabia
Titela Vilceanu	University of Craiova, Romania
Tomoki Yoshihisa	Osaka University, Japan
Tomoya Kawakami	NAIST, Japan
Toshihiro Yamauchi	Okayama University, Japan
Toshiya Takami	Oita University, Japan
Ugo Fiore	University of Naples "Frederico II", Italy
Venkatesha Prasad	Delft University of Technology, The Netherlands
Victor Kardeby	RISE Acreo, Sweden
Vlado Stankovski	University of Ljubljana, Slovenia
Walayat Hussain	University of Technology, Australia
Xingzhi Sun	IBM Research, Australia
Xu An Wang	Engineering University of CAPF, China
Xue Li	The University of Queensland, Australia
Yoshinobu Tamura	Tokyo City University, Japan
Yue Zhao	National Institutes of Health, USA
Yunbo Li	IMT Atlantique, France
Zasheen Hameed	Higher Colleges of Technology, United Arab Emirates
Zia Rehman	COMSATS University Islamabad, Pakistan

EIDWT-2021 Reviewers

Ali Khan Zahoor
Amato Flora
Amato Alba
BarolliAdmir
Barolli Leonard
Bista Bhed
Caballé Santi
Chellappan Sriram
Chen Hsing-Chung
Cui Baojiang
Di Martino Beniamino
Enokido Tomoya
Fun Li Kin
Gotoh Yusuke
Hussain Farookh
Hussain Omar
Javaid Nadeem
Ikeda Makoto
Ishida Tomoyuki
Kikuchi Hiroaki
Kolici Vladi
Koyama Akio
Kouki Samia

Kulla Elis
Leu Fang-Yie
Matsuo Keita
Koyama Akio
Kryvinska Natalia
Ogiela Lidia
Ogiela Marek
Okada Yoshihiro
Palmieri Francesco
Paruchuri Vamsi Krishna
Rahayu Wenny
Sato Fumiaki
Spaho Evjola
Sugawara Shinji
Takizawa Makoto
Taniar David
Terzo Olivier
Uehara Minoru
Venticinque Salvatore
Wang Xu An
Woungang Isaac
Xhafa Fatos

EIDWT-2021 Keynote Talks

Contract Tracing During Covid-19 Pandemic: An Australian Experience Synopsis

David Taniar

Monash University, Melbourne, Australia

Abstract. Contact tracing is the activity of retrieving historical activities and trips for a person where his presence at a specific location might affect other persons within a certain radius. Related to a contagious disease, an infected person might spread the pathogens to nearby people during close contact that can trigger a chain reaction of community transmission. The biggest problem in obtaining the historical activities in a contact tracing procedure is privacy and security issues. The privacy issue refers to private-related sensitive information that is not meant to be shared with anyone. However, during a contact tracing investigation, the authorities have the right to know every detail from a suspected patient. The security issue refers to the safety of the shared private information to the authority. Due to these issues, many patients are reluctant to share their past activities to the authority. This condition makes it even harder to obtain the right information from the patients. The next consequence is that the spreading of the diseases will be off the radar since contact tracing could not be done correctly. Several methods have been proposed to help contact tracing procedures. In general, there are two types of contact tracing methods: proximity-based and trajectory-based. While the proximity-based method lacks historical trips and suffers from multi-platforms communication issues, trajectory-based suffers from privacy issues. This speech will discuss these methods together with their pros and cons. In conclusion, a method that can preserve privacy and retain the details of the trip will also be explained in this session as an alternative method to support contact tracing.

Privacy Violation from Joint Attacks on Incremental Datasets

Juggapong Natwichai

Chiang Mai University, Chiang Mai, Thailand

Abstract. Data are continuously collected and grown; therefore, the privacy protection mechanisms designed for static data might not be able to cope with this situation effectively. In this talk, I will first present the possible privacy violations, attacks, which could occur, including a newly discovered type of violation and joint attack. After the attacks are formulated, then the characteristics of the privacy attacks are extracted in order to find approaches to preserve the privacy efficiently. Lastly, the preliminary experimental results will be presented.

Contents

Enhanced Focused Beam Routing
in Underwater Wireless Sensor Networks

Elis Kulla[1]([✉])[ID], Kengo Katayama[1], Keita Matsuo[2][ID], and Leonard Barolli[2][ID]

[1] Okayama University of Science (OUS), 1-1 Ridaicho,
Kita-ku, Okayama 700-0005, Japan
{kulla,katayama}@ice.ous.ac.jp
[2] Fukuoka Institute of Technology (FIT), 3-30-1 Wajiro-Higashi,
Higashi-Ku, Fukuoka 811-0295, Japan
{kt-matsuo,barolli}@fit.ac.jp

Abstract. Multi-hop transmissions in Underwater Wireless Sensor Networks (UWSN) have limited bandwidth and not stable links. Their objective consists in relaying data from each underwater device to the internet. Routing Protocols for UWSN are responsible for relaying data to the water surface, where an acoustic to electromagnetic converter uses well-established wireless communications to further relay them. Focused Beam Routing (FBR) is a well-known routing protocol for UWSN, which considers as forward candidates, only active neighbors which location is inside a cone specified by the direction from the forwarding node to the destination, an arbitrary angle and the communication distance.

In this paper, we propose an Enhanced FBR, which considers the depth of each candidate and limiting forwarding of data on not deeper nodes, and call it Depth-Based FBR (dFBR). We implement both FBR and its enhancement version (dFBR) in ONE Simulator, and compare their performances with Epidemic Routing (ER), a well-known routing protocols for Delay Tolerant Networks. The simulation results show that FBR and dFBR outperform ER in terms of Delivery Probability in our scenarios, where the buffer size is limited to a 30 MB memory.

Keywords: Underwater Wireless Sensor Networks · UWSN · Focused Beam Routing · FBR · ONE Simulator

1 Introduction

Underwater communications are shifting from military towards commercial applications [6]. They are becoming more and more popular in applications such as disaster detection/prevention, pollution monitoring in environmental systems, collection of scientific data in fields of biology and geology, mapping ocean floors in oceanography, coordinated navigation control and so on [1]. A typical application scenario includes several Autonomous Underwater Vehicles (AUVs), which are equipped with a wide range of sensors (environmental, chemical, navigation

L. Barolli et al. (Eds.): EIDWT 2021, LNDECT 65, pp. 1–9, 2021.
https://doi.org/10.1007/978-3-030-70639-5_1

and so on). They explore the seabed, use their sensors to collect data and submit the sensed data up to the sea surface, where air-water interface ships, boats or buoys are located (see Fig. 1). Furthermore, these air-water interfaces might also use electromagnetic waves to send the aggregated data back to monitoring centers or data processing centers.

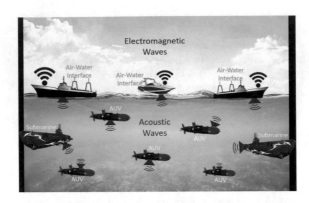

Fig. 1. Underwater acoustic communications and aerial wireless communications

The preferred communication media for underwater communications is acoustic waves in contrast to electromagnetic radio waves, which are widely used in the air. In fact, low frequency electromagnetic radio waves (30Hz–300Hz) can propagate for longer distances, but they require large antennae and high transmission power. Optical waves have better propagation, but they require directional coordination, which is almost impossible in underwater applications, where the devices are in constant movement [1].

Most of the research regarding the physical layer in underwater communications uses Phase Shift Keying (PSK) and Quadrature Amplitude Modulation (QAM) modulation techniques. With the increasing of processing capabilities of small devices, Orthogonal Frequency Division Modulation (OFDM) is also considered.

At the Medium Access Control (MAC) layer, Frequency Division Multiple Access (FDMA) does not work well with limited-band acoustic signals because they are affected by fading and multi-path [1]. Moreover, in acoustic communications there is a difference in delays between consecutive packets (jitter), which makes it very difficult to implement Time Division Multiple Access (TDMA) techniques. CDMA is so far the best solution, but newly designed techniques are yet to be developed, tested and implemented [2,5].

Because it is costly to implement network infrastructure in underwater environment, the network layer is mostly oriented towards adhoc architecture, where all participating nodes forward packets to other nodes, until the packets reach the intended destination. In fact, depending on the application, the adhoc architecture may consist of:

- Real-time data transmission, where participating nodes, either already know where to forward the received data, or the can find out in a very short amount of time. Such amount of time is usually considered as "the real time".
- Delay-tolerant data transmission, where different techniques are used to increase packet delivery ratio (PDR) even when there are no available routes to the intended destination. Delay-tolerant data transmissions, us the store-carry-forward paradigm, to transmit data from one node to another, until the data reach the destination.

Thus, existing adhoc and Delay Tolerant Network (DTN) Routing Protocols need to be redesigned, in order to deal with unstable links and high delay variance in underwater environment.

This paper's main objectives are: to introduce the implementation of FBR protocol, and compare its performance with Epidemic Routing (ER); and introduce an improved FBR, where the forwarding area is further narrowed down.

The remainder of the paper is organized as follows. In Sect. 2, we describe FBR, while comparing it to Epidemic Routing Protocol. In Sect. 3, we describe our simulation settings and discuss the simulation results. Then we conclude with conclusions and future works, in Sect. 4.

2   Focused Beam Routing

A more detailed description of FBR, can be found in [3]. Here we will shortly explain FBR in contrast to other routing paradigms in UWSN.

In UWSNs, the store-carry-forward paradigm is popular among routing protocols, hence Epidemic Routing (ER), which is visualized in Fig. 2(a), is the main forwarding technique. Here, the forwarding area is defined from the communicating distance, which in underwater environment is usually less than 100 m. One problem with ER protocol is that, the packets are copied to every possible relay node, creating a huge overhead in the network, which also impacts the network lifetime due to battery depletion. In fact in some applications, where nodes are scarce, the increased number of copies helps to increase the delivery probability, but in general, decreasing the overhead is beneficial for UWSN, because UWSN nodes also have limited storing resources.

Since the objective of routing protocols for UWSN in the majority of applications is to collect data to the modems in the water surface, and in order to decrease the number of copied packets, forcing nodes to strictly forward packets only to nodes that are closer to the surface has proven beneficial in terms of performance for these applications. This category of routing protocols is called depth-based routing (DBR) and is simplified in Fig. 2(b).

In order to furthermore decrease the overhead and energy consumption, some routing protocols, focus their forwarding area towards the surface data collector, and define it by a relatively small angle, as shown in Fig. 2(c). We implemented FBR as explained in the following.

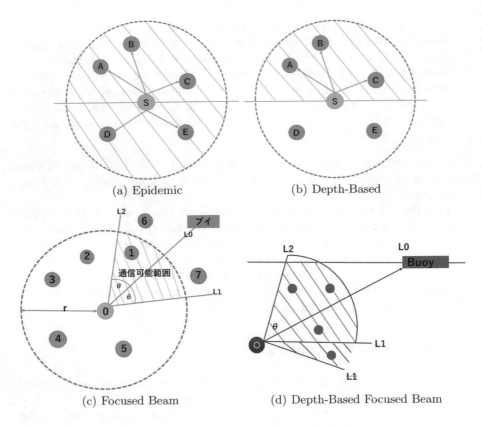

(a) Epidemic   (b) Depth-Based

(c) Focused Beam   (d) Depth-Based Focused Beam

Fig. 2. Underwater routing protocol paradigms

2.1   Implementation of FBR in The ONE Simulator

While there are a lot of routing protocol proposed and implemented around the world, we could not find an open implementation of FBR that could be easily editable and verifiable in different scenarios. Thus we implemented FBR in the well-known The One Simulator. The following assumptions are made, in order to simplify the implementation:

- The environment is considered with only 2 dimensions: the width on the horizontal axis and the depth in the vertical axis.
- Every node knows every other nodes' location in every moment, which in fact is very difficult to achieve in reality without a well-established infrastructure.
- Every participating node is mobile and moves based on the Random Waypoint Mobility Model, as implemented in The ONE.

Table 1. Simulation parameters' settings

Parameters	Value
Number of buoys	1
Number of nodes	$20, 40, 60$
Movement model	Random waypoint
Simulation time	43200 s (12 h)
Simulation area	500 m × 500 m
Communication distance	50 m
Buffer size	30 MB
Message size	500 kB–1000 kB
FBR's angle (θ)	$15°, 30°, 45°$

– There is only one buoy, and all transmissions are directed towards this buoy, which is located in the middle top of our 2D environment.

Then, while referring to Fig. 2(c), whenever a participating node (node O) receives a new message or contacts a new node, the following happens.

Direction towards the Buoy is calculated as the angle in degrees of line L_0.
The angle of transmission θ defines the transmission area on both sides of line L_0. The new transmission area is confined by lines L_1 and L_2.
Communication distance confines the transmission area, which, in the best case scenario, is an arc.
Define the transmission area (shaded area), which makes the transmissions directed towards the buoy.

2.2 Depth-Based FBR

During our simulation phase, we noticed that, in some rare cases when the transmitting nodes were near the surface and far from the buoy, the performance of FBR would decrease. We introduced a depth-based hybrid version of FBR (dFBR), where, each node further limits their forwarding area to nodes that are not deeper than itself, as shown in Fig. 2(d). In some scenarios, especially when the FBR angle is relatively small, the depth-based limitation does not apply, but results in general show slightly improved performance, compared to the traditional FBR.

3   Simulations

In order to analyse the enhanced protocol, we conducted simulations in different scenarios as shown in Table 1. We used The ONE Simulator [4], which has already implemented the epidemic routing protocol and the store-carry-forward mechanism. We limited the buffer size of UWSN nodes, in order to see the effect of overhead packets.

   In Fig. 3, we compare the delivery probability (DP) (number of delivered packets/number of sent packets) and average hop count (AHC) of the implemented FBR with ER for different number of nodes. The results show that ER has lower DP (see Fig. 3(a)), because nodes are forced to drop packets after their buffer has no more storage capacity. This can also be seen in Fig. 3(b), where we claim that packets are delivered in longer routes to the destination, because nodes in the optimal route have full buffers. We also noticed that, as the number of nodes increase, the delivery probability increases, because more routes become available to destination.

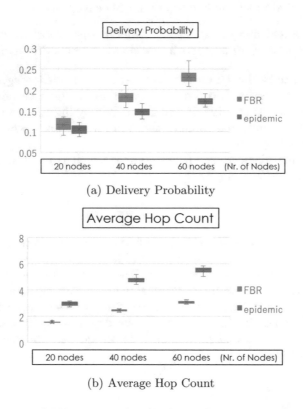

(a) Delivery Probability

(b) Average Hop Count

Fig. 3. Performance of FBR compared to Epidemic Routing, regarding the number of nodes.

In fact, ER is FBR with communication angle set to 180°. In order to investigate the effect of the communication angle in the performance of the network, we conducted the second simulations, with angles set to 15°, 30°, 45°. The results are shown in Fig. 4. When the number of nodes is small , the angle has almost no effect on delivery probability, but the average hop count increases when the angle increases. When the number of nodes is 40 and 60, the performance for both delivery probability and average hop count decreases, when the angle increases.

In Fig. 5, the performance of dFBR is compared to that of FBR, while the forwarding angle is changed from 15° to 45°. dFBR, slightly outperforms FBR in terms of DP, and we can clearly see that the AHC is smaller for dFBR. However, a thourough investigation is important to understand the effect of buffer size and other parameters.

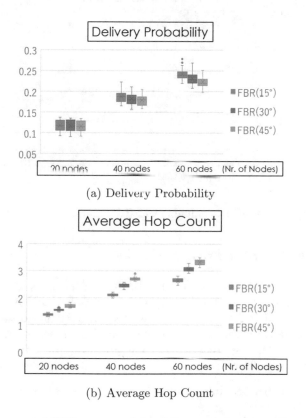

(a) Delivery Probability

(b) Average Hop Count

Fig. 4. Performance of FBR compared for different angles, regarding the number of nodes

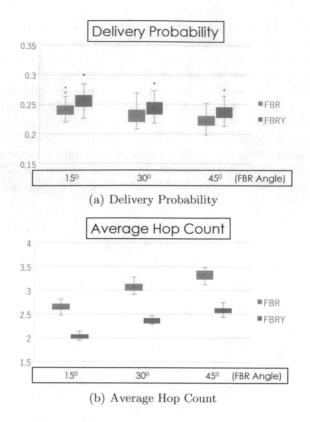

(a) Delivery Probability

(b) Average Hop Count

Fig. 5. Performance of dFBR compared to FBR, regarding the transmission angle.

4 Conclusions and Future Works

In this paper, we proposed an Enhanced FBR, which considers the depth of each candidate and limiting forwarding of data on not deeper nodes, and called it Depth-Based FBR (dFBR). We implemented both FBR and its enhancement version (dFBR) in ONE Simulator, and compared their performances with Epidemic Routing (ER), a well-known routing protocols for Delay Tolerant Networks. From the simulation results, we draw the following conclusions.

FBR shows better performance than ER, in scenarios where buffer size is relatively small.

When we increased the forwarding angle the performance of FBR decreased. The effect of the angle, shold be thouroughly investigated for all values af angle.

dFBR shows a slightly better performance than FBR. We still believe that FBR is suitable for a broad range of UWSN applications.

It still remains to be investigated the effect of forwarding angle in the energy consumption. This requires a deep analysis of MAC layer in UWSN and the acoustic spectrum.

References

1. Stojanovic, M.: Underwater acoustic communication. Wiley Encycl. Electr. Electron. Eng. (1999). https://doi.org/10.1002/047134608X.W5411
2. Akyildiz, I.F., Pompili, D., Melodia, T.: Underwater acoustic sensor networks: research challenges. Ad Hoc Netw. **3**(3), 257–279 (2005). https://doi.org/10.1016/j.adhoc.2005.01.004
3. Jornet, J.M., Stojanovic, M., Zorzi, M.: Focused beam routing protocol for underwater acoustic networks. In: Proceedings of the Third ACM International Workshop on Underwater Networks (WuWNeT 2008), pp. 75-82. Association for Computing Machinery, New York, NY, USA (2008)
4. Keränen, A., Ott, J., Kärkkäinen, T.: The ONE simulator for DTN protocol evaluation. In: Proceedings of the 2nd International Conference on Simulation Tools and Techniques. Italy, Rome (2009)
5. Yin, J., Du, P., Yang, G., Zhou, H.: Space-division multiple access for CDMA multiuser underwater acoustic communications. J. Syst. Eng. Electron. **26**(6), 1184–1190 (2015)
6. Potter, J., Alves, J., Green, D., Zappa, G., Nissen, I., McCoy, K.: The JANUS underwater communications standard. In: 2014 Underwater Communications and Networking (UComms), Sestri Levante, pp. 1–4 (2014)

A Hybrid Intelligent Simulation System for Node Placement in WMNs Considering Chi-Square Distribution of Mesh Clients and Different Router Replacement Methods

Seiji Ohara[1]([⊠]), Admir Barolli[2], Phudit Ampririt[1], Keita Matsuo[3], Leonard Barolli[3], and Makoto Takizawa[4]

[1] Graduate School of Engineering, Fukuoka Institute of Technology, 3-30-1 Wajiro-Higashi, Higashi-Ku, Fukuoka 811-0295, Japan
`seiji.ohara.19@gmail.com, iceattpon12@gmail.com`
[2] Department of Information Technology, Aleksander Moisiu University of Durres, L.1, Rruga e Currilave, Durres, Albania
`admir.barolli@gmail.com`
[3] Department of Information and Communication Engineering, Fukuoka Institute of Technology, 3-30-1 Wajiro-Higashi, Higashi-Ku, Fukuoka 811-0295, Japan
`{kt-matsuo,barolli}@fit.ac.jp`
[4] Department of Advanced Sciences, Faculty of Science and Engineering, Hosei University, 4-7-2 Kajino-Cho, Koganei-Shi, Tokyo 184-8584, Japan
`makoto.takizawa@computer.org`

Abstract. Wireless Mesh Networks (WMNs) are an important networking infrastructure and they have many advantages such as low cost and high-speed wireless Internet connectivity. However, they have some problems such as router placement, covering of mesh clients and load balancing. To deal with these problems, in our previous work, we implemented a Particle Swarm Optimization (PSO) based simulation system, called WMN-PSO, and a simulation system based on Genetic Algorithm (GA), called WMN-GA. Then, we implemented a hybrid simulation system based on PSO and distributed GA (DGA), called WMN-PSODGA. Moreover, we added in the fitness function a new parameter for the load balancing of the mesh routers called NCMCpR (Number of Covered Mesh Clients per Router). In this paper, we consider Chi-square distribution of mesh clients and five router replacement methods and carry out simulations using WMN-PSODGA system. The simulation results show that RDVM router replacement method has better performance than other methods.

1 Introduction

The wireless networks and devices can provide users access to information and communication anytime and anywhere [3,8–11,14,20,26,27,29,33]. Wireless Mesh Networks (WMNs) are gaining a lot of attention because of their low-cost nature that makes them attractive for providing wireless Internet connectivity.

L. Barolli et al. (Eds.): EIDWT 2021, LNDECT 65, pp. 10–23, 2021.
https://doi.org/10.1007/978-3-030-70639-5_2

A WMN is dynamically self-organized and self-configured, with the nodes in the network automatically establishing and maintaining mesh connectivity among itself (creating, in effect, an ad hoc network). This feature brings many advantages to WMN such as low up-front cost, easy network maintenance, robustness and reliable service coverage [1]. Moreover, such infrastructure can be used to deploy community networks, metropolitan area networks, municipal and corporative networks, and to support applications for urban areas, medical, transport and surveillance systems.

Mesh node placement in WMNs can be seen as a family of problems, which is shown (through graph theoretic approaches or placement problems, e.g. [6,15]) to be computationally hard to solve for most of the formulations [37].

We consider the version of the mesh router nodes placement problem in which we are given a grid area where to deploy a number of mesh router nodes and a number of mesh client nodes of fixed positions (of an arbitrary distribution) in the grid area. The objective is to find a location assignment for the mesh routers to the cells of the grid area that maximizes the network connectivity, client coverage and consider load balancing for each router. Network connectivity is measured by Size of Giant Component (SGC) of the resulting WMN graph, while the user coverage is simply the number of mesh client nodes that fall within the radio coverage of at least one mesh router node and is measured by Number of Covered Mesh Clients (NCMC). For load balancing, we added in the fitness function a new parameter called NCMCpR (Number of Covered Mesh Clients per Router).

Node placement problems are known to be computationally hard to solve [12, 13,38]. In previous works, some intelligent algorithms have been recently investigated for node placement problem [4,7,16,18,21–23,31,32].

In [24], we implemented a Particle Swarm Optimization (PSO) based simulation system, called WMN-PSO. Also, we implemented another simulation system based on Genetic Algorithm (GA), called WMN-GA [19], for solving node placement problem in WMNs. Then, we designed and implemented a hybrid simulation system based on PSO and distributed GA (DGA). We call this system WMN-PSODGA.

In this paper, we present the performance analysis of WMNs using WMN-PSODGA system considering Chi-square distribution of mesh clients and different router replacement methods.

The rest of the paper is organized as follows. We present our designed and implemented hybrid simulation system in Sect. 2. The simulation results are given in Sect. 3. Finally, we give conclusions and future work in Sect. 4.

2   Proposed and Implemented Simulation System

2.1   Particle Swarm Optimization

In PSO a number of simple entities (the particles) are placed in the search space of some problem or function and each evaluates the objective function at its current location. The objective function is often minimized and the exploration of the search space is not through evolution [17].

Each particle then determines its movement through the search space by combining some aspect of the history of its own current and best (best-fitness) locations with those of one or more members of the swarm, with some random perturbations. The next iteration takes place after all particles have been moved. Eventually the swarm as a whole, like a flock of birds collectively foraging for food, is likely to move close to an optimum of the fitness function.

Each individual in the particle swarm is composed of three $\mathcal{D}$-dimensional vectors, where $\mathcal{D}$ is the dimensionality of the search space. These are the current position $\vec{x}_i$, the previous best position $\vec{p}_i$ and the velocity $\vec{v}_i$.

The particle swarm is more than just a collection of particles. A particle by itself has almost no power to solve any problem; progress occurs only when the particles interact. Problem solving is a population-wide phenomenon, emerging from the individual behaviors of the particles through their interactions. In any case, populations are organized according to some sort of communication structure or topology, often thought of as a social network. The topology typically consists of bidirectional edges connecting pairs of particles, so that if j is in i's neighborhood, i is also in j's. Each particle communicates with some other particles and is affected by the best point found by any member of its topological neighborhood. This is just the vector $\vec{p}_i$ for that best neighbor, which we will denote with $\vec{p}_g$. The potential kinds of population "social networks" are hugely varied, but in practice certain types have been used more frequently. We show the pseudo code of PSO in Algorithm 1.

In the PSO process, the velocity of each particle is iteratively adjusted so that the particle stochastically oscillates around $\vec{p}_i$ and $\vec{p}_g$ locations.

2.2   Distributed Genetic Algorithm

Distributed Genetic Algorithm (DGA) has been used in various fields of science. DGA has shown their usefulness for the resolution of many computationally hard combinatorial optimization problems. We show the pseudo code of DGA in Algorithm 2.

Population of Individuals: Unlike local search techniques that construct a path in the solution space jumping from one solution to another one through local perturbations, DGA use a population of individuals giving thus the search a larger scope and chances to find better solutions. This feature is also known as "exploration" process in difference to "exploitation" process of local search methods.

Fitness: The determination of an appropriate fitness function, together with the chromosome encoding are crucial to the performance of DGA. Ideally we would construct objective functions with "certain regularities", i.e. objective functions that verify that for any two individuals which are close in the search space, their respective values in the objective functions are similar.

Algorithm 1. Pseudo code of PSO.

/* Initialize all parameters for PSO */
Computation maxtime:= Tp_{max}, $t := 0$;
Number of particle-patterns:= m, $2 \leq m \in N^1$;
Particle-patterns initial solution:= P_i^0;
Particle-patterns initial position:= x_{ij}^0;
Particles initial velocity:= v_{ij}^0;
PSO parameter:= ω, $0 < \omega \in R^1$;
PSO parameter:= C_1, $0 < C_1 \in R^1$;
PSO parameter:= C_2, $0 < C_2 \in R^1$;
/* Start PSO */
Evaluate(G^0, P^0);
while $t < Tp_{max}$ **do**
    /* Update velocities and positions */
    $v_{ij}^{t+1} = \omega \cdot v_{ij}^t$
        $+C_1 \cdot \mathrm{rand}() \cdot (best(P_{ij}^t) - x_{ij}^t)$
        $+C_2 \cdot \mathrm{rand}() \cdot (best(G^t) - x_{ij}^t)$;
    $x_{ij}^{t+1} = x_{ij}^t + v_{ij}^{t+1}$;
    /* if fitness value is increased, a new solution will be accepted. */
    Update_Solutions(G^t, P^t);
    $t = t + 1$;
end while
Update_Solutions(G^t, P^t);
return Best found pattern of particles as solution;

Selection: The selection of individuals to be crossed is another important aspect in DGA as it impacts on the convergence of the algorithm. Several selection schemes have been proposed in the literature for selection operators trying to cope with premature convergence of DGA. There are many selection methods in GA. In our system, we implement 2 selection methods: Random method and Roulette wheel method.

Crossover Operators: Use of crossover operators is one of the most important characteristics. Crossover operator is the means of DGA to transmit best genetic features of parents to offsprings during generations of the evolution process. Many methods for crossover operators have been proposed such as Blend Crossover (BLX-α), Unimodal Normal Distribution Crossover (UNDX), Simplex Crossover (SPX).

Mutation Operators: These operators intend to improve the individuals of a population by small local perturbations. They aim to provide a component of randomness in the neighborhood of the individuals of the population. In our system, we implemented two mutation methods: uniformly random mutation and boundary mutation.

Escaping from Local Optima: GA itself has the ability to avoid falling prematurely into local optima and can eventually escape from them during the search process. DGA has one more mechanism to escape from local optima by considering some islands. Each island computes GA for optimizing and they migrate its gene to provide the ability to avoid from local optima (See Fig. 1).

Convergence: The convergence of the algorithm is the mechanism of DGA to reach to good solutions. A premature convergence of the algorithm would cause that all individuals of the population be similar in their genetic features and thus the search would result ineffective and the algorithm getting stuck into local optima. Maintaining the diversity of the population is therefore very important to this family of evolutionary algorithms.

Algorithm 2. Pseudo code of DGA.

/* Initialize all parameters for DGA */
Computation maxtime:= Tg_{max}, $t := 0$;
Number of islands:= n, $1 \leq n \in N^1$;
initial solution:= P_i^0;
/* Start DGA */
Evaluate(G^0, P^0);
while $t < Tg_{max}$ **do**
   for all islands **do**
      Selection();
      Crossover();
      Mutation();
   end for
   $t = t + 1$;
end while
Update_Solutions(G^t, P^t);
return Best found pattern of particles as solution;

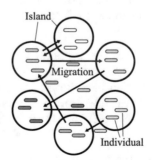

Fig. 1. Model of migration in DGA.

2.3 WMN-PSODGA Hybrid Simulation System

In this subsection, we present the initialization, particle-pattern, fitness function, and replacement methods. The pseudo code of our implemented system is shown in Algorithm 3. Also, our implemented simulation system uses Migration function as shown in Fig. 2. The Migration function swaps solutions among lands included in PSO part.

Algorithm 3. Pseudo code of WMN-PSODGA system.

Computation maxtime:= T_{max}, t := 0;
Initial solutions: P.
Initial global solutions: G.
/* Start PSODGA */
while $t < T_{max}$ **do**
   Subprocess(PSO);
   Subprocess(DGA);
   WaitSubprocesses();
   Evaluate(G^t, P^t)
   /* Migration() swaps solutions (see Fig. 2). */
   Migration();
   $t = t + 1$;
end while
Update_Solutions(G^t, P^t);
return Best found pattern of particles as solution;

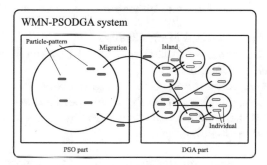

Fig. 2. Model of WMN-PSODGA migration.

Initialization

We decide the velocity of particles by a random process considering the area size. For instance, when the area size is $W \times H$, the velocity is decided randomly from $-\sqrt{W^2 + H^2}$ to $\sqrt{W^2 + H^2}$.

Particle-Pattern

A particle is a mesh router. A fitness value of a particle-pattern is computed by combination of mesh routers and mesh clients positions. In other words, each particle-pattern is a solution as shown is Fig. 3.

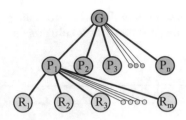

G: Global Solution
P: Particle-pattern
R: Mesh Router
n: Number of Particle-patterns
m: Number of Mesh Routers

Fig. 3. Relationship among global solution, particle-patterns, and mesh routers in PSO part.

Gene Coding

A gene describes a WMN. Each individual has its own combination of mesh nodes. In other words, each individual has a fitness value. Therefore, the combination of mesh nodes is a solution.

Fitness Function

The fitness function of WMN-PSODGA is used to evaluate the temporary solution of the router's placements. The fitness function is defined as:

$$Fitness = \alpha \times NCMC(x_{ij}, y_{ij}) + \beta \times SGC(x_{ij}, y_{ij}) + \gamma \times NCMCpR(x_{ij}, y_{ij}).$$

This function uses the following indicators.

- NCMC (Number of Covered Mesh Clients)
  The NCMC is the number of the clients covered by the SGC's routers.
- SGC (Size of Giant Component)
  The SGC is the maximum number of connected routers.
- NCMCpR (Number of Covered Mesh Clients per Router)
  The NCMCpR is the number of clients covered by each router. The NCMCpR indicator is used for load balancing.

WMN-PSODGA aims to maximize the value of the fitness function in order to optimize the placements of the routers using the above three indicators. Weight-coefficients of the fitness function are α, β, and γ for NCMC, SGC, and NCMCpR, respectively. Moreover, the weight-coefficients are implemented as $\alpha + \beta + \gamma = 1$.

Router Replacement Methods

A mesh router has x, y positions, and velocity. Mesh routers are moved based on velocities. There are many router replacement methods, such as:

Constriction Method (CM)
     CM is a method which PSO parameters are set to a week stable region ($\omega = 0.729$, $C_1 = C2 = 1.4955$) based on analysis of PSO by M. Clerc et al. [2, 5, 35].

Random Inertia Weight Method (RIWM)

In RIWM, the ω parameter is changing ramdomly from 0.5 to 1.0. The C_1 and C_2 are kept 2.0. The ω can be estimated by the week stable region. The average of ω is 0.75 [28,35].

Linearly Decreasing Inertia Weight Method (LDIWM)

In LDIWM, C_1 and C_2 are set to 2.0, constantly. On the other hand, the ω parameter is changed linearly from unstable region ($\omega = 0.9$) to stable region ($\omega = 0.4$) with increasing of iterations of computations [35,36].

Linearly Decreasing Vmax Method (LDVM)

In LDVM, PSO parameters are set to unstable region ($\omega = 0.9$, $C_1 = C_2 = 2.0$). A value of V_{max} which is maximum velocity of particles is considered. With increasing of iteration of computations, the V_{max} is kept decreasing linearly [30,34].

Rational Decrement of Vmax Method (RDVM)

In RDVM, PSO parameters are set to unstable region ($\omega = 0.9$, $C_1 = C_2 = 2.0$). The V_{max} is kept decreasing with the increasing of iterations as

$$V_{max}(x) = \sqrt{W^2 + H^2} \times \frac{T - x}{x}.$$

Where, W and H are the width and the height of the considered area, respectively. Also, T and x are the total number of iterations and a current number of iteration, respectively [25].

3   Simulation Results

In this section, we present the simulation results. Table 1 shows the common parameters for each simulation. Figure 4 shows the visualization results after the

Table 1. The common parameters for each simulation.

Parameters	Values
Distribution of mesh clients	Chi-square distribution
Number of mesh clients	48
Number of mesh routers	16
Radius of a mesh router	2.0–3.5
Number of GA islands	16
Number of migrations	200
Evolution steps	9
Selection method	Random method
Crossover method	UNDX
Mutation method	Uniform mutation
Crossover rate	0.8
Mutation rate	0.2
Area size	32.0×32.0

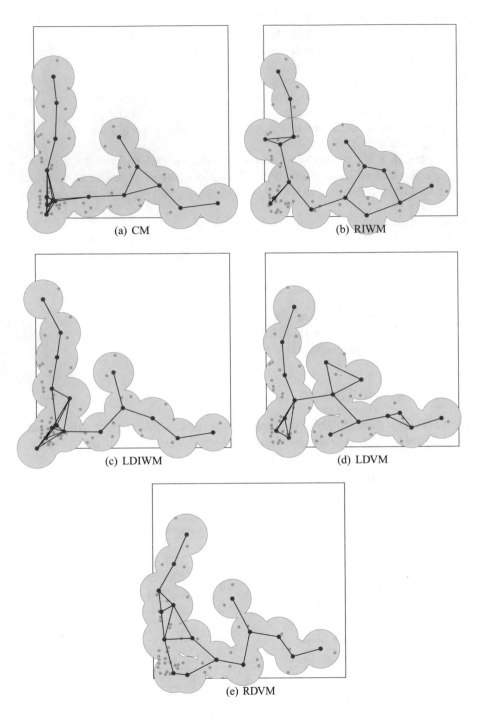

(a) CM

(b) RIWM

(c) LDIWM

(d) LDVM

(e) RDVM

Fig. 4. Visualization results after the optimization.

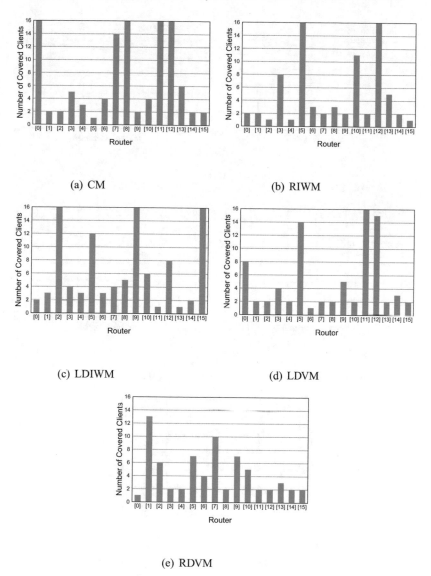

(a) CM

(b) RIWM

(c) LDIWM

(d) LDVM

(e) RDVM

Fig. 5. Number of covered clients by each router after the optimization.

optimization. While, Fig. 5 shows the number of covered clients by each router. In Fig. 6 are shown the transition of the standard deviations. The standard deviation is related to load balancing. When the standard deviation increased, the number of mesh clients for each router tends to be different. On the other hand, when the standard deviation decreased, the number of mesh clients for each router tends to go close to each other. The value of r in Fig. 6 means the correlation coefficient. In Fig. 6(a) and 6(d), the standard deviation is increased by increasing the number

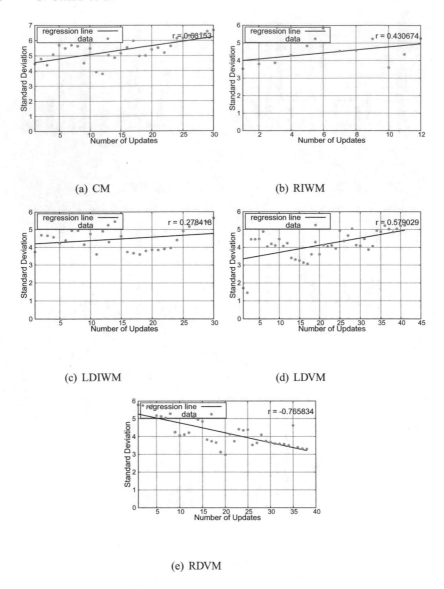

(a) CM

(b) RIWM

(c) LDIWM

(d) LDVM

(e) RDVM

Fig. 6. Standard deviation for five router replacement methods.

of updates. Figure 6(b) and Fig. 6(c) show that there is no correlation between the number of updates and the standard deviations. On the other hand, the standard deviation of Fig. 6(e) decreases by increasing the number of updates. Therefore, we conclude that RDVM has better behavior than other methods.

4   Conclusions

In this work, we evaluated the performance of WMNs using a hybrid simulation system based on PSO and DGA (called WMN-PSODGA). We considered Chi-square distribution of mesh clients and five router replacement methods for WMN-PSODGA. The simulation results show that RDVM router replacement method has better performance than other methods. In future work, we will consider other distributions of mesh clients.

References

1. Akyildiz, I.F., Wang, X., Wang, W.: Wireless mesh networks: a survey. Comput. Netw. **47**(4), 445–487 (2005)
2. Barolli, A., Sakamoto, S., Ozera, K., Ikeda, M., Barolli, L., Takizawa, M.: Performance evaluation of WMNs by WMN-PSOSA simulation system considering constriction and linearly decreasing Vmax methods. In: International Conference on P2P, Parallel, Grid, Cloud and Internet Computing, pp. 111–121. Springer (2017)
3. Barolli, A., Sakamoto, S., Barolli, L., Takizawa, M.: Performance analysis of simulation system based on particle swarm optimization and distributed genetic algorithm for WMNs considering different distributions of mesh clients. In: International Conference on Innovative Mobile and Internet Services in Ubiquitous Computing, Springer, pp. 32–45 (2018)
4. Barolli, A., Sakamoto, S., Ozera, K., Barolli, L., Kulla, E., Takizawa, M.: Design and implementation of a hybrid intelligent system based on particle swarm optimization and distributed genetic algorithm. In: International Conference on Emerging Internetworking, Data & Web Technologies, pp. 79–93. Springer (2018)
5. Clerc, M., Kennedy, J.: The particle swarm-explosion, stability, and convergence in a multidimensional complex space. IEEE Trans. Evol. Comput. **6**(1), 58–73 (2002)
6. Franklin, A.A., Murthy, C.S.R.: Node placement algorithm for deployment of two-tier wireless mesh networks. In: Proceeding of Global Telecommunications Conference, pp. 4823–4827 (2007)
7. Girgis, M.R., Mahmoud, T.M., Abdullatif, B.A., Rabie, A.M.: Solving the wireless mesh network design problem using genetic algorithm and simulated annealing optimization methods. Int. J. Comput. Appl. **96**(11), 1–10 (2014)
8. Goto, K., Sasaki, Y., Hara, T., Nishio, S.: Data gathering using mobile agents for reducing traffic in dense mobile wireless sensor networks. Mob. Inf. Syst. **9**(4), 295–314 (2013)
9. Inaba, T., Elmazi, D., Sakamoto, S., Oda, T., Ikeda, M., Barolli, L.: A secure-aware call admission control scheme for wireless cellular networks using fuzzy logic and its performance evaluation. J. Mob. Multimedia **11**(3&4), 213–222 (2015)
10. Inaba, T., Obukata, R., Sakamoto, S., Oda, T., Ikeda, M., Barolli, L.: Performance evaluation of a QoS-aware fuzzy-based CAC for LAN access. Int. J. Space-Based Situated Comput. **6**(4), 228–238 (2016)
11. Inaba, T., Sakamoto, S., Oda, T., Ikeda, M., Barolli, L.: A testbed for admission control in WLAN: a fuzzy approach and its performance evaluation. In: International Conference on Broadband and Wireless Computing, Communication and Applications, pp. 559–571. Springer (2016)
12. Lim, A., Rodrigues, B., Wang, F., Xu, Z.: k-center problems with minimum coverage. Theor. Comput. Sci. **332**(1–3), 1–17 (2005)

13. Maolin, T., et al.: Gateways placement in backbone wireless mesh networks. Int. J. Commun. Netw. Syst. Sci. **2**(1), 44–50 (2009)
14. Matsuo, K., Sakamoto, S., Oda, T., Barolli, A., Ikeda, M., Barolli, L.: Performance analysis of WMNs by WMN-GA simulation system for two WMN architectures and different TCP congestion-avoidance algorithms and client distributions. Int. J. Commun. Netw. Distrib. Syst. **20**(3), 335–351 (2018)
15. Muthaiah, S.N., Rosenberg, C.P.: Single gateway placement in wireless mesh networks. In: Proceeding of 8th International IEEE Symposium on Computer Networks, pp. 4754–4759 (2008)
16. Naka, S., Genji, T., Yura, T., Fukuyama, Y.: A hybrid particle swarm optimization for distribution state estimation. IEEE Trans. Power Syst. **18**(1), 60–68 (2003)
17. Poli, R., Kennedy, J., Blackwell, T.: Particle swarm optimization. Swarm Intell. **1**(1), 33–57 (2007)
18. Sakamoto, S., Kulla, E., Oda, T., Ikeda, M., Barolli, L., Xhafa, F.: A comparison study of simulated annealing and genetic algorithm for node placement problem in wireless mesh networks. J. Mob. Multimedia **9**(1–2), 101–110 (2013)
19. Sakamoto, S., Kulla, E., Oda, T., Ikeda, M., Barolli, L., Xhafa, F.: A comparison study of hill climbing, simulated annealing and genetic algorithm for node placement problem in WMNs. J. High Speed Netw. **20**(1), 55–66 (2014)
20. Sakamoto, S., Kulla, E., Oda, T., Ikeda, M., Barolli, L., Xhafa, F.: A simulation system for WMN based on SA: performance evaluation for different instances and starting temperature values. Int. J. Space-Based Situated Comput. **4**(3–4), 209–216 (2014)
21. Sakamoto, S., Kulla, E., Oda, T., Ikeda, M., Barolli, L., Xhafa, F.: Performance evaluation considering iterations per Phase and SA temperature in WMN-SA system. Mob. Inf. Syst. **10**(3), 321–330 (2014)
22. Sakamoto, S., Lala, A., Oda, T., Kolici, V., Barolli, L., Xhafa, F.: Application of WMN-SA simulation system for node placement in wireless mesh networks: a case study for a realistic scenario. Int. J. Mob. Comput. Multimedia Commun. (IJMCMC) **6**(2), 13–21 (2014)
23. Sakamoto, S., Oda, T., Ikeda, M., Barolli, L., Xhafa, F.: An Integrated simulation system considering WMN-PSO simulation system and network simulator 3. In: International Conference on Broadband and Wireless Computing, Communication and Applications, pp. 187–198. Springer (2016)
24. Sakamoto, S., Oda, T., Ikeda, M., Barolli, L., Xhafa, F.: Implementation and evaluation of a simulation system based on particle swarm optimisation for node placement problem in wireless mesh networks. Int. J. Commun. Netw. Distrib. Syst. **17**(1), 1–13 (2016)
25. Sakamoto, S., Oda, T., Ikeda, M., Barolli, L., Xhafa, F.: Implementation of a new replacement method in WMN-PSO simulation system and its performance evaluation. In: The 30th IEEE International Conference on Advanced Information Networking and Applications (AINA-2016), pp. 206–211 (2016)
26. Sakamoto, S., Obukata, R., Oda, T., Barolli, L., Ikeda, M., Barolli, A.: Performance analysis of two wireless mesh network architectures by WMN-SA and WMN-TS simulation systems. J. High Speed Netw. **23**(4), 311–322 (2017)
27. Sakamoto, S., Ozera, K., Barolli, A., Ikeda, M., Barolli, L., Takizawa, M.: Implementation of an intelligent hybrid simulation systems for WMNs based on particle swarm optimization and simulated annealing: performance evaluation for different replacement methods. Soft. Comput. **23**(9), 3029–3035 (2017)

28. Sakamoto, S., Ozera, K., Barolli, A., Ikeda, M., Barolli, L., Takizawa, M.: Performance evaluation of WMNs by WMN-PSOSA simulation system considering random inertia weight method and linearly decreasing Vmax method. In: International Conference on Broadband and Wireless Computing, Communication and Applications, pp. 114–124. Springer (2017)

29. Sakamoto, S., Ozera, K., Ikeda, M., Barolli, L.: Implementation of intelligent hybrid systems for node placement problem in WMNs considering particle swarm optimization, hill climbing and simulated annealing. Mob. Netw. Appl. **23**(1), 27–33 (2017)

30. Sakamoto, S., Ozera, K., Ikeda, M., Barolli, L.: Performance evaluation of WMNs by WMN-PSOSA simulation system considering constriction and linearly decreasing inertia weight methods. In: International Conference on Network-Based Information Systems, pp. 3–13. Springer (2017)

31. Sakamoto, S., Ozera, K., Oda, T., Ikeda, M., Barolli, L.: Performance evaluation of intelligent hybrid systems for node placement in wireless mesh networks: a comparison study of WMN-PSOHC and WMN-PSOSA. In: International Conference on Innovative Mobile and Internet Services in Ubiquitous Computing, pp. 16–26. Springer (2017)

32. Sakamoto, S., Ozera, K., Oda, T., Ikeda, M., Barolli, L.: Performance evaluation of WMN-PSOIIC and WMN-PSO simulation systems for node placement in wireless mesh networks: a comparison study. In: International Conference on Emerging Internetworking, Data & Web Technologies, pp. 64–74. Springer (2017)

33. Sakamoto, S., Ozera, K., Barolli, A., Barolli, L., Kolici, V., Takizawa, M.: Performance evaluation of WMN-PSOSA considering four different replacement methods. International Conference on Emerging Internetworking, pp. 51–64. Springer Data & Web Technologies, Berlin (2018)

34. Schutte, J.F., Groenwold, A.A.: A study of global optimization using particle swarms. J. Global Optim. **31**(1), 93–108 (2005)

35. Shi, Y.: Particle swarm optimization. IEEE Connections **2**(1), 8–13 (2004)

36. Shi, Y., Eberhart, R.C.: Parameter selection in particle swarm optimization. In: Evolutionary programming VII, pp. 591–600 (1998)

37. Vanhatupa, T., Hannikainen, M., Hamalainen, T.: Genetic algorithm to optimize node placement and configuration for WLAN planning. In: Proceeding of The 4th IEEE International Symposium on Wireless Communication Systems, pp. 612–616 (2007)

38. Wang, J., Xie, B., Cai, K., Agrawal, D.P.: Efficient mesh router placement in wireless mesh networks. In: Proceeding of IEEE Internatonal Conference on Mobile Adhoc and Sensor Systems (MASS-2007), pp. 1–9 (2007)

Effect of Slice Overloading Cost on Admission Control for 5G Wireless Networks: A Fuzzy-Based System and Its Performance Evaluation

Phudit Ampririt[1](✉), Seiji Ohara[1], Ermioni Qafzezi[1], Makoto Ikeda[2],
Leonard Barolli[2], and Makoto Takizawa[3]

[1] Graduate School of Engineering, Fukuoka Institute of Technology,
3-30-1 Wajiro-Higashi, Higashi-Ku, Fukuoka 811-0295, Japan
iceattpon12@gmail.com, seiji.ohara.19@gmail.com, eqafzezi@gmail.com
[2] Department of Information and Communication Engineering, Fukuoka Institute
of Technology, 3-30-1 Wajiro-Higashi, Higashi-Ku, Fukuoka 811-0295, Japan
makoto.ikd@acm.org, barolli@fit.ac.jp
[3] Department of Advanced Sciences, Faculty of Science and Engineering,
Hosei University, Kajino-Machi, Koganei-Shi, Tokyo 184-8584, Japan
makoto.takizawa@computer.org

Abstract. The Fifth Generation (5G) network is expected to be flexible
to satisfy user requirements and the Software-Defined Network (SDN)
with Network Slicing is a good approach for admission control. In this
paper, we consider the effect of Slice Overloading Cost on Admission
Control for 5G Wireless Networks. We present a Fuzzy-based system for
admission decision. We consider 4 input parameters: Quality of Service
(QoS), Slice Priority (SP), User Request Delay Time (URDT) and Slice
Overloading Cost (SOC) as a new parameter. From simulation results,
we can see that when QoS, SP, URDT parameters are increased, the AD
parameter is increased. But, when the SOC parameter is increased, the
AD parameter is decreased.

1 Introduction

Recently, the growth of wireless technologies and user's demand of services
are increasing rapidly. Especially in 5G networks, there will be billions of new
devices with unpredictable traffic pattern which provide high data rates. With
the appearance of Internet of Things (IoT), these devices will generate Big Data
to the Internet, which will cause to congest and deteriorate the QoS [1].

The 5G network will be expected to be better than 4G. The 5G network
will provide users with new experiences such as Ultra High Definition Television
(UHDT) on Internet and support a lot of IoT devices with long battery life and
high data rate on hotspot areas with high user density. In the 5G technology, the
routing and switching technologies aren't important anymore or coverage area is

L. Barolli et al. (Eds.): EIDWT 2021, LNDECT 65, pp. 24–35, 2021.
https://doi.org/10.1007/978-3-030-70639-5_3

shorter than 4G because it uses high frequency for facing higher device's volume for high user density [2-4].

Therefore, there are many research work that try to build systems which are suitable to 5G era. The SDN is one of them [5]. For example, the mobile handover mechanism with SDN is used for reducing the delay in handover processing and improve QoS. Also, by using SDN the QoS can be improved by applying Fuzzy Logic (FL) on SDN controller [6–8].

In our previous work [9, 10], we presented a Fuzzy-based system for admission control in 5G Wireless Networks considering three parameters: Grade of Service (GS), User Request Delay Time (URDT), Network Slice Size (NSS) and proposed a fuzzy-based scheme for evaluation of QoS in 5G wireless networks. In this paper, we considered Slice Overloading Cost (SOC) as a new parameter and present a Fuzzy-based system for admission control in 5G Wireless Networks.

The rest of the paper is organized as follows. In Sect. 2 is presented an overview of SDN. In Sect. 3, we present application of Fuzzy Logic for admission control. In Sect. 4, we describe the proposed fuzzy-based system and its implementation. In Sect. 5, we explain the simulation results. Finally, conclusions and future work are presented in Sect. 6.

2 Software-Defined Networks (SDNs)

The SDN is a new networking paradigm that decouples the data plane from control plane in the network. In traditional networks, the whole network is controlled by each network device. However, the traditional networks are hard to manage and control since they rely on physical infrastructure. Network devices must stay connected all the time when user wants to connect other networks. Those processes must be based on the setting of each device, making controlling the operation of the network difficult. Therefore, they have to be set up one by one. In contrast, the SDN is easy to manage and provide network software based services from a centralised control plane. The SDN control plane is managed by SDN controller or cooperating group of SDN controllers. The SDN structure is shown in Fig. 1 [11, 12].

- **Application Layer** builds an abstracted view of the network by collecting information from the controller for decision-making purposes. The types of applications are related to: network configuration and management, network monitoring, network troubleshooting, network policies and security.
- **Control Layer** receives instructions or requirements from the Application Layer and control the Infrastructure Layer by using intelligent logic.
- **Infrastructure Layer** receives orders from SDN controller and sends data among them.

The SDN can manage network systems while enabling new services. In congestion traffic situation, management system can be flexible, allowing users to easily control and adapt resources appropriately throughout the control plane.

Mobility management is easier and quicker in forwarding across different wireless technologies (e.g. 5G, 4G, Wifi and Wimax). Also, the handover procedure is simple and the delay can be decreased.

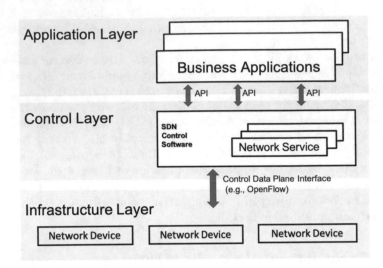

Fig. 1. Structure of SDN.

3   Outline of Fuzzy Logic

A Fuzzy Logic (FL) system is a nonlinear mapping of an input data vector into a scalar output, which is able to simultaneously handle numerical data and linguistic knowledge. The FL can deal with statements which may be true, false or intermediate truth-value. These statements are impossible to quantify using traditional mathematics. The FL system is used in many controlling applications such as aircraft control (Rockwell Corp.), Sendai subway operation (Hitachi), and TV picture adjustment (Sony) [13–15].

In Fig. 2 is shown Fuzzy Logic Controller (FLC) structure, which contains four components: fuzzifier, inference engine, fuzzy rule base and defuzzifier.

- **Fuzzifier** is needed for combining the crisp values with rules which are linguistic variables and have fuzzy sets associated with them.
- **The Rules** may be provided by expert or can be extracted from numerical data. In engineering case, the rules are expressed as a collection of IF-THEN statements.
- **The Inference Engine** infers fuzzy output by considering fuzzified input values and fuzzy rules.
- **The Defuzzifier** maps output set into crisp numbers.

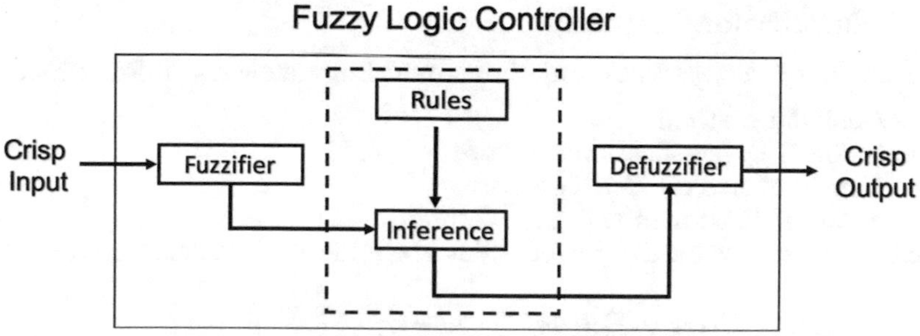

Fuzzy Logic Controller

Fig. 2. FLC structure.

3.1 Linguistic Variables

A concept that plays a central role in the application of FL is that of a linguistic variable. The linguistic variables may be viewed as a form of data compression. One linguistic variable may represent many numerical variables. It is suggestive to refer to this form of data compression as granulation.

The same effect can be achieved by conventional quantization, but in the case of quantization, the values are intervals, whereas in the case of granulation the values are overlapping fuzzy sets. The advantages of granulation over quantization are as follows:

- it is more general;
- it mimics the way in which humans interpret linguistic values;
- the transition from one linguistic value to a contiguous linguistic value is gradual rather than abrupt, resulting in continuity and robustness.

For example, let Temperature (T) be interpreted as a linguistic variable. It can be decomposed into a set of Terms: T (Temperature) = {Freezing, Cold, Warm, Hot, Blazing}. Each term is characterised by fuzzy sets which can be interpreted, for instance, "Freezing" as a temperature below 0 °C, "Cold" as a temperature close to 10 °C.

3.2 Fuzzy Control Rules

Rules are usually written in the form "IF x is S THEN y is T" where x and y are linguistic variables that are expressed by S and T, which are fuzzy sets. The x is a control (input) variable and y is the solution (output) variable. This rule is called Fuzzy control rule. The form "IF ... THEN" is called a conditional sentence. It consists of "IF" which is called the antecedent and "THEN" is called the consequent.

3.3 Defuzzificaion Method

There are many defuzzification methods, which are showing in following:

- The Centroid Method;
- Tsukamoto's Defuzzification Method;
- The Center of Are (COA) Method;
- The Mean of Maximum (MOM) Method;
- Defuzzification when Output of Rules are Function of Their Inputs.

4 Proposed Fuzzy-Based System

In this work, we use FL to implement the proposed system. In Fig. 3, we show the overview of our proposed system. Each evolve Base Station (eBS) will receive controlling order from SDN controller and they can communicate and send data with User Equipment (UE). On the other hand, the SDN controller will collect all the data about network traffic status and controlling eBS by using the proposed fuzzy-based system. The SDN controller will be a communicating bridge between eBS and 5G core network. The proposed system is called Fuzzy-based System for Admission Control (FBSAC) in 5G wireless networks. The structure of FBSAC is shown in Fig. 4. In this paper, the FBSAC system considers four input parameters: Quality of service (QoS), Slice Priority (SP), User Request Delay Time (URDT) and Slice Overloading Cost (SOC). The output parameter is Admission Decision (AD). We applied FL to evaluate the QoS, which is an input parameter of FBSAC system by considering three parameters: Slice Throughput (ST), Slice Delay (SD) and Slice Loss (SL).

Quality of Service (QoS): The QoS is an important parameter for admission control. A user with good QoS will have high priority to be accepted in the network.

Slice Priority (SP): In the case when there are different connection requests from different mobile devices, the system will accept with higher possibility the requests from high priority slices.

User Request Delay Time (URDT): The URDT is the waiting time of a user request in a buffer. When the user waiting time is long, the user's request will have to be served fast otherwise QoE will decrease.

Slice Overloading Cost (SOC): The SOC is the slice overloading cost. The slice with the lowest overloading cost value will have a high acceptance possibility for a new user.

Admission Decision (AD): By AD is decided whether to accept or reject a connection request.

The membership functions are shown in Fig. 5. We use triangular and trapezoidal membership functions because they are more suitable for real-time operations [16–19]. We show parameters and their term sets in Table 1. The Fuzzy Rule Base (FRB) is shown in Table 2 and has 135 rules. The control rules have the form: IF "condition" THEN "control action". For example, for Rule 1: "IF QoS is VB, SP is L, URDT is Sh and SOC is Sm THEN AD is AD1".

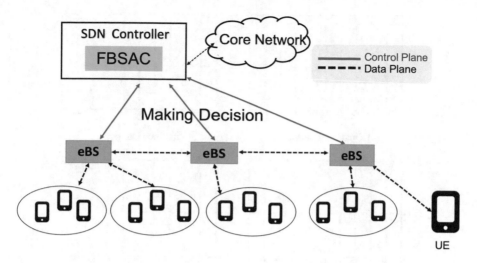

Fig. 3. Proposed system overview.

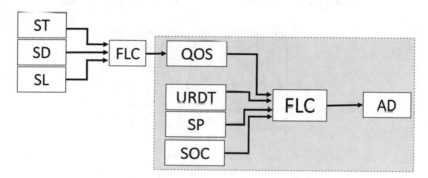

Fig. 4. Proposed system structure.

Table 1. Parameter and their term sets.

Parameters	Term set
Quality of Service (QoS)	Very Bad (VB), Bad (B), Moderate (Mo), Good (G), Very Good (VG)
Slice Priority (SP)	Low (L), Medium (M), High (H)
User Request Delay Time (URDT)	Short (Sh), Medium (Me), Long (Lo)
Slice Overloading Cost (SOC)	Small (Sm), Intermediate (In), Big (Bg)
Admission Decision (AD)	Admission Decision 1 (AD1), AD2, AD3, AD4, AD5, AD6, AD7, AD8, AD9

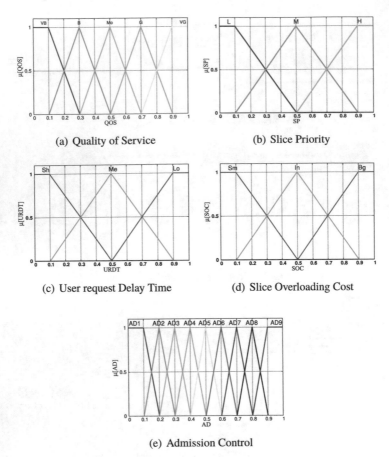

(a) Quality of Service

(b) Slice Priority

(c) User request Delay Time

(d) Slice Overloading Cost

(e) Admission Control

Fig. 5. Membership functions.

5   Simulation Results

In this section, we present the simulation result of our proposed system. The simulation results are shown in Fig. 6, Fig. 7 and Fig. 8. They show the relation of AD with SOC. We consider URDT, QoS and SP as constant parameters.

In Fig. 6(a), we consider the QoS value 0.1 and the SP value 0.1. When SOC is increased, we see that AD is decreased. For SOC 0.1, when URDT is increased from 0.1 to 0.5 and 0.5 to 0.9, the AD is increased by 12.25% and 20%, respectively.

We compare Fig. 6(a) with Fig. 6(b) to see how SP has affected AD. We change the SP value from 0.1 to 0.9. The AD is increasing by 51.38% when the SOC value is 0.3 and the URDT is 0.9. In Fig. 6(b), when we changed the SOC value from 0.3 to 0.7, the AD is decreased 20% when the URDT value is 0.5. This is because a higher SOC value means the slice is more overloaded and the network can't provide the service for a new user.

Table 2. Fuzzy rule base

Rule	QoS	SP	URDT	SOC	AD	Rule	QoS	SP	URDT	SOC	AD
1	VB	L	Sh	Sm	AD1	68	Mo	M	Me	In	AD5
2	VB	L	Sh	In	AD1	69	Mo	M	Me	Bg	AD3
3	VB	L	Sh	Bg	AD1	70	Mo	M	Lo	Sm	AD9
4	VB	L	Me	Sm	AD2	71	Mo	M	Lo	In	AD7
5	VB	L	Me	In	AD1	72	Mo	M	Lo	Bg	AD5
6	VB	L	Me	Bg	AD1	73	Mo	H	Sh	Sm	AD7
7	VB	L	Lo	Sm	AD4	74	Mo	H	Sh	In	AD5
8	VB	L	Lo	In	AD2	75	Mo	H	Sh	Bg	AD3
9	VB	L	Lo	Bg	AD1	76	Mo	H	Me	Sm	AD9
10	VB	M	Sh	Sm	AD2	77	Mo	H	Me	In	AD7
11	VB	M	Sh	In	AD1	78	Mo	H	Me	Bg	AD5
12	VB	M	Sh	Bg	AD1	79	Mo	H	Lo	Sm	AD9
13	VB	M	Me	Sm	AD4	80	Mo	H	Lo	In	AD9
14	VB	M	Me	In	AD2	81	Mo	H	Lo	Bg	AD8
15	VB	M	Me	Bg	AD1	82	G	L	Sh	Sm	AD3
16	VB	M	Lo	Sm	AD7	83	G	L	Sh	In	AD1
17	VB	M	Lo	In	AD4	84	G	L	Sh	Bg	AD1
18	VB	M	Lo	Bg	AD2	85	G	L	Me	Sm	AD5
19	VB	H	Sh	Sm	AD5	86	G	L	Me	In	AD3
20	VB	H	Sh	In	AD2	87	G	L	Me	Bg	AD2
21	VB	H	Sh	Bg	AD1	88	G	L	Lo	Sm	AD8
22	VB	H	Me	Sm	AD7	89	G	L	Lo	In	AD5
23	VB	H	Me	In	AD5	90	G	L	Lo	Bg	AD4
24	VB	H	Me	Bg	AD3	91	G	M	Sh	Sm	AD6
25	VB	H	Lo	Sm	AD9	92	G	M	Sh	In	AD3
26	VB	H	Lo	In	AD7	93	G	M	Sh	Bg	AD2
27	VB	H	Lo	Bg	AD5	94	G	M	Me	Sm	AD8
28	B	L	Sh	Sm	AD2	95	G	M	Me	In	AD6
29	B	L	Sh	In	AD1	96	G	M	Me	Bg	AD4
30	B	L	Sh	Bg	AD1	97	G	M	Lo	Sm	AD9
31	B	L	Me	Sm	AD4	98	G	M	Lo	In	AD8
32	B	L	Me	In	AD2	99	G	M	Lo	Bg	AD6
33	B	L	Me	Bg	AD1	100	G	H	Sh	Sm	AD8
34	B	L	Lo	Sm	AD6	101	G	H	Sh	In	AD6
35	B	L	Lo	In	AD4	102	G	H	Sh	Bg	AD4
36	B	L	Lo	Bg	AD2	103	G	H	Me	Sm	AD9
37	B	M	Sh	Sm	AD4	104	G	H	Me	In	AD8
38	B	M	Sh	In	AD2	105	G	H	Me	Bg	AD6
39	B	M	Sh	Bg	AD1	106	G	H	Lo	Sm	AD9
40	B	M	Me	Sm	AD6	107	G	H	Lo	In	AD9

(continued)

Table 2. (*continued*)

Rule	QoS	SP	URDT	SOC	AD	Rule	QoS	SP	URDT	SOC	AD
41	B	M	Me	In	AD4	108	G	H	Lo	Bg	AD8
42	B	M	Me	Bg	AD2	109	VG	L	Sh	Sm	AD5
43	B	M	Lo	Sm	AD8	110	VG	L	Sh	In	AD3
44	B	M	Lo	In	AD6	111	VG	L	Sh	Bg	AD2
45	B	M	Lo	Bg	AD4	112	VG	L	Me	Sm	AD8
46	B	M	Sh	Sm	AD6	113	VG	L	Me	In	AD5
47	B	H	Sh	In	AD4	114	VG	L	Me	Bg	AD3
48	B	H	Sh	Bg	AD2	115	VG	L	Lo	Sm	AD9
49	B	H	Me	Sm	AD8	116	VG	L	Lo	In	AD8
50	B	H	Me	In	AD6	117	VG	L	Lo	Bg	AD6
51	B	H	Me	Bg	AD4	118	VG	M	Sh	Sm	AD8
52	B	H	Lo	Sm	AD9	119	VG	M	Sh	In	AD5
53	B	H	Lo	In	AD8	120	VG	M	Sh	Bg	AD4
54	B	H	Lo	Bg	AD7	121	VG	M	Me	Sm	AD9
55	Mo	L	Sh	Sm	AD2	122	VG	M	Me	In	AD8
56	Mo	L	Sh	In	AD1	123	VG	M	Me	Bg	AD6
57	Mo	L	Sh	Bg	AD1	124	VG	M	Lo	Sm	AD9
58	Mo	L	Me	Sm	AD4	125	VG	M	Lo	In	AD9
59	Mo	L	Me	In	AD2	126	VG	M	Lo	Bg	AD8
60	Mo	L	Me	Bg	AD1	127	VG	H	Sh	Sm	AD9
61	Mo	L	Lo	Sm	AD7	128	VG	H	Sh	In	AD8
62	Mo	L	Lo	In	AD4	129	VG	H	Sh	Bg	AD6
63	Mo	L	Lo	Bg	AD3	130	VG	H	Me	Sm	AD9
64	Mo	M	Sh	Sm	AD5	131	VG	H	Me	In	AD9
65	Mo	M	Sh	In	AD2	132	VG	H	Me	Bg	AD8
66	Mo	M	Sh	Bg	AD1	133	VG	H	Lo	Sm	AD9
67	Mo	M	Me	Sm	AD7	134	VG	H	Lo	In	AD9
						135	VG	H	Lo	Bg	AD9

In Fig. 7, we increase the value of QoS to 0.5. The AD value in Fig. 7 is higher than in Fig. 6. When the SP increases from 0.1 to 0.9, the AD increases 50% when SOC value is 0.5 and the URDT value is 0.5. In Fig. 7(b), when URDT is 0.5, all AD values are higher than 0.5. This means that the system accepts all requests from new users.

In Fig. 8, we increase the value of QoS to 0.9. We see that the AD value is increased much more compared with the results of Fig. 7 and Fig. 8.

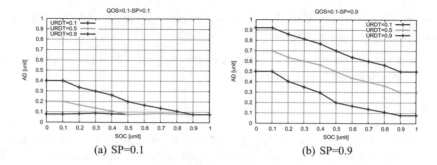

Fig. 6. Simulation results for QoS = 0.1.

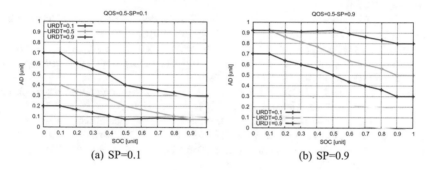

Fig. 7. Simulation results for QoS = 0.5.

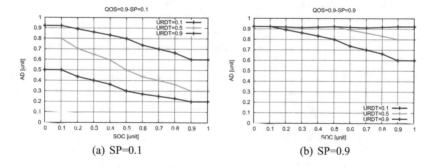

Fig. 8. Simulation results for QoS = 0.9.

6 Conclusions and Future Work

In this paper, we considered SOC as a new parameter and presented a Fuzzy-based system for Admission Control System in 5G Wireless Networks. From simulation results, we conclude as follows.

- When SOC value is increased, the AD value is decreased. This means that the acceptance possibility will be low.

- When QoS, SP and URDT value are increased, the AD value is increased. Thus, the acceptance possibility will be high.

In the future, we will consider other parameters and make extensive simulations to evaluate the proposed system.

References

1. Navarro-Ortiz, J., Romero-Diaz, P., Sendra, S., Ameigeiras, P., Ramos-Munoz, J.J., Lopez-Soler, J.M.: A survey on 5G usage scenarios and traffic models. IEEE Commun. Surv. Tutor. **22**, 905–929 (2020)
2. Hossain, S.: 5G wireless communication systems. Am. J. Eng. Res. (AJER) **2**(10), 344–353 (2013)
3. Giordani, M., Mezzavilla, M., Zorzi, M.: Initial access in 5G mmWave cellular networks. IEEE Commun. Mag. **54**(11), 40–47 (2016)
4. Kamil, I.A., Ogundoyin, S.O.: Lightweight privacy-preserving power injection and communication over vehicular networks and 5G smart grid slice with provable security. Internet Things **8**, 100–116 (2019)
5. Hossain, E., Hasan, M.: 5G cellular: key enabling technologies and research challenges. IEEE Instrum. Measur. Mag. **18**(3), 11–21 (2015)
6. Yao, D., Su, X., Liu, B., Zeng, J.: A mobile handover mechanism based on fuzzy logic and MPTCP protocol under SDN architecture*. In: 18th International Symposium on Communications and Information Technologies (ISCIT-2018), pp. 141–146, September 2018
7. Lee, J., Yoo, Y.: Handover cell selection using user mobility information in a 5G SDN-based network. In: 2017 Ninth International Conference on Ubiquitous and Future Networks (ICUFN 2017), pp. 697–702, July 2017
8. Moravejosharieh, A., Ahmadi, K., Ahmad, S.: A fuzzy logic approach to increase quality of service in software defined networking. In: 2018 International Conference on Advances in Computing,Communication Control and Networking (ICACCCN 2018), pp. 68–73, October 2018
9. Ampririt, P., Ohara, S., Liu, Y., Ikeda, M., Maeda, H., Barolli, L.: A fuzzy-based system for admission control in 5G wireless networks considering software-defined network approach. In: International Conference on Emerging Internetworking, Data & Web Technologies, pp. 73–81. Springer (2020)
10. Ampririt, P., Ohara, S., Qafzezi, E., Ikeda, M., Barolli, L., Takizawa, M.: Integration of software-defined network and fuzzy logic approaches for admission control in 5G wireless networks: a fuzzy-based scheme for QoS evaluation. In: Barolli, L., Takizawa,, M., Enokido, T., Chen, H.C., Matsuo, K. (eds.) Advances on Broad-Band Wireless Computing, Communication and Applications, pp. 386–396. Springer, Cham (2021)
11. Li, L.E., Mao, Z.M., Rexford, J.: Toward software-defined cellular networks. In: 2012 European Workshop on Software Defined Networking, pp. 7–12, October 2012
12. Mousa, M., Bahaa-Eldin, A.M., Sobh, M.: Software defined networking concepts and challenges. In: 2016 11th International Conference on Computer Engineering & Systems (ICCES), pp. 79–90. IEEE (2016)
13. Jantzen, J.: Tutorial on fuzzy logic. Technical report, Department of Automation, Technical University of Denmark (1998)
14. Mendel, J.M.: Fuzzy logic systems for engineering: a tutorial. Proc. IEEE **83**(3), 345–377 (1995)

15. Zadeh, L.A.: Fuzzy logic. Computer **21**, 83–93 (1988)
16. Norp, T.: 5G requirements and key performance indicators. J. ICT Stand. **6**(1), 15–30 (2018)
17. Parvez, I., Rahmati, A., Guvenc, I., Sarwat, A.I., Dai, H.: A survey on low latency towards 5G: ran, core network and caching solutions. IEEE Commun. Surv. Tutor. **20**(4), 3098–3130 (2018)
18. Kim, Y., Park, J., Kwon, D., Lim, H.: Buffer management of virtualized network slices for quality-of-service satisfaction. In: 2018 IEEE Conference on Network Function Virtualization and Software Defined Networks (NFV-SDN), pp. 1–4 (2018)
19. Barolli, L., Koyama, A., Yamada, T., Yokoyama, S.: An integrated CAC and routing strategy for high-speed large-scale networks using cooperative agents. IPSJ J. **42**(2), 222–233 (2001)

Implementation of a Simulation System for Optimal Number of MOAP Robots Using Elbow and Silhouette Theories in WMNs

Keita Matsuo[1(✉)], Kenshiro Mitsugi[2], Atushi Toyama[2], and Leonard Barolli[1]

[1] Department of Information and Communication Engineering, Fukuoka Institute of Technology (FIT), 3-30-1 Wajiro-Higashi, Higashi-Ku, Fukuoka 811-0295, Japan
{kt-matsuo,barolli}@fit.ac.jp
[2] Graduate School of Engineering, Fukuoka Institute of Technology (FIT), 3-30-1 Wajiro-Higashi, Higashi-Ku, Fukuoka 811-0295, Japan
{mgm20108,mgm20105}@bene.fit.ac.jp

Abstract. Recently, various communication technologies have been developed in order to satisfy the requirements of many users. Especially, mobile communication technology continues to develop rapidly and Wireless Mesh Networks (WMNs) are attracting attention from many researchers in order to provide cost efficient broadband wireless connectivity. The main issue of WMNs is to improve network connectivity and stability in terms of user coverage. In our previous work, we presented Moving Omnidirectional Access Point (MOAP) robot. The MOAP robot should move omnidirectionaly in the real space to provide good communication and stability for WMNs. For this reason, we need to find optimal number of MOAP robots. In this paper, we implemented a simulation system to find optimal number of MOAP robots considering elbow and silhouette theories for WMNs in order to achieve a good communication environment.

1 Introduction

Recently, communication technologies have been developed in order to satisfy the requirements of many users. Especially, mobile communication technologies continue to develop rapidly and has facilitated the use of laptops, tablets and smart phones in public spaces [4]. In addition, Wireless Mesh Networks (WMNs) [1] are becoming on important network infrastructure. These networks are made up of wireless nodes organized in a mesh topology, where mesh routers are interconnected by wireless links and provide Internet connectivity to mesh clients.

WMNs are attracting attention from many researchers in order to provide cost efficient broadband wireless connectivity. The main issue of WMNs is to improve network connectivity and stability in terms of user coverage. This problem is very closely related to the family of node placement problems in

L. Barolli et al. (Eds.): EIDWT 2021, LNDECT 65, pp. 36–47, 2021.
https://doi.org/10.1007/978-3-030-70639-5_4

WMNs [5,8,10]. In these papers is assumed that routers move by themselves or by using network simulator moving models.

In our research, we consider a moving robot as network device. In order to realize a moving access point, we implemented a moving omnidirectional access point robot (called MOAP robot). It is important that the MOAP robot moves to an accurate position in order to have a good connectivity. Thus, the MOAP robot can provide good communication and stability for WMNs. In this work, we implement a simulation system for MOAP robot by considering elbow and silhouette theories to decide the optimal number of MOAP robots.

The rest of this paper is structured as follows. In Sect. 2, we introduce the related work. In Sect. 3, we present our implemented moving omnidirectional access point robot. In Sect. 4, we use elbow and silhouette theory to decide optimal number of MOAP robots. In Sect. 5, we describe the implemented simulation system. Finally, conclusions and future work are given in Sect. 6.

2   Related Work

Different techniques are developed to solve the problem of moving robots position. One of important research area is indoor position detection, because the outdoor position can be detected easily by using GPS (Global Positioning System). However, in the case of indoor environment, we can not use GPS. So, it is difficult to find the target position.

Asahara et al. [2] proposed to improve the accuracy of the self position estimation of a mobile robot. A robot measures a distance to an object in the mobile environment by using a range sensor. Then, the self position estimation unit estimates a self position of the mobile robot based on the selected map data and range data obtained by the range sensor. Wang et al. [11] proposed the ROS (Robot Operating System) platform. They designed a WiFi indoor initialize positioning system by triangulation algorithm. The test results show that the WiFi indoor initialize position system combined with AMCL (Adaptive Monte Carlo Localization) algorithm can be accurately positioned and has high commercial value. Nguyen et al. [9] proposed a low speed vehicle localization using WiFi fingerprinting. In general, these researches rely on GPS in fusion with other sensors to track vehicle in outdoor environment. However, as indoor environment such as car park is also an important scenario for vehicle navigation, the lack of GPS poses a serious problem. They used an ensemble classification method together with a motion model in order to deal with the issue. Experiments show that proposed method is capable of imitating GPS behavior on vehicle tracking. Ban et al. [3] proposed indoor positioning method integrating pedestrian Dead Reckoning with magnetic field and WiFi fingerprints. Their proposed method needs WiFi and magnetic field fingerprints, which are created by measuring in advance the WiFi radio waves and the magnetic field in the target map. The proposed method estimates positions by comparing the pedestrian sensor and fingerprint values using particle filters. Matsuo et al. [6,7] implemented and evaluated a small size omnidirectional wheelchair.

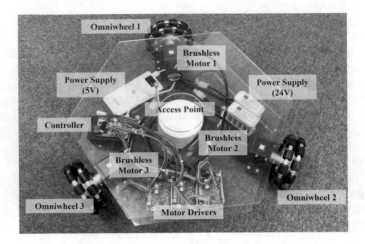

Fig. 1. Implemented MOAP robot.

3   Moving Omnidirectional Access Point Robot

In this section, we describe the implemented MOAP (Moving Omnidirection Access Point) robot. We show the implemented MOAP robot in Fig. 1. The MOAP robot can move omnidirectionaly keeping the same direction and can provide access points for network devices. In order to realize our proposed MOAP robot, we used omniwheels which can rotate omnidirectionaly in front, back, left and right. The movement of the MOAP robot is shown in Fig. 2. We would like to control the MOAP robot to move accurately in order to offer a good environment for communication.

3.1   Overview of MOAP Robot

Our implemented MOAP robot has 3 omniwheels, 3 brushless motors, 3 motor drivers and a controller. The MOAP robot requires 24 V battery to move and 5 V battery for the controller. We show the specification of MOAP robot in Table 1.

3.2   Control System

We designed the control system for operation of MOAP robot, which is shown in Fig. 3. We are using brushless motors as main motor to move the robot, because the motor can be controlled by PWM (Pulse Width Modulation). We used Rasberry Pi as a controller. However, the controller has only 2 PWM hardware generators. But, we need to use 3 generators, so we decided to use the software generator to get a square wave for the PWM. As software generator, we use the Pigpio which can generate better signal than other software generators and make PWM signals with 32 lines. Figure 4 shows the square signal generated by Pigpio.

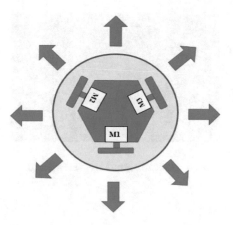

Fig. 2. Movement of our implemented MOAP robot.

Table 1. Specification of MOAP robot.

Item	Specification
Length	490.0 [mm]
Width	530.0 [mm]
Height	125.0 [mm]
Brushless motor	BLHM015K-50 (orientalmotor corporation)
Motor driver	BLH2D15-KD (orientalmotor corporation)
Controller	Raspberry Pi 3 Model B+
Power supply	DC24 V battery
PWM driver	Pigpio (the driver can generate PWM signal with 32 line)

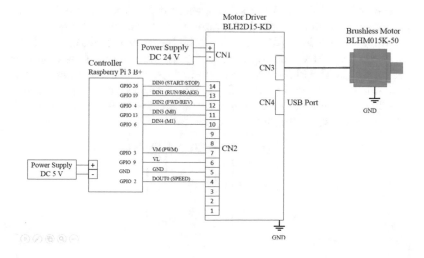

Fig. 3. Control system for MOAP robot.

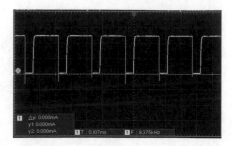

Fig. 4. Square signal by using Pigpio.

3.3 Kinematics

For the control of the MOAP robot are needed the robot's rotation degrees, movement speed and direction.

Let us consider the movement of the robot in 2 dimensional space. In Fig. 5, we show the kinematics of MOAP robot. In this figure, there are 3 onmiwheels which are placed 120° with each other. The omniwheels can move in clockwise and counter clockwise directions, we decided clockwise is positive rotation as shown in the figure. We consider the speed for each omniwheel M1, M2 and M3, respectively.

As shown in Fig. 5, the axis of the MOAP robot are x, y and the speed is $v = (\dot{x}, \dot{y})$ and the rotating speed is $\dot{\theta}$. In this case, the moving speed of the MOAP robot can be expressed by Eq. (1).

$$V = (\dot{x}, \dot{y}, \dot{\theta}) \tag{1}$$

Based on the Eq. (1), the speed of each omniwheel can be decided. By considering the control value of the motor speed ratio of each omniwheel as linear and synthesising the vector speed of 3 omniwheels, we can get Eq. (2) by using Reverse Kinematics, where (d) is the distance between the center and the omniwheels. Then, from the rotating speed of each omniwheel based on Forward Kinematics, we get the MOAP robot's moving speed. If we calculate the inverse matrix of Eq. (2), we get Eq. (3). Thus, when the MOAP robot moves in all directions (omnidirectional movement), the speed for each motor (theoretically) is calculated as shown in Table 2.

$$\begin{vmatrix} M_1 \\ M_2 \\ M_3 \end{vmatrix} = \begin{vmatrix} 1 & 0 & d \\ -\frac{1}{2} & -\frac{\sqrt{3}}{2} & d \\ -\frac{1}{2} & \frac{\sqrt{3}}{2} & d \end{vmatrix} \begin{vmatrix} \dot{x} \\ \dot{y} \\ \dot{\theta} \end{vmatrix} \tag{2}$$

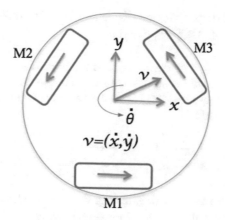

Fig. 5. The kinematics of MOAP robot.

Table 2. Motor speed ratio.

Direction	Motor speed ratio		
(degrees)	Motor1	Motor2	Motor3
0	0.00	−0.87	0.87
30	0.50	−1.00	0.50
60	0.87	−0.87	0.00
90	1.00	−0.50	−0.50
120	0.87	0.00	−0.87
150	0.50	0.50	−1.00
180	0.00	0.87	−0.87
210	−0.50	1.00	−0.50
240	−0.87	0.87	0.00
270	−1.00	0.50	0.50
300	−0.87	0.00	0.87
330	−0.50	−0.50	1.00
360	0.00	−0.87	0.87

$$
\begin{vmatrix} \dot{x} \\ \dot{y} \\ \dot{\theta} \end{vmatrix} = \begin{vmatrix} \frac{2}{3} & -\frac{1}{3} & -\frac{1}{3} \\ 0 & -\frac{1}{\sqrt{3}} & \frac{1}{\sqrt{3}} \\ \frac{1}{3d} & \frac{1}{3d} & \frac{1}{3d} \end{vmatrix} \begin{vmatrix} M_1 \\ M_2 \\ M_3 \end{vmatrix} \tag{3}
$$

4   Elbow and Silhouette Theory to Decide Optimal Number of MOAP Robots

Elbow and Silhouette theory considers K-means clustering. We show K-means function in Eq. (4). In this case, C_i means i th cluster and x_{ij} is j th of i th data. K is the number of clusters. Ideal clustering is achieved when the value of Eq. (4) will be minimized.

$$\underset{C_1,\dots,CK}{min}\{\sum_{k=1}^{K}\frac{1}{|C_k|}\sum_{i,i'\in C_k}\sum_{j=1}^{p}(x_{ij}-x_{i'j})^2\} \tag{4}$$

We show the K-means clustering in Fig. 6, the dots show clients. In Fig. 6(a), we deployed 150 clients in random way on 2D space (100 m × 100 m). After that, we used K-means clustering as shown in Fig. 6(b). We consider that the centroids can communicate with each-other for this scenario. In order to decide the number of optimal clusters, we used Elbow and Silhouette theory.

4.1   Elbow Theory

Elbow theory uses the distance between centroid and clients (see Fig. 6(b)). In Eq. (5), *All_Distance* means total distance between each centroid and clients in the cluster. If there is only one cluster, that *All_Distance* value is maximum. When the cluster number increases the *All_Distance* value will be decreased. The relation between *All_Distance* and the number of clusters is shown in Fig. 7. From this figure, we can see that the optimal number of clusters is 3. This is the elbow value.

$$All\_Distance = \sum_{k=1}^{K}\sum_{i,i'\in C_k}\sum_{j=1}^{p}(x_{ij}-x_{i'j})^2 \tag{5}$$

4.2   Silhouette Theory

Silhouette theory use the value of Silhouette coefficient. We show the equation of Silhouette coefficient in Eq. (6), where $a^{(i)}$ is the average distance between i th client and other clients in the same cluster and $b^{(i)}$ is the average distance between i th client and other clients in the nearest cluster. The, $s^{(i)}$ shows the degree of success or failure of clustering. The value range for $s^{(i)}$ is -1 to 1. If $s^{(i)}$ value is near to 1 the clients in the same cluster are very close to each other. When $s^{(i)}$ value is 0, the clients are located on the border between clusters. Also, when the value of $s^{(i)}$ is negative, the client does not belong to an appropriate cluster.

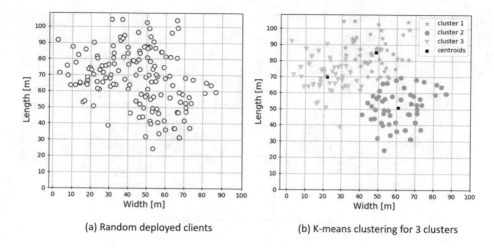

(a) Random deployed clients   (b) K-means clustering for 3 clusters

Fig. 6. Simulation with 3 clusters.

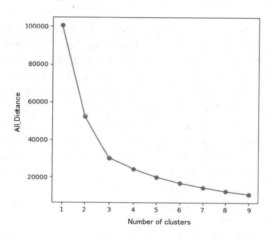

Fig. 7. Relation between All_Distance and number of clusters (for 3 clusters).

We show the K-means clustering and Silhouette coefficient with 8 clusters in Figs. 8 and 9. In Fig. 8(a), we deployed 800 clients on the 2D space (150 m × 120 m) randomly. After that we clustered the clients by 8 clusters using K-means clustering in order to analyze the clusters using the Silhouette theory (see Fig. 8(b)).

In Fig. 9 are shown silhouette coefficients of 8 clusters, where the vertical line means average value. The thickness of each clusters shows the number of clients in the cluster. The silhouette coefficient is high when the thickness are almost the same.

$$s^{(i)} = \frac{b^{(i)} - a^{(i)}}{max\{a^{(i)}, b^{(i)}\}} \tag{6}$$

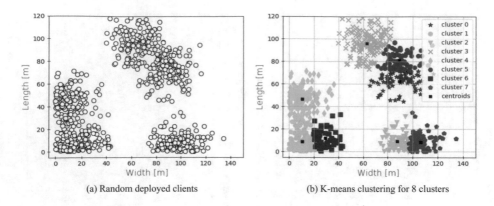

(a) Random deployed clients   (b) K-means clustering for 8 clusters

Fig. 8. Simulation with 8 clusters.

5 Implemented Simulation System

We implemented a simulation system considering elbow and silhouette theories to deploy the MOAP robots. The simulation system is shown in Fig. 10. The left side of the figure shows the operation screen, while the right side shows the zoomed part of the figure. The upper part shows the input parameters of the simulation system, while lower part shows the results. Moreover, the system can show some results in graphical form (see Fig. 11).

In input parameters part, there are some parameters such as the number of MOAP robots, number of clients, TX power, minimum RSSI, frequency, and so on. In results part are shown the communication radius for each Access Point, number of client in each cluster, cover ratio of clients for each cluster, cover ratio of all clients and so on.

In order to get the elbow value more clearly, we calculate the elbow value by using difference of All_Distance values for each point. Figure 12 shows the difference of All_Distance values ($\Delta 1$ to $\Delta 6$) for each clusters. For example, when the cluster number is 2, the elbow value can be calculated by $\Delta 1/\Delta 2$. Also, we show elbow value for 2 to 6 number of clusters in Table 3. So, the elbow point is 4, because in this case the elbow value is maximum. The simulation system shows the elbow point and silhouette coefficient values, which are used to decided the optimal number of MOAP robots.

In many cases, when the elbow point is very clear, we can use elbow theory to decided the optimal number of clusters. But, in the case when it is difficult to decided the elbow point, the system uses the silhouette coefficient.

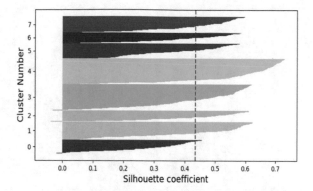

Fig. 9. Silhouette coefficient with 8 clusters.

Table 3. Elbow values for 2 to 6 number of clusters.

Number of clusters	Elbow value	Equation
2	2.056	$\Delta1/\Delta2$
3	1.560	$\Delta2/\Delta3$
4	3.264	$\Delta3/\Delta4$
5	1.729	$\Delta4/\Delta5$
6	1.480	$\Delta5/\Delta6$

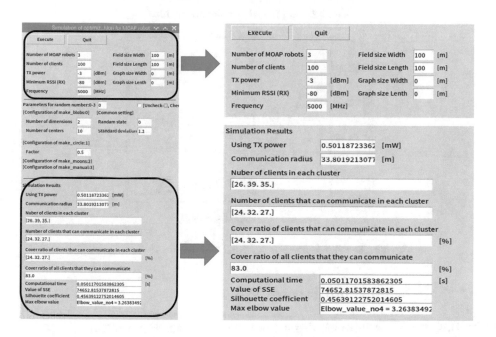

Fig. 10. Operation screen of implemented simulation system.

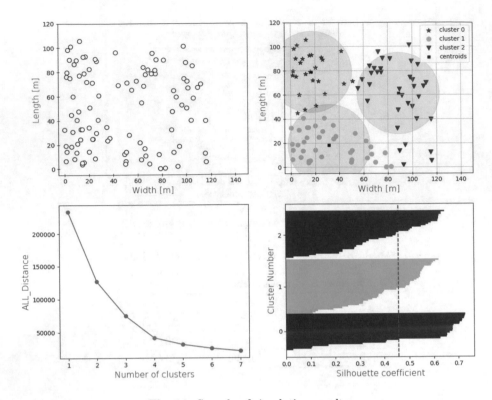

Fig. 11. Sample of simulation result.

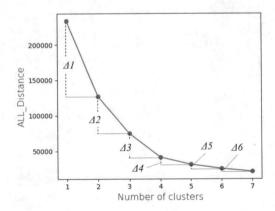

Fig. 12. The different value of All_Distance for each clusters.

6 Conclusions and Future Work

In this paper, we introduced our implemented simulation system for optimal number of MOAP robots. We showed some of the previous works and discussed the related problems and issues. Then, we presented in details the kinematics and the control methodology for MOAP robot. In addition, we proposed elbow and silhouette theories to determine the number of MOAP for good communication environment in WMNs. The implemented simulation system could find good points to deploy the number of MOAP robots. The simulation results show that the silhouette theory and elbow theory can help to decide the optimal number of MOAP robots.

In the future work, we would like to add other functions to the simulation system and carry out extensive simulations to evaluate the proposed system.

References

1. Akyildiz, I.F., Wang, X., Wang, W.: Wireless mesh networks: a survey. Comput. Netw. **47**(4), 445–487 (2005)
2. Asahara, Y., Mima, K., Yabushita, H.: Autonomous mobile robot, self position estimation method, environmental map generation method, environmental map generation apparatus, and data structure for environmental map, US Patent 9,239,580, 19 January 2016
3. Ban, R., Kaji, K., Hiroi, K., Kawaguchi, N.: Indoor positioning method integrating pedestrian dead reckoning with magnetic field and WiFi fingerprints. In: 2015 Eighth International Conference on Mobile Computing and Ubiquitous Networking (ICMU), pp. 167–172, January 2015
4. Hamamoto, R., Takano, C., Obata, H., Ishida, K., Murase, T.: An access point selection mechanism based on cooperation of access points and users movement. In: 2015 IFIP/IEEE International Symposium on Integrated Network Management (IM), pp. 926–929, May 2015
5. Maolin, T.: Gateways placement in backbone wireless mesh networks. Int. J. Commun. Netw. Syst. Sci. **2**(01), 44–50 (2009)
6. Matsuo, K., Barolli, L.: Design and implementation of an omnidirectional wheelchair: control system and its applications. In: Proceedings of the 9th International Conference on Broadband and Wireless Computing, Communication and Applications (BWCCA 2014), pp. 532–535 (2014)
7. Matsuo, K., Liu, Y., Elmazi, D., Barolli, L., Uchida, K.: Implementation and evaluation of a small size omnidirectional wheelchair. In: Proceedings of the IEEE 29th International Conference on Advanced Information Networking and Applications Workshops (WAINA 2015), pp. 49–53 (2015)
8. Muthaiah, S.N., Rosenberg, C.: Single gateway placement in wireless mesh networks. In: Proceedings of ISCN, vol. 8, pp. 4754–4759 (2008)
9. Nguyen, D., Recalde, M.E.V., Nashashibi, F.: Low speed vehicle localization using WiFi fingerprinting. In: 2016 14th International Conference on Control, Automation, Robotics and Vision (ICARCV), pp. 1–5, November 2016
10. Oda, T., Barolli, A., Spaho, E., Xhafa, F., Barolli, L., Takizawa, M.: Performance evaluation of WMN using WMN-GA system for different mutation operators. In: 2011 14th International Conference on Network-Based Information Systems, pp. 400–406, September 2011
11. Wang, T., Zhao, L., Jia, Y., Wang, J.: WiFi initial position estimate methods for autonomous robots. In: 2018 WRC Symposium on Advanced Robotics and Automation (WRC SARA). pp. 165–171, August 2018

A Fuzzy-Based Approach for Reducing Transmitted Data Considering Data Difference Parameter in Resilient WSNs

Daisuke Nishii[1], Makoto Ikeda[2(✉)], and Leonard Barolli[2]

[1] Graduate School of Engineering, Fukuoka Institute of Technology,
3-30-1 Wajiro-higashi, Higashi-ku, Fukuoka 811-0295, Japan
mgm20106@bene.fit.ac.jp
[2] Department of Information and Communication Engineering,
Fukuoka Institute of Technology, 3-30-1 Wajiro-higashi,
Higashi-ku, Fukuoka 811-0295, Japan
makoto.ikd@acm.org, barolli@fit.ac.jp

Abstract. Resilient Wireless Sensor Networks (WSNs) can collect data for long-term operation even if the network condition is unstable due to the disaster situation. In this paper, we propose an intelligent transmission control system based on fuzzy logic in resilient WSNs. From the evaluation results, we found that our proposed system can reduce the transaction and control the transmission interval for various conditions.

Keywords: Resilient Wireless Sensor Network · Fuzzy logic · WSNs

1 Introduction

Recently in Japan, some location-based statistical information services on mobile phones have attracted attention. From user's location data of mobile phone, we are able to see the density of users per area on the website. These sensed data can be linked with the information of the mobile phone carrier to see the approximate number of each generation and gender in each specific area.

Different sensors are widely scattered around our daily life, and mobile devices are one of the most important sensors. In several fields of science, smart solutions based on sensors are providing us with the possibility to make decisions that are even better than human decisions [4,7,11,13,15].

Resilient Wireless Sensor Network (WSN) is one of the key factors of the Sustainable Development Goals (SDGs) [2,5,14,17]. In [12], the authors present a restoration scheme of fuzzy-based distribution system for emergency situation. In [10], we proposed a fuzzy-based approach for transmission control of sensed data for disaster situation. We considered three input parameters.

In this paper, we propose a fuzzy-based transmission control system in resilient WSNs. We consider not only real-time sensed data but also observed data released by Japan Meteorological Agency (JMA). For evaluation, we use

© The Author(s), under exclusive license to Springer Nature Switzerland AG 2021
L. Barolli et al. (Eds.): EIDWT 2021, LNDECT 65, pp. 48–57, 2021.
https://doi.org/10.1007/978-3-030-70639-5_5

distance to receiver, data priority, data difference and data change amount as input parameters.

The structure of the paper is as follows. In Sect. 2, we describe resilient WSNs. In Sect. 3, we present the design of simulation system for resilient WSNs. In Sect. 4, we provide the evaluation results. Finally, conclusions and future work are given in Sect. 5.

2   Resilient WSNs

In this section, we present the overview of WSNs and distributed edge devices.

2.1   Overview of Resilient WSNs

WSNs are able to monitor physical phenomena, processing the sensed data, making decisions based on the sensed data, and completing the appropriate tasks when required. To reduce the transactions and resources, sensor node does not send every bit of sensed data within the limits of available resources [1]. When important data has to be sent, the sensor nodes can send that data back to the sink, which controls the task from a distance, or they may send the data to a cluster head that can conduct the operation independently of the sink node.

2.2   Distributed Edge Devices

In this work, we explain an integrated mobile device that serves as an edge. The edge is composed of a sensing element and a communication element. We assume that many edge devices are utilized in the measurement. The sensing element is assumed to be implemented in Raspberry Pi. We assume a sensor to measure parameters as Distance to Receiver (DR), Data Priority (DP), Data Difference (DD) and Data Change Amount (DCA). All sensing data are measured by network tool equipped with the wireless network interface.

We implemented a fuzzy logic simulation system written in Python. The edge controls the transmission interval to cluster head or sink. In this way, our intelligent system acts to five different levels of responsibility to improve the transactions and resources within the database.

3   Proposed Fuzzy-Based Transmission Control System

Fuzzy sets and fuzzy logic [18] have been developed to deal with vagueness and uncertainty in a inference process of a intelligent system such as knowledge-based system, logical control system and so on [3,6,8,16].

The proposed system consists of a Fuzzy Logic Controller (FLC). The FLC basic elements are the fuzzifier, inference engine, fuzzy rule base and defuzzifier. We use triangular and trapezoidal membership functions for FLC, because they are suitable for real-time operation [9]. We use DR, DP, DD, DCA input parameters for FLC. The fuzzy membership functions to conduct in resilient WSNs are

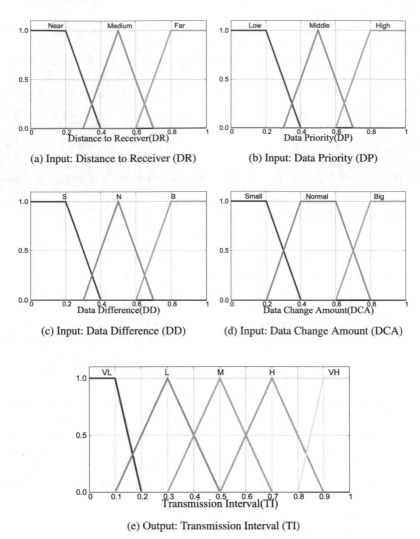

(a) Input: Distance to Receiver (DR)

(b) Input: Data Priority (DP)

(c) Input: Data Difference (DD)

(d) Input: Data Change Amount (DCA)

(e) Output: Transmission Interval (TI)

Fig. 1. Fuzzy membership function.

shown in Fig. 1. The output linguistic parameter is the Transmission Interval (TI). The Fuzzy Rule Base (FRB) is shown in Table 1.

Distance to Receiver (DR): The edge has a network interface controller, which connects to a wireless radio-based computer network. We use the Received Signal Strength Indicator (RSSI) or GPS module to measure the DR. We consider three levels of DR for different distances.

Table 1. Fuzzy rule base.

Rule	DR	DP	DD	DCA	TI	Rule	DR	DP	DD	DCA	TI
1	Near	High	B	Small	H	42	Medium	Middle	N	Big	M
2	Near	High	B	Normal	H	43	Medium	Middle	S	Small	L
3	Near	High	B	Big	VH	44	Medium	Middle	S	Normal	L
4	Near	High	N	Small	M	45	Medium	Middle	S	Big	M
5	Near	High	N	Normal	H	46	Medium	Low	B	Small	L
6	Near	High	N	Big	VH	47	Medium	Low	B	Normal	M
7	Near	High	S	Small	M	48	Medium	Low	B	Big	H
8	Near	High	S	Normal	M	49	Medium	Low	N	Small	VL
9	Near	High	S	Big	H	50	Medium	Low	N	Normal	L
10	Near	Middle	B	Small	H	51	Medium	Low	N	Big	M
11	Near	Middle	B	Normal	H	52	Medium	Low	S	Small	VL
12	Near	Middle	B	Big	VH	53	Medium	Low	S	Normal	VL
13	Near	Middle	N	Small	M	54	Medium	Low	S	Big	L
14	Near	Middle	N	Normal	M	55	Far	High	B	Small	M
15	Near	Middle	N	Big	H	56	Far	High	B	Normal	M
16	Near	Middle	S	Small	L	57	Far	High	B	Big	H
17	Near	Middle	S	Normal	M	58	Far	High	N	Small	L
18	Near	Middle	S	Big	H	59	Far	High	N	Normal	L
19	Near	Low	B	Small	M	60	Far	High	N	Big	M
20	Near	Low	B	Normal	M	61	Far	High	S	Small	L
21	Near	Low	B	Big	H	62	Far	High	S	Normal	L
22	Near	Low	N	Small	L	63	Far	High	S	Big	M
23	Near	Low	N	Normal	L	64	Far	Middle	B	Small	L
24	Near	Low	N	Big	M	65	Far	Middle	B	Normal	M
25	Near	Low	S	Small	L	66	Far	Middle	B	Big	H
26	Near	Low	S	Normal	L	67	Far	Middle	N	Small	VL
27	Near	Low	S	Big	M	68	Far	Middle	N	Normal	L
28	Medium	High	B	Small	M	69	Far	Middle	N	Big	M
29	Medium	High	B	Normal	H	70	Far	Middle	S	Small	VL
30	Medium	High	B	Big	VH	71	Far	Middle	S	Normal	VL
31	Medium	High	N	Small	M	72	Far	Middle	S	Big	L
32	Medium	High	N	Normal	M	73	Far	Low	B	Small	L
33	Medium	High	N	Big	H	74	Far	Low	B	Normal	L
34	Medium	High	S	Small	L	75	Far	Low	B	Big	M
35	Medium	High	S	Normal	M	76	Far	Low	N	Small	VL
36	Medium	High	S	Big	H	77	Far	Low	N	Normal	VL
37	Medium	Middle	B	Small	M	78	Far	Low	N	Big	L
38	Medium	Middle	B	Normal	M	79	Far	Low	S	Small	VL
39	Medium	Middle	B	Big	H	80	Far	Low	S	Normal	VL
40	Medium	Middle	N	Small	L	81	Far	Low	S	Big	L
41	Medium	Middle	N	Normal	L						

Data Priority (DP): We consider the DP to support various situations. A sink or cluster head can ask for different priorities of data depending on the situation. For example, the cluster may increase the priority of the required data.

Data Difference (DD): The DD indicates the difference between our sensed data and the observed data released by JMA. The difference is handled as a unit of 0 to 1. For example, if there is no difference, the system calculates 0. The DD has three different levels, which can be interpreted as: Small (S), Normal (N) and Big (B).

Data Change Amount (DCA): The DCA is used to consider time sequence data. We consider three levels of DCA. For example, sometimes there is almost no change from the previous data at time of updating. The proposed system will help to increase the transmission interval at this time.

Transmission Interval (TI): Our system can control the data transmission interval to selected sink or cluster head depending on the situation. The transmission interval has five different levels, which can be interpreted as: Very Low (VL), Low (L), Middle (M), High (H), Very High (VH).

4   Evaluation Results

We present the simulation results of our fuzzy-based transmission control system for resilient WSNs from Fig. 2 to Fig. 4. The S-DD in the figures indicates that DD is small, N-DD indicates that DD is normal and B-DD indicates that DD is big.

We show in Fig. 2 the simulation results when DR is near. The output TI increases with increase of DCA regardless the data priority. After 0.6 unit of DCA, the output TI increases one unit higher. Then, when DP becomes more than Middle, TI is mostly middle or higher. On the other hand, N-DD and S-DD have the same result when DP is low as shown in Fig. 2(a). This phenomenon can also be seen in Fig. 3(b), Fig. 4(a) and Fig. 4(c).

The simulation results of TI when DR is medium are shown in Fig. 3. Compared with results when DR is near, the output results change widely according to the DP. When the DR is Far (see Fig. 4), the SI results for S-DD and N-DD are less than or equal to M.

These results show that the transmission interval is shorter for each DP when DR is near. We discuss four cases resulted in VH in the FRB of Table 1. These cases follow the rule numbers 3, 6, 12 and 30. We can see that when the DCA becomes more than 0.8, the output TI is 0.9. Thus, the transmission interval in our proposed system is set to VH at this case.

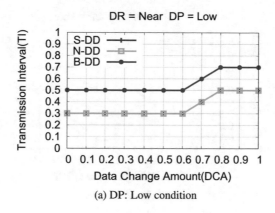

(a) DP: Low condition

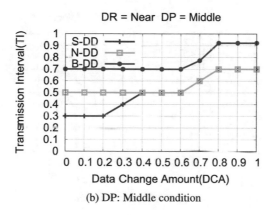

(b) DP: Middle condition

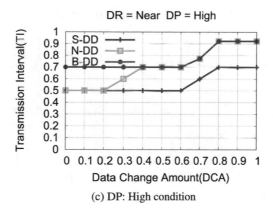

(c) DP: High condition

Fig. 2. Simulation results when DR is Near.

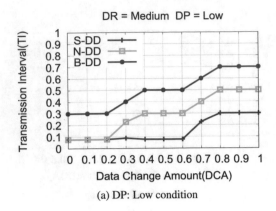

(a) DP: Low condition

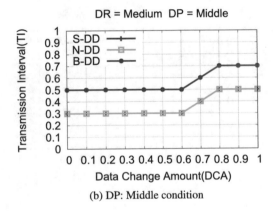

(b) DP: Middle condition

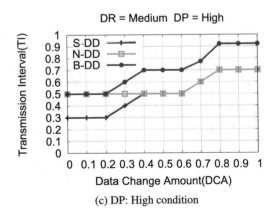

(c) DP: High condition

Fig. 3. Simulation results when DR is Medium.

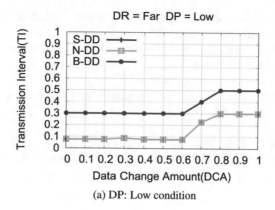

(a) DP: Low condition

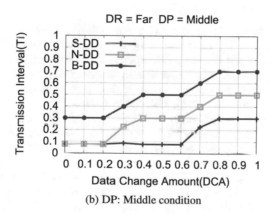

(b) DP: Middle condition

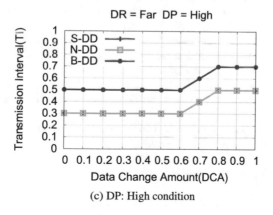

(c) DP: High condition

Fig. 4. Simulation results when DR is Far.

5   Conclusions

In this paper, we evaluated the performance of the proposed fuzzy-based transmission control system for resilient WSNs. We considered four input parameters to select the level of transmission interval. From the evaluation results, we found that our proposed simulation system can reduce the transaction and control the transmission interval for various conditions.

In the future work, we would like to implement the testbed to evaluate the proposed system in real life.

Acknowledgments. This work has been partially funded by the research project from Comprehensive Research Organization at Fukuoka Institute of Technology (FIT), Japan.

References

1. Aazam, M., Huh, E.N.: Fog computing and smart gateway based communication for cloud of things. In: Proceedings of the International Conference on Future Internet of Things and Cloud (FiCloud-2014), pp. 464–470, August 2014
2. Al-Fuqaha, A., Guizani, M., Mohammadi, M., Aledhari, M., Ayyash, M.: Internet of Things: a survey on enabling technologies, protocols, and applications. IEEE Commun. Surv. Tutorials **17**(4), 2347–2376 (2015)
3. Balan, K., Manuel, M.P., Faied, M., Krishnan, M., Santora, M.: A fuzzy based accessibility model for disaster environment. In: Proceedings of the IEEE International Conference on Robotics and Automation (ICRA-2019), pp. 2304–2310, May 2019
4. Chimatapu, R., Hagras, H., Kern, M., Owusu, G.: Hybrid deep learning type-2 fuzzy logic systems for explainable AI. In: Proceedings of the IEEE International Conference on Fuzzy Systems (FUZZ-IEEE-2020), pp. 1–6, July 2020
5. Guo, Z., Li, G., Zhou, M., Feng, W.: Resilient configuration approach of integrated community energy system considering integrated demand response under uncertainty. IEEE Access **7**, 87513–87533 (2019)
6. Gupta, I., Riordan, D., Sampalli, S.: Cluster-head election using fuzzy logic for wireless sensor networks. In: Proceedings of the 3rd Annual Communication Networks and Services Research Conference (CNSR-2005), pp. 255–260 (2005)
7. Jammeh, E.A., Fleury, M., Wagner, C., Hagras, H., Ghanbari, M.: Interval type-2 fuzzy logic congestion control for video streaming across IP networks. IEEE Trans. Fuzzy Syst. **17**(5), 1123–1142 (2009)
8. Li, T.S., Chang, S.J., Tong, W.: Fuzzy target tracking control of autonomous mobile robots by using infrared sensors. IEEE Trans. Fuzzy Syst. **12**(4), 491–501 (2004)
9. Mendel, J.M.: Fuzzy logic systems for engineering: a tutorial. Proc. IEEE **83**(3), 345–377 (1995)
10. Nishii, D., Ikeda, M., Barolli, L.: A fuzzy-based approach for transmission control of sensory data in resilient wireless sensor networks during disaster situation. In: Proceedings of the 15th International Conference on Broadband and Wireless Computing, Communication and Applications (BWCCA-2020), pp. 296–303, October 2020

11. Petrakis, E.G.M., Sotiriadis, S., Soultanopoulos, T., Renta, P.T., Buyya, R., Bessis, N.: Internet of Things as a service (iTaaS): challenges and solutions for management of sensor data on the cloud and the fog. Internet Things **3–4**, 156–174 (2018)
12. Reddy, G.H., Chakrapani, P., Goswami, A.K., Choudhury, N.B.D.: Fuzzy based approach for restoration of distribution system during post natural disasters. IEEE Access **6**, 3448–3458 (2018)
13. Ruan, J., Jiang, H., Li, X., Shi, Y., Chan, F.T.S., Rao, W.: A granular GA-SVM predictor for big data in agricultural cyber-physical systems. IEEE Trans. Industr. Inf. **15**(12), 6510–6521 (2019)
14. Schmitt, S., Will, H., Aschenbrenner, B., Hillebrandt, T., Kyas, M.: A reference system for indoor localization testbeds. In: Proceedings of the International Conference on Indoor Positioning and Indoor Navigation (IPIN-2012), Sydney, Australia, pp. 1–8, November 2012
15. Silver, D., Schrittwieser, J., Simonyan, K., Antonoglou, I., Huang, A., Guez, A., Hubert, T., Baker, L., Lai, M., Bolton, A., Chen, Y., Lillicrap, T., Hui, F., Sifre, L., van den Driessche, G., Graepel, T., Hassabis, D.: Mastering the game of go without human knowledge. Nature **550**, 354–359 (2017)
16. Su, X., Wu, L., Shi, P.: Sensor networks with random link failures: distributed filtering for T-S fuzzy systems. IEEE Trans. Industr. Inf. **9**(3), 1739–1750 (2013)
17. Sung, J.Y., Guo, L., Grinter, R.E., Christensen, H.I.: My Roomba is Rambo: intimate home appliances. In: Proceedings of the 9th International Conference on Ubiquitous Computing (UbiComp-2007), Seoul, South Korea, pp. 145–162, September 2007
18. Zadeh, L.: Fuzzy logic, neural networks, and soft computing. ACM Commun. **37**, 77–84 (1994)

An Energy-Efficient Migration Algorithm for Virtual Machines to Reduce the Number of Migrations

Naomichi Noaki[1][✉], Takumi Saito[1], Dilawaer Duolikun[1], Tomoya Enokido[2], and Makoto Takizawa[1]

[1] Hosei University, Tokyo, Japan
{naomichi.noaki.2k,takumi.saito.3j}@stu.hosei.ac.jp,
dilewerdolikun@gmail.com, makoto.takizawa@computer.org
[2] Rissho University, Tokyo, Japan
eno@ris.ac.jp

Abstract. It is critical to reduce electric energy consumed in information systems, especially servers in clusters to realize green society. By making a virtual machine migrate from a host server to a guest server, the energy consumption of the host server can be reduced since application processes on the virtual machine leave the host server. On the other hand, the guest server consumes more energy to additionally perform the application processes. In the MDMG (Maximum Energy Consumption Difference by Virtual Machine Migration) algorithm previously proposed, virtual machines frequently migrate among servers while the total energy consumption of servers can be reduced. The more frequently virtual machines migrate, the longer time is spent to make the virtual machine migrate. In this paper, we propose an RWM (Reducing additionally Wasted Migrations) algorithm to more reduce energy consumption of servers and the execution time of application processes by reducing the number of migrations. In the evaluation, we show the energy consumption of servers and the average execution time of application processes can be reduced in the RWM algorithm compared with the MDMG algorithm.

Keywords: Live migration of virtual machines · Energy-efficient virtual machine migration · Power consumption model · RWM algorithm

1 Introduction

It is critical to reduce electric energy consumed by information systems to realize green societies by decreasing carbon dioxide emission on the earth. Information systems are composed of clients and clusters of servers. An application on a client issues a request to a cluster of servers. Here, one host sever is selected where an application process to handle the request is performed. Since servers consume more energy than clients, we have to reduce the energy consumption of servers.

L. Barolli et al. (Eds.): EIDWT 2021, LNDECT 65, pp. 58–70, 2021.
https://doi.org/10.1007/978-3-030-70639-5_6

The SPC (Simple Power Consumption) [9] and MLPC (Multi-Level Power Consumption) [11,12] models are proposed to show the total electric power [W] to be consumed by a whole server to perform application processes. Here, the models are macro-level ones where power consumption of each hardware component is not considered. By using the models, the energy consumption of a server and the execution time of application processes can be estimated. In order to reduce the energy consumption of servers in a cluster, a host server, which is expected to consume less energy, is first selected to perform an application process [5,7,8,10–12]. Then, the process gets active, i.e. starts on the server. After that, the server might consume more electric energy than estimated, e.g. due to congestion. In one approach, active processes on a server migrate to another guest server if the energy consumption can be reduced. Virtual machine technologies are widely used to virtualize computation resources like CPUs in a cluster [3]. Especially, in the live migration, a virtual machine can migrate from a host server to a guest server without terminating active processes on the virtual machine. It takes longer time to perform a process on a virtual machine if the virtual machine migrates. The migration time of a virtual machine is measured and the MDMG (Maximum energy consumption Difference by virtual machine MiGration) algorithm is proposed in our previous paper [13]. Here, the total energy consumption of servers and the average execution time of processes are reduced compared with non-migration algorithms. However, virtual machines frequently migrate among servers. While a virtual machine migrates from a host server to a guest server, processes on the virtual machine are suspended. According to our measurement [13], the migration time is 0.405 [s].

In this paper, we newly propose an RWM (Reducing the number of Wasted Migrations) algorithm to select a host server to perform a new process and make virtual machines migrate to more reduce the energy consumption of servers by decreasing the number of migrations. Here, the number of active processes on each server is tried to be kept proportional to the computation rate of the server. In the evaluation, we show the total energy consumption of servers and the average execution time of application processes can be reduced in the RWM algorithm.

In Sect. 2, we present the system model. In Sect. 3, we discuss energy consumption of servers to make virtual machine migrate. In Sect. 4, we propose the RWM algorithm. In Sect. 5, we evaluate the RWM algorithm.

2   System Model

A cluster S is composed of servers $s_1, \ldots, s_m$ ($m \geq 1$) which support a set V of virtual machines. Each server s_t supports applications with a set V_t of virtual machines. A server s_t supporting a virtual machine vm_k is a *host* server of vm_k. Let $h(vm_k)$ stand for a host server which hosts a virtual machine vm_k. A client issues a request to a cluster S [Fig. 1]. A virtual machine vm_k on a server s_t is selected and the request is sent to the virtual machine vm_k. An application process p_i to handle the request is created and performed on the virtual machine vm_k. In this paper, a term *process* means an application process. An *active* process is a process being performed on a virtual machine. Let PV_k indicate a set of active processes on a virtual machine vm_k. Let nv_k show the number of $|PV_k|$ active processes on a virtual machine vm_k. A virtual machine vm_k is *larger* than a virtual machine vm_h ($vm_k > vm_h$) iff $nv_k > nv_h$.

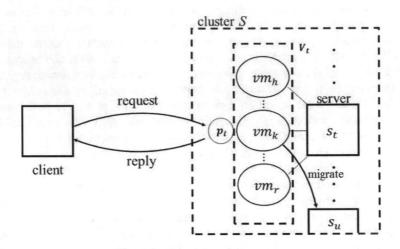

Fig. 1. Virtual machines in a cluster.

A virtual machine can migrate from a host server to another guest server in the live manner [3], where processes on the virtual machine do not terminate while suspended. In our experiment [13], the migration time of a virtual machine between a pair of homogeneous PCs interconnected in a 10 Gbps LAN is 0.405 [s].

A server s_t is composed of np_t (≥ 1) homogeneous CPUs. Each CPU [1] includes nc_t (≥ 1) homogeneous cores. Each core supports the same number ct_t of threads. A server s_t supports processes with totally nt_t ($= np_t \cdot nc_t \cdot ct_t$) threads. Each process is at a time performed on one thread. $CP_t(\tau)$ is a set of active processes on a server s_t at time τ. Here, the electric power $NE_t(n)$ [W] consumed by a server s_t to concurrently perform n (≥ 0) processes is given as follows [10,12]:

[Power consumption for n processes]

$$NE_t(n) = \begin{cases} minE_t \text{ if } n = 0. \\ minE_t + n \cdot (bE_t + cE_t + tE_t) \text{ if } 1 \leq n \leq np_t. \\ minE_t + np_t \cdot bE_t + n \cdot (cE_t + tE_t) \\ \quad \text{if } np_t < n \leq nc_t \cdot np_t. \\ minE_t + np_t \cdot (bE_t + nc_t \cdot cE_t) + n \cdot tE_t \\ \quad \text{if } nc_t \cdot np_t < n < nt_t. \\ maxE_t \text{ if } n \geq nt_t. \end{cases} \quad (1)$$

The electric power consumption $E_t(\tau)$ [W] of a server s_t at time τ is assumed to be $NE_t(|CP_t(\tau)|)$ in this paper.

Let $minT_{ti}$ show the minimum execution time of a process p_i, which is only performed on a server s_t without any other process. Let $minT_i$ be a minimum one of $minT_{1i}, \ldots, minT_{mi}$. A server s_f where $minT_{fi} = minT_i$ is *fastest* in a cluster S. The computation rate TCR_t of a server s_t is $minT_i/minT_{ti}$ (≤ 1) for any process p_i. The maximum computation rate $XSCR_t$ $(\leq nt_t)$ of a server s_t is $nt_t \cdot TCR_t$. The maximum computation rate $maxPCR_{ti}$ of a process p_i on a server s_t is TCR_t.

If n (≥ 1) processes are concurrently performed on a thread, each process is assumed to be fairly allocated with the same computation rate TCR_t/n. The process computation rate $NPR_t(n)$ $(\leq TCR_t)$ [vs/sec] of each of n active process on a server s_t is defined as follows [4,6,11]:

[MLCM (Multi-Level Computation with Multiple CPUs) model]

$$NPR_t(n) = \begin{cases} nt_t \cdot TCR_t \, / \, n & \text{if } n > nt_t. \\ TCR_t \text{ if } n \leq nt_t. \end{cases} \tag{2}$$

The computation rate $PR_t(\tau)$ of each actve process at time τ is $NPR_t(|CP_t(\tau)|)$. If a process p_i on a server s_t starts at time st and ends at time et, $\sum_{\tau=st}^{et} PR_t(\tau) = minT_i$. Thus, $minT_i$ shows the amount of computation of a process p_i.

Each process p_i is performed on a server s_t as follows:

[Computation model of a process p_i]

1. At time τ a process p_i starts, the computation residue RP_i is $minT_i$;
2. At each time τ, $RP_i = RP_i - NPR_t(|CP_t(\tau)|)$, if $RP_i \leq 0$, p_i terminates;

The server computation rate $NSR_t(n)$ of a server s_t to perform n processes is $n \cdot NPR_t(n)$. $NSR_t(n)$ shows the throughput of a server s_t.

Suppose a new process p_i starts on a server s_t where n_t active processes are performed. The execution time $ET_t(n_t, p_i)$ means time by which every active process and the pre process p_i to terminate. $EE_t(n_t, p_i)$ shows energy to be consumed by a server s_t for $ET_t(n_t, p_i)$ time units from current time τ. $EE_t(n_t, p_i)$ and $ET_t(n_t, p_i)$ are obtained as E and T as follows:

The execution time $ETN_t(R, n)$ for the total computation residue R of n processes on a server s_t is given as follows:

$$ETN_t(R, n) = R/NSR_t(n). \tag{3}$$

3   Energy Consumption Model of Migration

First, we consider case that no virtual machine migrates from a host server s_t to a guest server s_u. Here, a pair of execution time NET_t and NET_u of the servers s_t nad s_u are $ET_t(n_t, \_)$ and $ET_u(n_u, \_)$, respectively. The servers s_t

---

Algorithm 1: algorithm

---

1 **input** : s_t = server ;
2   P_t = set of active processes on s_t;
3   p_i = new process ;
4   τ = current time ;
5 **output** : E = energy consumption of s_t ;
6   T = execution time of s_t ;
7 $t = \tau$; $E = 0$;
8 $RP_i = minT_i$; $P_t = P_t \cup \{p_i\}$; $n = |P_t|$;
9 **while** ($P_t \neq \phi$) **do**
10   **for** each process p_j in p_t **do**
11          $RP_j = RP_j - NPR_t(n)$;
12   **if** $RP_j \leq 0$, $P_t = P_t - \{p_j\}$; /* p_j terminates */
13   **for end** ;
14      $E = E + NE_t(n)$;
15      $t = t + 1$;
16 **while end**;
17 $T = t - \tau + 1$

---

and s_u consume energy, $EE_t(RS_t, \_)$ and $EE_u(RS_t, \_)$, respectively. Suppose $NET_t < NET_u$. Here, the server s_t just consumes power $minE_t$ [W] from time NET_t to time NET_u. Hence, the servers s_t and s_u totally consume the energy TEE_{tu} :

$$TEE_{tu} = EE_t(RS_t, \_) + EE_u(RS_u, \_) + NTEE_{tu}. \tag{4}$$

$$NTEE_{tu} = \begin{cases} (NET_t - NET_u) \cdot minE_t & \text{if } NET_t \geq NET_u. \\ (NET_u - NET_t) \cdot minE_u & \text{otherwise.} \end{cases} \tag{5}$$

Next, we consider case a virtual machine vm_k migrates from a host server s_t to a guest server s_u. Suppose the virtual machine vm_k starts migrating from the host server s_t to the guest server s_u at time τ. The execution time NET_t of the host server s_t decreases to $MET_t = ETN_t(RS_t - RV_k, n_t - nv_k) = (RS_t - RV_k)/NSR_t(n_t - nv_k)$ since the virtual machine vm_k leaves the server s_t. The virtual machine vm_k restarts on the server s_u at time $\tau + mt$. Here, mt is the migration time of the virtual machine. Every process on vm_k is suspended, i.e. is not performed during the migration from time MET_t to $\tau + mt$. The computation residue RS_u of the guest server s_u at time τ is reduced to $\alpha_u \cdot RS_u$ at time $\tau + mt$ since the processes are performed on s_u. If $NET_u \geq mt$, α_u is $(NET_u - mt)/NET_u$ (≤ 1) [Fig. 2]. Otherwise, $\alpha_u = 0$, i.e. every process terminates on s_u before $\tau + mt$.

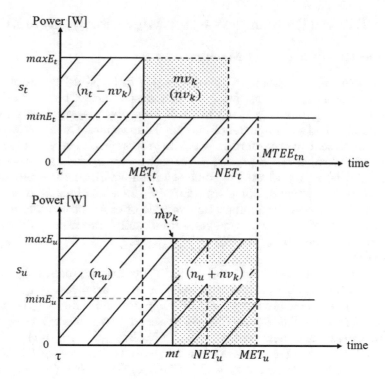

Fig. 2. Migration of a virtual machine mv_k ($mt \leq NET_u$ and $MET_t < MET_u$).

The execution time MET_u [time unit (tu)] of the guest server s_u is as follows:

$$MET_u = \begin{cases} ETN_u((1 - \alpha_u) \cdot RS_u, \ n_u) + \\ \quad ETN_u(\alpha_u \cdot RS_u + RV_k, \ n_u + nv_k) & \text{if } NET_u \geq mt. \\ mt + ETN_u(RV_k, \ nv_k) & \text{otherwise.} \end{cases} \quad (6)$$

The total energy MEE_{tu}^k [W·tu] consumed by the servers s_t and s_u is as follows:

$$MEE_{tu}^k = \begin{cases} mt \cdot EE_u((1 - \alpha_u) \cdot RS_u, \ \_) + \\ \quad ETN_u(\alpha_u \cdot RS_u + RV_k, \ n_u + nv_k) \cdot NE_u(n_u + nv_k) + \\ \quad MTEE_{tu} & \text{if } NET_u \geq mt. \\ EE_u(RS_u, \ \_) + (mt - MET_u) \cdot minE_u + \\ \quad ETN_u(RV_k, \ nv_k) + MTEE_{tu} & \text{otherwise.} \end{cases} \quad (7)$$

$$MTEE_{tu} = \begin{cases} (MET_t - MET_u) \cdot minE_t & \text{if } MET_t \geq MET_u. \\ (MET_u - MET_t) \cdot minE_u & \text{otherwise.} \end{cases} \quad (8)$$

4   An RWM (Reducing Wasted Migrations) Algorithm

4.1   Selection of a Virtual Machine

Let P be a set of processes p_1, ..., p_n $(n \geq 1)$ to be issued by clients and S be a set of servers s_1, ..., s_m $(m \geq 1)$ in a cluster. Let V be a set of virtual machines vm_1, ..., vm_l $(l \geq 1)$ in the set S. Each server s_t supports one active virtual machine and some number of idle virtual machines. On an *active* virtual machine, processes can be performed while no process on an *idle* virtual machine.

First, if a client issues a process p_i, a host server s_t and an active virtual machine vm_k on s_t are selected in some selection algorithm. The process p_i is performed on the virtual machine vm_k. Let PS_t be a set of active processes on a server s_t. RP_i shows the computation residue of each active process p_i. RS_t and RV_t are the total computation residues of active processes on a server s_t and a virtual machine vm_k, i.e. $RS_t = \sum_{p_i \in PS_t} RP_i$ and $RV_k = \sum_{p_i \in PV_k} RP_i$, respectively.

Let n_t and nv_k be numbers $|PS_t|$ and $|PV_k|$ of active processes on a server s_t and a virtual machine vm_k, respectively. Suppose a process p_i newly starts on a server s_t. The new process p_i has the computation residue RP_i $(= minT_i)$. The execution time $ET_t(RS_t, p_i)$ [tu] of a server s_t to perform both all the active processes and a new process p_i is $ETN_t(RS_t + RP_i, n_t + 1) = (RS_t + RP_i)/NSR_t(n_t + 1) = (RS_t + minT_i)/NSR_t(n_t + 1)$. The energy $EE_t(RS_t, p_i)$ to be consumed by a server s_t is $ET_t(RS_t, p_i)$ [tu] $\cdot$ $NE_t(n_t + 1)$ [W] $= ETN_t(RS_t, minT_i) \cdot NE_t(n_t + 1)$.

For a new process p_i, a host server s_t is selected where the energy consumption $EE_t(RS_t, p_i)$ is smallest on the cluster S in our previous MDMG algorithm [13]. Then, a smallest active virtual machine vm_h is selected on the server s_t.

Let NP be a set of new processes which start at current time τ. In our previous algorithms [13,14], a server is selected for each process independently of the other processes in the set NP. In this paper, servers are selected for new processes in the set NP so that the total energy consumption of the servers can be reduced to perform all the processes in the set NP.

In order to reduce the number of migrations of virtual machines, we take the following strategies:

1. Processes in the set NP are distributed to servers so that the number n_t of active processes of each server s_t is proportional to the server computation rate SCR_t.
2. If a selected server s_t supports no active virtual machine, one idle virtual machine is activated.
3. Processes in the set NP are selected in an order of the computation residue. In the previous algorithms [13], processes in the set P are selected in an order of process identifier.

Servers are selected in the Algorithm 2.

Algorithm 2: Server selection in the RWM algorithm

---

1 **input** : τ = current time;
2   S = set of servers ;
3   NP = set of new processes which start at time τ ;
4 **output** : $PV = \{\langle p_i, vm_k, s_t \rangle \mid s_t =$ host server, $vm_k =$ virtual machine on s_t, $p_i \in NP \}$;
5 RS_t = total residue of active processes on each server s_t ;
6 RP = total residue of active processes on the servers in S, i.e. $\sum_{s_t \in S} RS_t$;
7 NRP = total residue of the new processes, $\sum_{p_i \in NP} minT_i$;
8 $TXSCR$ = total server computation rate of the servers in S, i.e. $\sum_{s_t \in S} XSCR_t$;
9 $SR_t = XSCR_t \, / \, TXSCR$ for each server s_t in S ;
10 $D_t = (RP + NRP) \cdot SR_t$ - RS_t for each server s_t in S ;
11 $X = NP$; $PV = \phi$;
12 **while** $X \neq \phi$ **do**
13   p_i = process in X whose $minT_i$ is the largest ;
14   s_t = server where $D_t > 0$ is largest in S ;
15   vm_k = active virtual machine on s_t whose nv_t is the smallest ;
16      $PV = PV \cup \{\langle p_i, vm_k, s_t \rangle\}$;
17      $X = X$ - $\{p_i\}$; $RS_t = RS_t + minT_i$;
18 **while end**;

In the EVMS [14] and MDMG [13] algorithms, virtual machines on servers migrate to some servers which support more computation rates. In this paper, we take the following strategies to make virtual machines migrate at each time unit:

1. Every md time units, all the servers hosting active virtual machines are checked to decide which virtual machines to migrate to which servers.
2. Once a virtual machine vm_k migrates from a host server s_t to a guest server s_u, the servers s_t and s_u are not selected as a guest server of another virtual machine.
3. A virtual machine vm_k migrates from a host server s_t to a guest server s_u only if $|(n_u + nv_k)/(n_t - nv_k) - SCR_u/SCR_t|$ is smaller than some constant ε.
4. For each tuple $\langle vm_k, s_t, s_u \rangle$ in the set RS obtained in the migration algorithm, a virtual machine vm_k migrates from a host server s_t to a guest server s_u.

The migration algorithm is shown in Algorithm 3.

Algorithm 3: Migration in the RWM algorithm

---

1 **input** : S = set of servers ;
2   NP = set of new processes which start at time τ ;
3 **output** : $VS = \{\langle vm_k, s_t, s_u \rangle \mid s_t$ = host server, s_u = guest server, vm_k = virtual machine on $s_t\}$;
4 $HS = S$; $GS = S$; $VS = \phi$; $X = NP$;
5 RS_t = total residue of active processes on each server s_t ;
6 RP = total residue of active processes on all servers in S, i.e. $\sum_{s_t \in S} RS_t$;
7 NRP = total residue of the new processes, $\sum_{p_i \in NP} minT_i$;
8 $TXSCR$ = total server computation rate of the servers in S, i.e. $\sum_{s_t \in S} XSCR_t$;
9 $SR_t = XSCR_t / TXSCR$ for each server s_t in S ;
10 $D_t = (RP + NRP) \cdot SR_t - RS_t$ for each server s_t in S ;
11 **while** $HS \neq \phi$ **do**
12    $\quad s_t$ = server in HS where $D_t \leq 0$ is smallest ;
13    $\quad vm_k$ = active virtual machine on s_t whose n_t is the smallest ;
14    $\quad HS = HS - \{s_t\}$; $GS = GS - \{s_t\}$;
15    $\quad$ **if** $GS \neq \phi$ **then**
16      $\quad\quad s_u$ = server where $D_u \geq 0$ is largest ;
17      $\quad\quad GS = GS - \{s_u\}$
18    $\quad$ **if end** ;
19    $\quad VS = VS \cup \{\langle vm_h, s_t, s_u \rangle\}$;
20 **while end** ;

---

5   Evaluation

We evaluate the RWM algorithm to select a host virtual machine to perform a process issued by a client and make virtual machines migrate in terms of the total energy consumption of servers and the average execution time of processes. We consider a cluster composed of four servers $s_1, \ldots, s_4 (m = 4)$ as shown in Table 1. The fastest thread computation rate TCR_1 of the server s_1 is one, i.e. $TCR_1 = 1$. For the other servers s_2, s_3, and s_4, $TCR_2 = 0.8$, $TCR_3 = 0.6$, and $TCR_4 = 0.4$. The performance and energy parameters of the servers are shown in Table 1. For example, the server s_1 supports the maximum server computation rate $XSCR_1 = 16$ by sixteen threads where the maximum power consumption $maxE_1$ is 230 [W] and the minimum power consumption $minE_1$ is 150 [W]. The server s_4 supports $XSCR_4 = 3.2$ by eight threads where $maxE_4 = 77$ and $minE_4 = 40$. The servers s_2 and s_3 support the same number, twelve threads while $maxE_2 > maxE_1$ and $minE_2 > minE_1$. The server s_3 is more energy-efficient than the server s_2.

Each server s_t supports two virtual machines in the cluster S. One time unit means one step of the simulation. There are totally eight virtual machines. The migration time mt is one [tu] which shows 0.41 [s] based on the measurement [13]. Each server s_t is checked every four time units, i.e. $md = 4$ if an active virtual machine on the server s_t is to migrate to another server.

Let P be a set of processes $p_1, \ldots, p_n$ ($n \geq 1$) to be issued. For each process p_i in the set P, the starting time $stime_i$ and the minimum execution time $minT_i$ are randomly taken where $0 < stime_i \leq xtime$ and $10 \leq minT_i \leq 50$. Here, $xtime = 100$ [tu]. At each time τ, if there is a process p_i whose $stime_i$ is τ, one server s_t is selected by a selection algorithm. The process p_i is added to the set P_t of the selected server s_t, i.e. $P_t = P_t \cup \{p_i\}$. For each server s_t, active processes in the set P_t are performed. The energy variable E_t is incremented by the power consumption $NE_t(|P_t|)$. If $|P_t| = \phi$, E_t is just incremented by $minE_t$. If $|P_t| > 0$, the variable T_t is incremented by one [time unit] since the server s_t is active. The variable T_t shows how long the server s_t is active. For each process p_i in the set P_t, the computation residue R_i of the process p_i is decremented by the process computation rate $NPR_t(n_t)$. If $R_i \leq 0$, the process p_i terminates, i.e. $P_t = P_t - \{p_i\}$ and $P = P - \{p_i\}$. Until the set P gets empty, the steps are iterated. The variables E_t and T_t give the total energy consumption [W · tu] and execution time [tu] of each server respectively.

In the evaluation, we consider a non-migration (NMG) algorithm [15] and the MDMG algorithm [13] in addition to the RWM algorithm proposed in this paper. In the NMG algorithm, each server s_t supports only one active virtual machine which just stays on the server s_t, i.e. does not migrate. For each process p_i, a server s_t is selected in the same way as the RWM algorithm but no virtual machine migrates. In the MDMG algorithm, a server s_t is selected for each process p_i whose $EE_t(RS_t, p_i)$ is minimum. A virtual machine migrates from a host server s_t to s_u so as to mostly reduce the energy consumption.

The simulator is implemented in SQL on a Sybase [2] database. Information on servers, virtual machines, and processes are stored in tables of the database and the tables are manipulated in SQL.

Figure 3 shows the total energy consumption of the servers $s_1, \ldots, s_4$ for number n of processes. The total energy consumption of the servers of the MDMG algorithm is smaller than the NMG algorithm. The total energy consumption of the servers can be reduced in the MDMG algorithm.

Figure 4 shows the average execution time of n processes. The average execution time of the RWM algorithm is almost the same as the MDMG algorithm and is shorter than the NMG algorithm.

Figure 5 shows the number of migrations of virtual machines. Compared with the MDMG algorithm, the number of migrations of virtual machines in the RWM algorithm is reduced.

Table 1. Parameters of servers.

	np_t	nc_t	nt_t	TCR	$XSCR$	$minE[W]$	$maxE[W]$	$pE[W]$	$cE[W]$	$tE[W]$
s_1	1	8	16	1.0	16.0	150.0	270.0	40.0	8.0	1.0
s_2	1	6	12	0.8	9.6	128.0	200.0	30.0	5.0	1.0
s_3	1	6	12	0.6	7.2	80.0	130.0	20.0	3.0	1.0
s_4	1	4	8	0.4	3.2	40.0	67.0	15.0	2.0	0.5

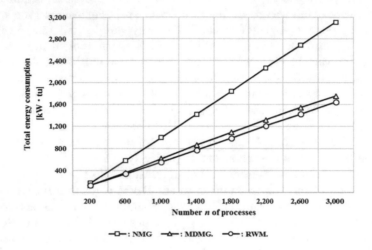

Fig. 3. Total energy consumption of servers.

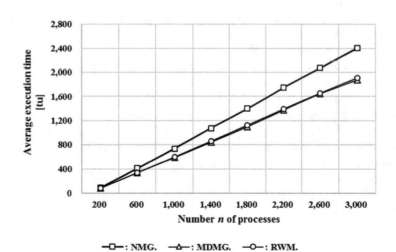

Fig. 4. Average execution time of processes.

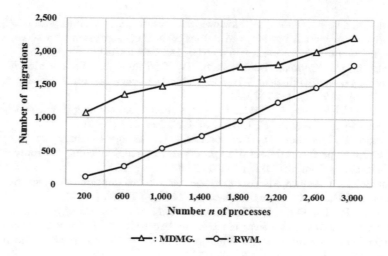

Fig. 5. Number of migrations of virtual machines.

6 Concluding Remarks

In this paper, we discussed the virtual machine migration approach to reducing the energy consumption of servers. In our previous algorithms, virtual machines frequently migrate among servers. Since it takes time to make a virtual machine migrate [13], the number of migrations of virtual machines should be reduced. In this paper, we proposed the RWM algorithm to make a virtual machine migrate from a host server to a guest server. Here, virtual machines migrate so that the number of active processes on each server is proportional to the server consumption rate and the number of migrations of virtual machines is decreased. In the evaluation, we showed the energy consumption of servers and the average execution time of processes can be reduced in the RWM algorithm compared with the migration MDMG and the non-migration NMG algorithms. The number of migrations of virtual machines can be decreased in the RWM algorithm compared with the MDMG algorithm.

References

1. Intel xeon processor 5600 series: The next generation of intelligent server processors, white paper (2010). http://www.intel.com/content/www/us/en/processors/xeon/xeon-5600-brief.html
2. Sybase (2014). http://www.cultofmac.com/167829/sybasesap-afaria-offers-ios-and-pcmanagement-options-mobile-management-month/
3. A virtualization infrastructure for the linux kernel (kernel-based virtual machine). https://en.wikipedia.org/wiki/Kernel-basedVirtualMachine
4. Duolikun, D., Aikebaier, A., Enokido, T., Takizawa, M.: Energy-aware passive replication of processes. Int. J. Mob. Multimedia 9(1,2), 53–65 (2013)

5. Duolikun, D., Enokido, T., Takizawa, M.: Static and dynamic group migration algorithms of virtual machines to reduce energy consumption of a server cluster. Trans. Comput. Collective Intell. **11610**, 144–166 (2019)
6. Duolikun, D., Kataoka, H., Enokido, T., Takizawa, M.: Simple algorithms for selecting an energy-efficient server in a cluster of servers. Int. J. Commun. Netw. Distrib. Syst. **21**(1), 1–25 (2018)
7. Enokido, T., Ailixier, A., Takizawa, M.: A model for reducing power consumption in peer-to-peer systems. IEEE Syst. J. **4**(2), 221–229 (2010)
8. Enokido, T., Duolikun, D., Takizawa, M.: The energy consumption laxity-based algorithm to perform computation processes in virtual machine environments. Int. J. Grid Utiliy Comput. **10**(5), 545–555 (2019)
9. Enokido, T., Takizawa, M.: An integrated power consumption model for distributed systems. IEEE Trans. Industr. Electron. **15**(4), 366–385 (2012)
10. Kataoka, H., Duolikun, D., Enokido, T., Takizawa, M.: Energy-efficient virtualisation of threads in a server cluster. In: Proceedings of the 10th International Conference on Broadband and Wireless Computing, Communication and Applications (BWCCA-2015), pp. 288–295 (2015)
11. Kataoka, H., Nakamura, S., Duolikun, D., Enokido, T., Takizawa, M.: Multi-level power consumption model and energy-aware server selection algorithm. Int. J. Grid Utility Comput. (IJGUC) **8**(3), 201–210 (2017)
12. Kataoka, H., Sawada, A., Duolikun, D., Enokido, T., Takizawa, M.: Energy-aware server selection algorithm in a scalable cluster. In: Proceedings of IEEE the 30th International Conference on Advanced Information Networking and Applications (AINA-2016), pp. 565–572 (2016)
13. Noaki, N., Saito, T., Duolikun, D., Enokido, T., Takizawa, M.: An energy-efficient algorithm for virtual machines to migrate considering migration time. In: The 15th International Conference on Broadband and Wireless Computing, Communication and Applications (BWCCA-2020) (2020)
14. Noaki, N., Saito, T., Duolikun, D., Enokido, T., Takizawa, M.: Energy-efficient migration of virtual machine. In: The 23rd International Conference on Network-Based Information Systems (NBiS-2020) (2020)
15. Noguchi, K., Saito, T., Duolikun, D., Enokido, T., Takizawa, M.: An algorithm to select a server to minimize the total energy consumption of a cluster. In: 15th International Conference on P2P, Parallel, Grid, Cloud and Internet Computing (3PGCIC-2020) (2020)

Design and Implementing of the Dynamic Tree-Based Fog Computing (DTBFC) Model to Realize the Energy-Efficient IoT

Keigo Mukae[1]([envelope]), Takumi Saito[1]([envelope]), Shigenari Nakamura[2]([envelope]),
Tomoya Enokido[3]([envelope]), and Makoto Takizawa[1]([envelope])

[1] Hosei University, Tokyo, Japan
{keigo.mukae.2d,takumi.saito.3j}@stu.hosei.ac.jp,
makoto.takizawa@computer.org
[2] Tokyo Metropolitan Industrial Technology Research Institute, Tokyo, Japan
nakamura.shigenari@gmail.com
[3] Rissho University, Tokyo, Japan
eno@ris.ac.jp

Abstract. The IoT (Internet of Things) consumes huge amount of electric energy since millions to billions of device nodes are interconnected. In order to decrease the energy consumption of the IoT, the TBFC (Tree-based Fog Computing) model is proposed in our previous studies. Here, fog nodes are hierarchically structured where a root node shows a cloud of servers and a leaf node indicates an edge node which communicates with sensors and actuators. Each fog node supports a subsequence of subprocesses of an application process. As a volume of sensor data increases, some fog node gets overloaded. In the DTBFC (Dynamic TBFC) model, the tree structure is dynamically changed by splitting and replicating fog nodes so that the energy consumption of fog nodes can be reduced. In this paper, we newly propose a DFC (Dynamic FC) algorithm to dynamically change the tree structure of fog nodes to reduce the energy consumption and execution time of the fog nodes in the DTBFC model. We also implement the DFC algorithm in a node.

Keywords: IoT · TBFC (Tree-Based Fog Computing) model ·
DTBFC (Dynamic TBFC) model · DFC (Dynamic FC) algorithm

1   Introduction

In the IoT (Internet of Things) [7], not only computers like servers and clients but also millions to billions of devices supporting sensors and actuators are interconnected in networks. The FC (Fog Computing) model [15] is proposed to reduce the communication and processing traffic to handle sensor data in the IoT. Here, subprocesses of an application process to handle sensor data are distributed to not only servers but also fog nodes to improve the performance and responsibility.

© The Author(s), under exclusive license to Springer Nature Switzerland AG 2021
L. Barolli et al. (Eds.): EIDWT 2021, LNDECT 65, pp. 71–81, 2021.
https://doi.org/10.1007/978-3-030-70639-5_7

In the TBFC (Tree-Based FC) model [5,6,8–13], fog nodes are hierarchically structured in a tree, where a root node is a server in clouds and a leaf node is an edge node which communicates with sensors and actuators. Each fog node receives input data from child nodes, calculates output data on the input data, and sends the output data to a parent fog node. Thus, a child-parent relation of fog nodes shows the output-input relation of data. Here, the tree structure of the fog nodes is static, i.e. not changed even if the amount of sensor data is so changed that some fog nodes are overloaded or underloaded.

In the DTBFC (Dynamic TBFC) model [14], fog nodes are dynamically connected and disconnected in a tree as the traffic of each node increases and decreases, respectively, so that the energy consumption and execution time of fog nodes can be reduced. In this paper, an application process is assumed to be a sequence of subprocesses. An edge subprocess receives input data from sensors and sends output data calculated on the input data to a succeeding subprocess. Thus, each subprocess receives input data from a preceding subprocess and sends output data to a succeeding subprocess. As the amount of sensor data increases, a fog node f is more heavily loaded and consumes more energy. The fog node f is splitted or replicated to a pair of nodes f and f_1 to reduce energy consumption and execution time. In the splitting way, some postfix subsequence of subprocesses of a node f is supported by a new node f_1 and the node f supports only the prefix of subprocesses. In the replication way, the new fog node f_1 supports the same subsequence of the subprocesses and is the same parent node as the node f. In this paper, we propose a DFC (Dynamic FC) algorithm where the tree structure is dynamically changed by splitting and replicating fog nodes A DFC algorithm is implemented. Each fog node consumes energy even if the node is idle. If a fog node f is splitted and replicated to multiple fog nodes, the energy consumption of each node can be smaller than the node f. On the other hand, each fog node gets idle and just consumes energy to wait for data from the child nodes. We measure time just each fog nodes waits for data from child nodes.

In Sect. 2, we propose the DTBFC model. In Sect. 3, we discuss the energy consumption and execution time of a fog node. In Sect. 5, we implement the DFC algorithm to split and replicate fog nodes.

2   Dynamic Tree-Based Fog Computing (DTBFC) Model

In the IoT, an application process handles data from sensors and issues actions to actuators. In this paper, an application process P is assumed to be a sequence of subprocesses $\langle p_1, ..., p_m \rangle$ $(m \geq 1)$, where p_1 is an edge and p_m is a root. If $i < j$, p_i precedes p_j, i.e. p_i is performed prior to p_j. Here, the edge subprocess p_1 receives input data from sensors. The subprocess p_1 obtains output data d_1 by calculating on the input data from the sensors. Then, the subprocess p_1 sends the output data d_1 to the succeeding subprocess p_2. Thus, each subprocess p_i receives input data d_1 from a preceding subprocess p_{i-1} and sends output data d_i to a succeeding subprocess p_{i+1}. Let $|d|$ show the size of data d. The output ratio α_i of the subprocess p_i is $|d_{i+1}|$ / $|d_i|$. A sequence $P' = \langle p_i, p_{i+1}, ..., p_j \rangle$

$(1 \leq i \leq j \leq m)$ is a *subsequence* of a sequence $P = \langle p_1, ..., p_i, ..., p_j, ..., p_m \rangle$ $(P' \sqsubset P)$. Subsequences $\langle p_1, ..., p_i \rangle$ and $\langle p_j, ..., p_m \rangle$ $(1 \leq i, j \leq m)$ are a *prefix* and *postfix* of a sequence $P = \langle p_1, ..., p_m \rangle$, respectively. For a pair of sequence $P_1 = \langle p_i, ..., p_j \rangle$ and $P_2 = \langle p_k, ..., p_l \rangle$, a concatenation $P_1 | P_2$ is a sequence $\langle p_i, ..., p_j, p_k, ..., p_l \rangle$.

In the DTBFC model [14], a root node f has child nodes $f_1, ..., f_c$ $(c \geq 1)$. Each node f_i has also child nodes $f_{i1}, ..., f_{i,c_i}$ $(c_i \geq 1)$. Thus, a fog node f_R has child nodes $f_{R1}, ..., f_{R,c_R}$ $(c_R \geq 1)$ where the label R is a sequence of numbers. The label R of a node f_R shows a path from the root node f to the node f_R. Here, f_R is a parent node of each node f_{Ri}. $C(f_R)$ is a set of child nodes and $pt(f_R)$ shows a parent node of a node f_R. Each node f_R supports a subsequence $SP(f_R)$ $(\sqsubset P)$ of subprocesses of an application process P. The output ratio β_R of a node f_R is $\Pi_{p_i \in SP(f_R)} \alpha_i$. For each edge node f_R where $R = i_1 \, i_2 \, ... \, i_h$ $(1 \leq h \leq m)$, a concatenation $SP(f_{i_1}) | SP(f_{i_1 i_2}) | ... | SP(f_{i_1 i_2 ... i_h})$ is a subprocesses sequence $\langle p_1, ..., p_m \rangle$ of the application process P.

All the subprocesses $p_1, ..., p_m$ of an application process P are initially supported by one fog node f, which is a root node, e.g. servers in a cloud [1]. A sensor sends sensor data to the root node f, where output data is calculated on the sensor data by the subprocesses $p_1, ..., p_m$. As the amount of sensor data increases, the execution time and energy consumption of the node f increase and network traffic to the fog node f also increases. In order to increase the performance and reduce the energy consumption, a prefix $\langle p_1, ..., p_l \rangle$ and postfix $\langle p_{l+1}, ..., p_m \rangle$ of the subprocesses supported by the node f for some l $(\leq m)$ are distributed to the nodes f_1 and f, respectively, in a splitting way as shown in Fig. 1(1). Here, the node f is a parent node of the node f_1. The fog node f_1 receives all the input data from the child nodes and calculates output data d_1 on the input data by the subprocesses $\langle p_1, ..., p_l \rangle$. The fog node f_1 then sends the output data d_1 to the node f. The fog node f supports subprocesses $\langle p_{l+1}, ..., p_m \rangle$ to calculate on the data d_1. The processing load of the node f thus decreases since the subprocesses $\langle p_1, ..., p_l \rangle$ are not performed. If the amount of input data of the nodes f and f_1 decreases, the node f_1 is disconnected from the tree and the node f supports all the subprocesses $\langle p_1, ..., p_l, p_{l+1}, ..., p_m \rangle$.

Next, if the fog node f_1 gets too heavily loaded, a replica f_2 of the node f_1 is additionally connected to the parent fog node f. A replica f_2 supports the same subsequence $\langle p_1, ..., p_l \rangle$ of the subprocesses as the node f_1 as shown in Fig. 1(2). The nodes f_1 and f_2 are new child nodes of the node f. Each of the child nodes in $C(f)$ sends the output data to one of the nodes f_1 and f_2. Thus, the nodes f_1 and f_2 receive smaller amount of sensor data than the node f_1 before replication. If the amount of sensor data to be sent to the nodes f_1 and f_2 decreases so that one node can calculate on all the sensor data, one node, say f_2 is disconnected and every sensor data is sent to the node f_1.

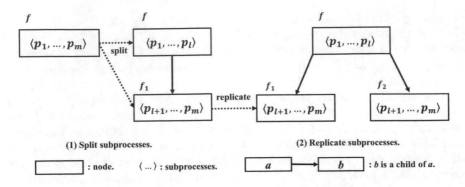

Fig. 1. Splitting and replication of subprocesses.

3　Computation and Energy Consumption Models of a Fog Node

3.1　Computation Model

An application process P is a sequence $\langle p_1, ..., p_m \rangle$ $(m \geq 1)$ of subprocesses. That is, the subprocesses are sequentially performed in an order of $p_1, ..., p_m$.. A fog node f_R receives input data $D_R = \{d_{R1}, ..., d_{R,c_R}\}$ of size i_R $(= |D_R|)$ from the child nodes $f_{R1}, .., f_{R,c_R}(c_R \geq 1)$. The fog node f_R calculates the output data d_R of size o_R $(= |d_R|)$ on the input data D_R by the subprocesses $SP(f_R)$. Here, let $SP(f_R)$ be a subsequence $\langle p_{s_R}, p_{s_R+1}, ..., p_{e_R} \rangle$ $(\sqsubseteq P)$ $(1 \leq s_R \leq e_R \leq m$) of subprocesses supported by a fog node f_R. The output ratio β_R of the node f_R is $\alpha_{s_R}...\alpha_{e_R}$. When α_k is the output ratio of each subprocess p_k in $SP(f_R)$. Then, the fog node f_R sends the output data d_R to a parent node $pt(f_R)$, where the size o_R of the output data d_R is $\beta_R \cdot i_R$.

Let CR_R show the computation rate of a fog node f_R. The residue rd_R of input data in f_R is $|id_R|$ on receipt of the input data id_R. For each time unit, the residue rd_R is decremented by the computation rate CR_R. If $rd_R \leq 0$, the output data od_R is sent to the parent node. CR shows the computation rate of the root node.

$TC_R(SQ, x)$ shows the execution time of the subprocesses SQ to calculate of data of size x on a fog node f_R. $TC(\langle p_i \rangle, x)$ and $TC_R(\langle p_i \rangle, x)$ show the execution time [sec] of a subprocess p_i to calculate on input data of size x by a root node f and a fog node f_R, respectively. $TC(\langle p_i \rangle, x)/TC_R(\langle p_i \rangle, x)$ is CR_R/CR. The execution time $TC(\langle p_i \rangle, x)$ of each subprocess p_i on a root node f to calculate on input data of size x is $ct_i \cdot C_i(x)$ where ct_i is a constant and $C_i(x)$ is x or x^2. $C_i(x)$ shows the computation complexity of the subprocess p_i. The size of the output data which the subprocesses p_i calculates on the input data of size x is $\alpha_i \cdot x$ where α_i is the output ratio of the subprocess p_i. $TC_R(SQ, x)$ by $SQ = \langle f_{s_R}, f_{s_R+1}, ..., f_{e_R} \rangle$ is as follows:

$$TC_R(SQ, \; x) = TC_R(f_{s_R}, x)$$
$$+ TC_R(f_{e_R+1}, \alpha_{s_R} x) + \ldots + TC_R(f_{e_R}, \alpha_{s_R} \ldots \alpha_{e_R-1} x). \tag{1}$$

The output ratio β_R of the node f_R with the subprocesses $SQ = \langle f_{s_R}, f_{s_R+1}, \ldots, f_{e_R} \rangle$ is $\alpha_{s_R} \alpha_{s_R+1} \ldots \alpha_{e_R}$. The size of the output data d_{e_R} of the subprocess p_{e_R}, i.e. the size of the output data d_R of the node f_R is $\beta_R \cdot x$.

A pair of the execution time $TI_{Ri}(x)$ and $TO_R(x)$ of a node f_R to receive from a child fog node f_{Ri} and send data of size x to the parent node are proportional to the data size x, i.e. $TI_{Ri}(x) = rc_{Ri} + rt_{Ri} \cdot x$ and $TO_R(x) = sc_R + st_R \cdot x$. Here, rc_{Ri}, rt_{Ri}, sc_R, and st_R are constants. A node f_R receives input data $d_{R1}, \ldots, d_{R,c_R}$ from child nodes $f_{R1}, \ldots, f_{R,c_R}$, respectively, where $x_i = |d_{Ri}|$ for $i = 1, \ldots, c_R$ and $x = |D_R| = x_1 + \cdots + x_{c_R}$. We assume $TI_{Ri}(x) = TI_R(x)$, i.e. $rc_{Ri} = rc_R$ and $rt_{Ri} = rt_R$ for every fog node f_{Ri}. It takes $TTI_R(x) = TI_R(x_1) + \cdots + TI_R(x_{c_R}) = c_R \cdot rc_R + rt_R \cdot x$ [sec] to receive the input data $d_{R1}, \ldots, d_{R,c_R}$ [14].

Here, if f_R is a root, $\delta_R = 0$, else $\delta_R = 1$. The total execution time $TT_R(SQ, x)$ [sec] of a fog node f_R with subprocesses $SQ(\sqsubset P)$ for input data of size x is given as $TT_R(SQ, x) = TTI_R(x) + TC_R(SQ, x) + \delta_R \cdot TO_R(\beta_R \cdot x)$.

3.2   Energy Consumption Model

Next, we consider electric energy to be consumed by each fog node f_R. Let $EI_R(x)$, $EC_R(SQ, \; x)$, and $EO_R(x)$ show electric energy [J] consumed by a fog node f_R supporting subprocesses $SQ \; (= SP(f_R))$ to receive, calculate on, and send data of size x, respectively. In this paper, we assume each fog node f_R follows the SPC (Simple Power Consumption) model [2–4]. Here, a fog node f_R consumes the power $maxE_R$ [W] to calculate on data of size x for $TC_R(SQ, \; x)$ [sec]. Hence, a node f_R consumes the energy $EC_R(SQ, x)$ [J] to perform subprocesses $SQ \; (\sqsubset P)$ on input data of size $x \; (> 0)$.

$$EC_R(SQ, x) = maxE_R \cdot TC_R(SQ, \; x). \tag{2}$$

A pair of the power consumption PI_R and PO_R [W] of a fog node f_R to receive and send data are $re_R \cdot maxE_R$ and $se_R \cdot maxE_R$, respectively, where $re_R \; (\leq 1)$ and $se_R \; (\leq 1)$ are constants. Here, the energy consumption $EI_R(x)$ and $EO_R(x)$ [J] of a fog node f_R, are given as follows:

$$EI_R(x) = PI_R \cdot TTI_R(x) = re_R \cdot maxE_R \cdot (c_R \cdot rc_R + rt_R \cdot x). \tag{3}$$
$$EO_R(x) = PO_R \cdot TO_R(x) = se_R \cdot maxE_R \cdot (sc_R + st_R \cdot x). \tag{4}$$

A fog node f_R supporting subprocesses $SQ(\sqsubset P)$ consumes energy $EF_R(SQ, x)$ to receive and calculate on input data $d_{R1}, \ldots, d_{R,c_R}$ of size $x_1, \ldots, x_{c_R}$, where $x = x_1 + \cdots + x_{c_R}$ and send the output data d_R of size $\beta_R \cdot x$ to a parent node $pt(f_R)$:

$$EF_R(SQ, x) = EI_R(x) + EC_R(SQ, x) + \delta_R \cdot EO_R(\beta_R \cdot x)$$
$$= \{(re_R \cdot (c_R \cdot rc_R + rt_R \cdot x)) + EC_R(SQ, x)$$
$$+ \delta_R \cdot se_R \cdot (sc_R + st_R \cdot \beta_R \cdot x)\} \cdot maxE_R. \tag{5}$$

Here, $\delta_R = 0$ if f_R is a root node, else $\delta_R = 1$.

4   Dynamic TBFC (DTBFC) Model

4.1   Ways to Split and Replicate Fog Nodes

First, suppose a fog node f_R supporting subprocesses $SP(f_R) = \langle p_{s_R}, p_{s_R+1}, ..., p_{e_R} \rangle$ has initially child nodes $cf_{R1}, ..., cf_{R,q_R}$ ($q_R \geq 0$) and a parent node $f_{R'}$ ($= pt(f_R)$). Let $CF(f_R)$ be a set $\{cf_{R1}, ..., cf_{R,q_R}\}$ ($q_R \geq 0$) of child fog nodes of a fog node f_R. The fog node f_R receives input data $d_{R1}, ..., d_{R,q_R}$ from child nodes $cf_{R1}, ..., cf_{R.q_R}$, respectively, and calculates output data d_R on the input data. Let x_i be the size of the input data d_{Ri}. Here, x is the total size $|D_R|$ of input data D_R, i.e. $x = x_1 + ... + x_{q_R}$.

In order to improve the throughput of a fog node f_R, one fog node f_{R1} is connected to the fog node f_R as a child node and the prefix $\langle p_{s_R}, ..., p_l \rangle (s_R \leq l \leq e_R)$ of $SP(f_R)$ moves to the node f_{R1} and the fog node f_R supports only the postfix $\langle p_{l+1}, ..., p_{e_R} \rangle$. Since some subprocesses supported by a fog node f_R are moved to f_{R1}. The execution time and energy consumption of the nodef_R can be reduced. This is a splitting way of a fog node f_R (Fig. 2).

Next, suppose the fog node f_{R1} is congested, e.g. the execution time gets longer. Here, new fog nodes $f_{R2}, ..., f_{R,c_R}(c_R \geq 2)$ are connected to the fog node f_R as child nodes. The subprocesses $p_{s_R}, ..., p_{e_R}$ supported by the node f_{R1} are copied to the nodes $f_{R2}, ..., f_{R,c_R}$. Every fog node f_{Rj} supports the same subprocesses ($j = 2, ..., c_R$) as the node f_{R1}. Let $CSP(f_R)$ show subprocesses $SP(f_{R1})$. Let B_R be the output ratio of the subprocesses $CSP(f_R)$, i.e. $\beta_{Ri} = B_R$ for every node f_{Ri}. Each child node cf_{Ri} in the set $CF(f_R)$ is connected to one of the nodes $f_{R1}, ..., f_{R,c_R}$ and each node f_{Rj} has at least one child node cf_{Ri}. Input data from child nodes in $CF(f_R)$ are distributed and in parallel processed on the nodes $f_{R1}, ..., f_{R,c_R}$. Thus, the execution time of each fog node f_{Ri} can be reduced. This is a replicating way of a fog node f_{Ri}.

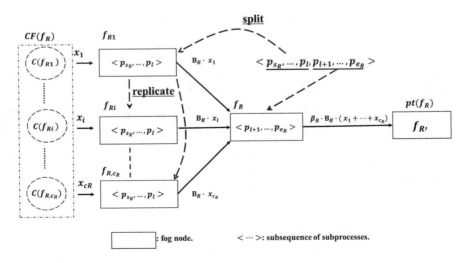

Fig. 2. Splitting and replication of fog node.

4.2   Energy Consumption and Execution Time

First, we consider simple case each edge node receives one sensor data. Each fog node f_{Ri} receives input data ID_{Ri} from the child nodes and calculates the output data od_{Ri} on the input data ID_{Ri} by subprocesses $SP(f_{Ri}) = CSP(f_R) = \langle p_{s_R}, ..., p_l \rangle$. The parent node f_R of the node f_{Ri} supports subprocesses $SP(f_R) = \langle p_{l+1}, ..., p_{e_R} \rangle$.

We consider how much electric energy a fog node f_R and the child fog nodes $f_{R1}, ..., f_{R,c_R}$ consume after each child node f_{Ri} receives input data from the child nodes $C(f_{Ri})$ until sending output data.

First, we assume every child fog node f_{Ri} receives the input data id_{Ri} from the child nodes at the same time τ. Each time a child node f_{Ri} sends the output data od_{Ri}, the node f_R receives od_{Ri} as input data id_{Ri}. The fog node f_R starts calculating output data od_R on a collection of the input data $id_{R1}, ..., id_{R,c_R}$ on receipt of all the input data. We also assume the communication delay time between every pair of a parent node and a child node is zero. This means, on finishing sending the output data od_{Ri}, the parent node f_R starts receiving od_{Ri} as input

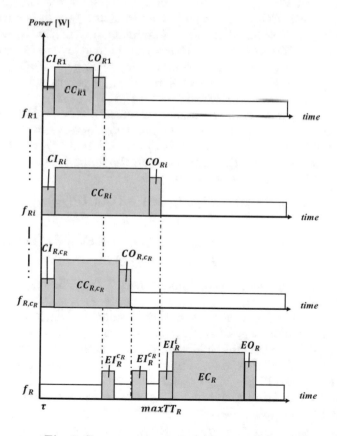

Fig. 3. Energy consumption of subprocesses.

data id_{Ri}. We consider each child node f_{Ri} which receives the input data ID_{Ri} of size x_i from the child nodes. The execution time $TT_{Ri}(CSP(f_R), x_i)$ of the child node f_{Ri} is $TI_{Ri}(x_i) + TC_{Ri}(CSP(f_R), x_i) + TO_{Ri}(B_R \cdot x_i)$ [sec]. The child node f_{Ri} consumes energy $EF_{Ri}(CSP(f_R), x_i) = EI_{Ri}(x_i) + EC_{Ri}(CSP(f_R), x_i) + TO_{Ri}(B_R \cdot x_i)$ [J]. Each time a fog node f_{Ri} sends the output data $od_{Ri}(= id_{Ri})$ of size $B_R \cdot x_i$ to the parent node f_R, the parent node f_R receives the input data id_{Ri}.

Let $SP(f_R)$ be a sequence $\langle p_{s_R}, p_{s_R+1}, ..., p_{e_R} \rangle$ of subprocesses supported by a fog node f_R. Let RP be a subsequence $\langle p_{s_R}, ..., p_l \rangle$ and TP be a subsequence $\langle p_{l+1}, ..., p_{e_R} \rangle$ for $s_R \leq l \leq e_R$, i.e. $SP(f_R) = RP|TP$. The fog node f_R is splitted and replicated to $f_{R1}, ..., f_{R,c_R}(c_R \geq 0)$ where f_R supports the subsequence TP and each f_{Ri} supports the sequence $TP(= CSP(f_R))$. For a sequence $RP = \langle p_{s_R}, ..., p_l \rangle$, $b(TP)$ stands for the output ratio $B_R = \alpha_{s_R}...\alpha_{e_R}$ of the subprocesses TP. Here, the nodes $f_R, f_{R1}, ..., f_{R,c_R}$ consume the energy:

$$TE_R(l, c_R, x) = EF_R(RP, b(TP) \cdot x) + \Sigma_{i=1}^{c_R} EF_{Ri}(TP, x_i). \tag{6}$$

Here, x_i is the size of input data of each child node f_{Ri} and $x = x_1 + ... + x_{e_R}$. In Fig. 3, CI_{Ri}, CC_{Ri}, and CO_{Ri} show the energy $EI_{Ri}(x_i)$, $EC_{Ri}(TP, x_i)$, and $EO_{Ri}(B_R \cdot x_i)$ of a child node f_{Ri}, respectively, where $TP = CSP(f_R)$. Let $maxTT_R$ be a maximum one of the execution time $TT_{R1}(TP, x_1), ..., TT_{R,c_R}(TP, x_{c_R})$. Here, the parent node f_R spends time $PTI_{Ri} = TI_R(B_R \cdot x_i)$ [sec] and consumes energy $PEI_R = EI_R(B_R \cdot x_i)$ [J] for each child node f_{Ri}. From time τ to $\tau + maxTT_R$, a parent node f_R consumes energy $maxE_R \cdot (maxTT_R - (PTI_{Ri} + ... + PTI_{R,c_R})) + (PEI_{R1} + ... + PEI_{R,c_R})$. This means, the parent node f_R starts calculating output data od_R on the input data $id_{R1}, ..., id_{R,c_R}$ at time $\tau + maxTT_R$. Here, $x = x_1 + ... + x_{c_R}$. Then, it takes $PET_R(SP(f_R), B_R \cdot x)$ [sec] and the node f_R consumes energy $EF_R(SP(f_R), B_R \cdot x)$ to do the calculation on the input data of size $B_R \cdot x$ and transmit the output data of size $\beta_R \cdot B_R \cdot x$:

$$PET_R(SP(f_R), B_R \cdot x) = TI_R(B_R \cdot x_l) + TC_R(SP(f_R), B_R \cdot x)$$
$$+ TO_R(\beta_R \cdot B_R \cdot x). \tag{7}$$
$$EF_R(SP(f_R), B_R \cdot x) = \Sigma_{i=1}^{c_R} EI_R(B_R \cdot x_i) + EC_R(SP(f_R), B_R \cdot x)$$
$$+ EO_R(\beta_R \cdot B_R \cdot x). \tag{8}$$

Here, $f_{Rk}(1 \leq k \leq c_R)$ shows a child node where $TT_{Rk}(CSP(f_R), x_k)$ is longest in the child nodes..

Each child node f_{Ri} consumes not only energy $EF_{Ri}(CSP(f_R), \cdot x_i)$ to handle the input data id_{Ri} of size x_i but also energy $minE_{Ri} \cdot (maxTT_R - TT_{Ri}(CSP(f_R), x_i))$ where the child node f_{Ri} does nothing. The time $maxTT_R - TT_{Ri}(CSP(f_R), x_i)$ is *idle time* of a node f_{Ri}. The parent node f_R also spends idle time $IE_R = maxTT_R - \Sigma_{i=1}^{c_R} TI_{Ri}(B_R \cdot x_i)$[sec].

The idle time IE_{Ri} of a child node f_{Ri} and IE_R of a parent node f_R are given as follows:

$$IE_{Ri} = minE_{Ri} \cdot (maxTT_R - TT_{Ri}(CSP(f_R), x_i) + PET_R(SP(f_R), B_R \cdot x)). \tag{9}$$
$$IE_R = minE_R \cdot (maxTT_R - \Sigma_{i=1}^{c_R} TI_R(B_R \cdot x_i) + TI_R(B_R \cdot x_l)). \tag{10}$$

The parent node f_R supporting subprocesses $SP(f_R) = \langle f_{l+1}, ..., f_{e_R} \rangle$ and the child nodes $f_{R1}, ..., f_{R,c_R}$ supporting subprocesses $CSP(f_R) = \langle f_{s_R}, ..., f_l \rangle$ totally consume energy $SE_R(l, c_R, x)[J]$ for time units $ST_R(l, c_R, x)[\text{sec}]$:

$$ST_R(l, c_R, x) = maxTT_R + TI_R(x_l) + TC_R(SP(f_R), \ B_R \cdot x) + TO_R(\beta_R \cdot B_R \cdot x). \quad (11)$$

$$SE_R(l, c_R, x) = \Sigma_{i=1}^{c_R}(EF_{Ri}(CSP(f_R), x_i) + IE_{Ri}) + EF_R(SP(f_R), B_R \cdot x) + IE_R. \quad (12)$$

On receipt of input data id_{Ri} from every child node f_{Ri}, if a fog node f_R still does the calculation, every input data id_{Ri} is enqueued in a receipt queue RQ_{Ri}. Let DT be a DTBFC tree of a set F of fog nodes. A DFC (Dynamic FC) algorithm is shown in Algorithm 1. First, we select a fog node f_R where the length of the receipt queue RQ_R is longest. Next, we find l ($s_R \leq l \leq e_R$) where $SE_R(l, c_R, x)$ is minimum for total size of input data of f_R. Then, we find the numbers l and c_R where $SE_R(l, c_R, x)$ is minimum. The prefix $\langle f_{s_R}, ..., f_l \rangle$ is supported by a fog node f_R and the postfix $\langle f_{l+1}, ..., f_{e_R} \rangle$ is supported by child fog nodes $f_{R1}, ..., f_{R,c_R}$.

Algorithm 1: [DFC algrithm]

---

1 **input**: DT = tree of fog nodes;
2 **output**: f_R = fog node in DT where $SP(f_R) = \langle p_{s_R}, ..., p_{e_R} \rangle$;
3            $f_{R1}, ..., f_{R,c_R}$ = new child nodes of f_R;
4 **select** a fog node f_R where a receipt queue RQ_R is the longest;
5 CF = set $\{cf_{R1}, ..., cf_{R,q_R}\}$ of child nodes of f_R;
6 $x = |D_R|$; /* size of input data of f_R */
7 **select** l such that $SE_R(l, 1, x)$ is smallest ($s_R \leq l \leq e_R$);
8 **select** c_R such that $SE_R(l, c_R, x)$ is smallest ($1 \leq c_R \leq q_R(= |CR|)$);
9 $SP(f_R) = \langle p_{l+1}, ..., p_{e_R} \rangle$;
10 $SP(f_{Ri}) = \langle p_{s_R}, ..., p_l \rangle$ for $i = 1, ..., c_R$;
11 **for** each node cf_{Rj} in CF, a node f_{Rk} such that $k = (j \mod c_R)$ is taken as a
           parent node of cf_{Rj};

---

5   Implementation of a DFC Algorithm

Each fog node f_R is implemented in C language, which is composed of three modules I_R, C_R, and O_R [10]. Here, I_R is an input module which receives input data from child nodes. C_R is a computation module by which output data is calculated on the input data. O_R is an output module which sends the output data to the parent node. The computation module C_R starts only if I_R receives input data from every child node. Input data sent by a child node is stored in the receipt queue RQ_R.

6   Concluding Remarks

In this paper, we discussed the DTBFC (Dynamic Tree Based Fog Computing) model. A fog node is splitted and replicated to multiple fog nodes to reduce the total energy consumption of a fog a fog node, as the traffic of the fog node increases in the tree. We made clear the execution time and energy consumption of fog nodes. By using the model, we propose the DFC algorithm to split and replicate nodes so that the energy consumption of nodes can be reduced.

References

1. Creeger, M.: Cloud computing: an overview. Queue **7**(5), 3–4 (2009)
2. Enokido, T., Ailixier, A., Takizawa, M.: A model for reducing power consumption in peer-to-peer systems. IEEE Syst. J. **4**(2), 221–229 (2010)
3. Enokido, T., Ailixier, A., Takizawa, M.: Process allocation algorithms for saving power consumption in peer-to-peer systems. IEEE Trans. Industr. Electron. **58**(6), 2097–2105 (2011)
4. Enokido, T., Ailixier, A., Takizawa, M.: An extended simple power consumption model for selecting a server to perform computation type processes in digital ecosystems. IEEE Trans. Industr. Electron. **10**(2), 1627–1636 (2014)
5. Guo, Y., Oma, R., Nakamura, S., Duolikun, D., Enokido, T., Takizawa, M.: Data and subprocess transmission on the edge node of TWTBFC model. In: Proceedings of the 11-th International Conference on Intelligent Networking and Collaborative Systems (INCoS 2019), pp. 80–90 (2019)
6. Guo, Y., Oma, R., Nakamura, S., Duolikun, D., Enokido, T., Takizawa, M.: Evaluation of a two-way tree-based fog computing (TWTBFC) model. In: Proceedings of the 13th International Conference on Innovative Mobile and Internet Services in Ubiquitous Computing (IMIS 2019), pp. 72–81 (2019)
7. Hanes, D., Salgueiro, G., Grossetete, P., Barton, R., Henry, J.: IoT Fundamentals: Networking Technologies, Protocols, and Use Cases for the Internet of Things. Cisco Press (2018)
8. Oma, R., Nakamura, S., Duolikun, D., Enokido, T., Takizawa, M.: An energy-efficient model for fog computing in the internet of things (IoT). Internet Things **1–2**, 14–26 (2018)
9. Oma, R., Nakamura, S., Duolikun, D., Enokido, T., Takizawa, M.: Energy-efficient recovery algorithm in the fault-tolerant tree-based fog computing (FTBFc) model. In: Proceedings of the 33rd International Conference on Advanced Information Networking and Applications (AINA-2019), pp. 132–143 (2019)
10. Oma, R., Nakamura, S., Duolikun, D., Enokido, T., Takizawa, M.: Evaluation of data and subprocess transmission strategies in the tree-based fog computing (TBFC) model. In: Proceedings of the 22nd International Conference on Network-Based Information Systems (NBiS 2019), pp. 15–26 (2019)
11. Oma, R., Nakamura, S., Duolikun, D., Enokido, T., Takizawa, M.: A fault-tolerant tree-based fog computing model. Int. J. Web Grid Serv. (IJWGS) **15**(3), 219–239 (2019)
12. Oma, R., Nakamura, S., Duolikun, D., Enokido, T., Takizawa, M.: Subprocess transmission strategies for recovering from faults in the tree-based fog computing (TBFC) model. In: Proceedings of the 13th International Conference on Complex, Intelligent, and Software Intensive Systems (CISIS 2019), pp. 50–61 (2019)

13. Oma, R., Nakamura, S., Enokido, T., Takizawa, M.: A nodes selection algorithm for fault recovery in the GTBFC model. In: Proceedings of the 14th International Conference on Broad-Band Wireless Computing, Communication and Applications (BWCCA 2019), pp. 81–92 (2019)
14. Oma, R., Nakamura, S., Enokido, T., Takizawa, M.: A dynamic tree-based fog computing (DTBFC) model for the energy-efficient IoT. In: Proceedings of the 8th International Conference on Emerging Internet, Data and Web Technologies (EIDWT 2020), pp. 24–34 (2020)
15. Rahmani, A., Liljeberg, P., Preden, J.S., Jantsch, A.: Fog Computing in the Internet of Things. Springer, Cham (2018)

An Algorithm to Select an Energy-Efficient Server for an Application Process in a Cluster

Kaiya Noguchi[1(✉)], Takumi Saito[1(✉)], Dilawaer Duolikun[1(✉)],
Tomoya Enokido[2(✉)], and Makoto Takizawa[1(✉)]

[1] Hosei University, Tokyo, Japan
{kaiya.noguchi.9w,takumi.saito.3j}@stu.hosei.ac.jp,
dilewerdolkun@gmail.com, makoto.takizawa@computer.org
[2] Rissho University, Tokyo, Japan
eno@ris.ac.jp

Abstract. We have to decrease electric energy consumption of information systems, especially servers to realize green societies. Information systems are composed of servers. One server is selected to perform an application process issued by a client so as to not only increase the performance but also reduce the energy consumption of the servers. In our previous studies, the ESECS algorithm is proposed to select a host server to perform an application process. Here, only the energy consumption of each server is compared with the other servers. In this paper, we take into consideration the total energy consumption of the other servers in addition to a host server. We propose an MTES (Minimization of Total Energy consumption of Servers) algorithm to select a host server where the total energy consumption of all the servers can be minimized in this paper. In the evaluation, we show the total energy consumption of servers and the average execution time of processes can be reduced in the MTES algorithm compared with other algorithms.

Keywords: Green information systems · MTES algorithm · Power consumption model · Computation model

1 Introduction

In information systems, servers in clusters consume huge amount of electric energy compared with clients and devices. It is critical to reduce the energy consumption of information systems, especially servers in clusters to realize green societies by reducing carbon dioxide emission on the earth. For each new application process issued by a client, one host server is selected in the cluster where the application process is to be performed. In our previous studies [4,9–11], algorithms to select an energy-efficient host server for each new application process are proposed. In order to select an energy-efficient host server, we need a model to give the power consumption of each server to perform application processes.

L. Barolli et al. (Eds.): EIDWT 2021, LNDECT 65, pp. 82–92, 2021.
https://doi.org/10.1007/978-3-030-70639-5_8

A pair of the SPC (Simple Power Consumption) and SC (Simple Computation) models [5,6] and a pair of the MLPC (Multi-Level Power Consumption) and MLC (Multi-Level Computation) models [10,11] are proposed to give the power consumption [W] of a server and the execution time of each application process on the server, respectively. The models are *macro-level* ones [5] where the total power consumption of a whole server to perform application processes is considered without being conscious of the power consumption of each hardware component like CPU.

In the ESECS (ESEC Selection) algorithm [12] to select a host server, we consider not only current active application processes but also possible application processes to be issued after current time to estimate the energy consumption of each server and the execution time of each application process. In the ESECS algorithm, the energy to be consumed by each host server is compared with the other servers. While the application process is performed on a host server, another server consumes power even if no process is performed. In the TEA algorithm [11], the total energy consumption of not only a host server but also the other servers are taken into consideration. In this paper, we newly propose an MTES (Minimization of Total Energy consumption of Servers) algorithm to select a server to perform an application process p_i. Here, we estimate termination time ET_t every active process to terminate on each server s_t. We also estimate time NT_t every active process and the new process p_i to terminate on each server s_t. In the TEA algorithm, the total energy TEE_t to be consumed by all the servers from current time to time NT_t for each server s_t is obtained and a server s_t whose TEE_t is minimum is selected as a host server of the new processes p_i. In the MTES algorithm, for each server s_t, time TT_t every process to terminate on every server is obtained, i.e. longest one of NT_t and ET_u for every server s_u. Then, the total energy MEE_t to be consumed by all the servers from current time to time TT_t is calculated. A server s_t whose MEE_t is minimum is selected to perform the new process p_i.

In the evaluation, we show the total energy consumption of servers and the average execution time of application processes can be more reduced in the MTES algorithm than the ESECS algorithm and other algorithms.

In Sect. 2, we present a system model and the power consumption and computation models. In Sect. 3, we propose the MTES algorithm to select servers to energy-efficiently perform application processes. In Sect. 5, we evaluate the MTES algorithm.

2   Power Consumption and Computation Models

A cluster S is composed of servers $s_1, \cdots, s_m$ ($m \geq 1$). A client c_i issues a request q_i to a load balancer L. The load balancer L selects a host server s_t in the cluster S, so that the total energy consumption of the servers $s_1, \cdots, s_m$ can be reduced and the average execution time of application processes on the servers can be shortened and forwards the request q_i to the server s_t. An application process p_i is created to handle the request q_i and performed on the sever s_t. The server s_t sends a reply r_i to the client c_i.

In this paper, a term *process* stands for an application process to be performed on a server, which only consumes CPU resources of a server, i.e. computation process [5]. A process p_i is *active* on a server s_t if and only if (iff) the process p_i is performed on the server s_t. Otherwise, the process p_i is *idle*. $CP_t(\tau)$ is a set of active processes on a server s_t at time τ.

A server s_t is equipped with $np_t(\geq 1)$ homogeneous CPUs [1] each of which includes $nc_t(\geq 1)$ homogeneous cores. Each core supports the same number ct_t of threads. Thus, a server s_t supports processes with $nt_t(= np_t \cdot nc_t \cdot ct_t)$ threads. Each process is at a time performed on one thread. Here, the electric power $NE_t(n)$ [W] of a server s_t to concurrently perform $n(\geq 0)$ processes is given in the MLPC (Multi-Level Power Consumption) model [9–11]:

[MLPC Model]

$$
NE_t(n) = \begin{cases}
minE_t & \text{if } n = 0. \\
minE_t + n \cdot (bE_t + cE_t + tE_t) & \text{if } 1 \leq n \leq np_t. \\
minE_t + np_t \cdot bE_t + n \cdot (cE_t + tE_t) & \text{if } np_t < n \leq nc_t \cdot np_t. \\
minE_t + np_t \cdot (bE_t + nc_t \cdot cE_t) + n \cdot tE_t & \text{if } nc_t \cdot np_t < n < nt_t. \\
maxE_t & \text{if } n \geq nt_t.
\end{cases} \tag{1}
$$

The power consumed by a server s_t increases by bE_t, cE_t, and tE_t [W] as a CPU, core, and thread are activated, respectively.

The electric power $E_t(\tau)$ [W] consumed by a server s_t at time τ is assumed to be $NE_t(|CP_t(\tau)|)$ in our approach. That is, the electric power consumption of a server s_t depends on the number n of active processes. A server s_t consumes electric energy $\sum_{\tau=st}^{et} E_t(\tau)$ [W · time unit (tu)] from time st to et.

Let $minT_{ti}$ show the minimum execution time [tu] of a process p_i, i.e. only the process p_i is performed on a thread of a server s_t without any other process. Let $minT_i$ be a minimum one of $minE_{1i}$, $\cdots$, $minT_{mi}$. That is, $minT_i$ is $minT_{fi}$ of a server s_f which supports the fastest thread. The *computation amount* of a process p_i is defined to be $minT_i$ [5–8]. In most server applications, well-known processes, i.e. transactions are performed. Here, we can get the minimum execution time $minT_{ti}$ of each process p_i on each server s_t.

The thread computation rate TCR_t of a server s_t is $minT_i/minT_{ti}$ (≤ 1). The maximum computation rate $XSCR_t$ of a server s_t is $nt_t \cdot TCR_t$. If a process p_i is only performed on the server s_t without any other process on a server s_t, the process p_i is performed at the maximum rate $XPCR_t$ which is TCR_t. We assume every active process p_i is fairly allocated with the same computation rate on each server. The process computation rate $NPR_t(n)$ of each active process on a server s_t where n active processes are performed is defined in the MLC (Multi-Level Computation) model [2,4,10]:

[MLC (Multi-Level Computation) Model]

$$
NPR_t(n) = \begin{cases}
nt_t \cdot TCR_t/n & \text{if } n > nt_t. \\
TCR_t & \text{if } n \leq nt_t.
\end{cases} \tag{2}
$$

The computation rate $PR_{ti}(\tau)$ of each process p_i at time τ is assumed to be $NPR_t(n)$ on a server s_t where $n = |CP_t(\tau)|$. The computation rate $NSR_t(n)$ of a server s_t is $n \cdot NPR_t(n)$ where n processes are active. If a process p_i starts on a server s_t at time st and terminates at time et, $\sum_{\tau=st}^{et} NPR_t(|CP_t(\tau)|) = minT_i$. Thus, $minT_i$ shows the amount of computation of each process p_i.

A process p_i is modeled to be performed on a server s_t as follows [6–8, 10]:

[Computation Model of a Process p_i]

1. At time τ a process p_i starts, the computation residue RP_i of a process p_i is $minT_i$;
2. At each time τ, $RP_i = RP_i - NPR_t(|CP_t(\tau)|)$;
3. If $RP_i \leq 0$, the process p_i terminates at time τ.

The total computation residue RS_t of active processes on a server s_t is $\sum_{p_j \in CP_t(\tau)} RP_j$. Let n_t be the total number $|CP_t(\tau)|$ of active processes performed on a server s_t at time τ. The execution time $NET_t(n_t)$ [tu] of n_t active processes on a server s_t is given as follows;

$$NET_t(n_t) = \sum_{p_j \in CP_t(\tau)} R_j / SCR_t(n_t). \tag{3}$$

A server s_t consumes electric power $NE_t(n_t)$ [W] for time $NET_t(n_t)$ [tu]. Hence, the energy consumption $NEE_t(n_t)$ of a server s_t to perform active processes is given as follows;

$$NEE_t(n_t) = NET_t(n_t) \cdot NE_t(n_t). \tag{4}$$

Figure 1 shows the estimation model of energy consumption of a server s_t at time τ. Here, EE_t gives the estimated energy consumption of a server s_t to perform every active processes and $etime_i$ is the estimated termination time of each current active process p_i on a server s_t.

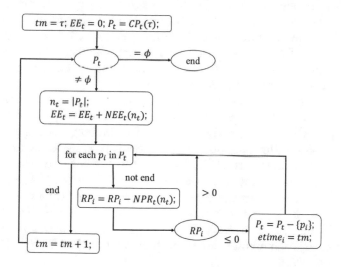

Fig. 1. Estimation model.

3 An MTES Algorithm to Select a Host Server

A client issues a request to a load balancer L of a cluster S of servers $s_1, \cdots, s_m$ $(m \geq 1)$. The load balancer L selects a server s_t in the cluster S. Then, a process p_i to handle the request is created and performed on the server s_t. Let P be a set $\{p_1, \cdots, p_n\}(n \geq 1)$ of processes to be performed on the servers in the cluster S. $minT$ is the average minimum execution time of all the processes.

Suppose a new process p_i is issued to a server s_t at time τ. Then, not only current active processes $CP_t(\tau)$ but also the new process p_i are performed on the server s_t. Let n_t be the number $|CP_t(\tau)|$ of active processes. RP_i shows the computation residue of each process p_i. If p_i is a new process, $RP_i = minT_i$. RS_t is the total computation residue of all the current active processes on the server s_t. Here, (n_t+1) processes with the total computation residue $(\sum_{p_j \in CP_t(\tau)} RP_j + minT_i)$ are performed on the server s_t at time τ. Hence, the execution time $TET_t(n_t, p_i)$ [tu] of the server s_t to perform $(n_t + 1)$ processes, i.e. n_t current active processes and one new process p_i is given as follows;

$$TET_t(n_t, p_i) = (\sum_{p_j \in CP_t(\tau)} RP_j + minT_i)/SCR_t(n_t + 1). \tag{5}$$

The server s_t consumes the electric power $NE_t(n_t + 1)$ [W] for $TET_t(n_t, p_i)$ time units [tu]. Hence, the server s_t consumes energy $TEE_t(n_t, p_i)$ [W·tu]:

$$TEE_t(n_t, p_i) = TET_t(n_t, p_i) \cdot NE_t(n_t + 1). \tag{6}$$

In the ESECS algorithm [12], only the energy consumption $TEE_t(n_t, p_i)$ of each server s_t to perform a new process p_i is considered. That is, a server s_t where $TEE_t(n_t, p_i)$ is minimum is selected as a host server of a new process p_i in a cluster. Here, another server s_u also consumes energy while the process p_i is performed on the host server s_t. For example, a server s_u consumes the minimum power $minE_u$ [W] even if no process is performed, i.e. s_u is idle. In the paper [3], the TEA (Totally Energy Aware) server selection algorithm is proposed, where the total energy consumption of not only a host server of a new process p_i but also the other servers are taken into consideration.

Suppose a cluster S is composed of three servers s_t, s_u, and s_v as shown in Fig. 2. Here, ET_t, ET_u, and ET_v show time $NET_t(n_t)$, $NET_u(n_u)$, and $NET_v(n_v)$ [tu] when all the active processes to terminate on the servers s_t, s_u, and s_v, respectively. Here, suppose a new process p_i is performed on the server s_t. NT_t shows termination time $TET_t(n_t, p_i)$ [tu] of not only every active process but also the new process p_i on the server s_t. For each server s_x in the cluster S, $TE_x(\gamma)$ is the energy consumption [W · tu] for time τ to time $\tau + \gamma$, i.e. $TE_x(\gamma) = \sum_{y=\tau}^{\tau+\gamma} E_x(y)$. TE shows the total energy to be consumed by the servers in the cluster S, i.e. $TE = \sum_{s_x \in S} TE_x(\gamma)$. In this paper, a server s_t where $TE_t(\gamma)$ is minimum for some time units γ is selected. In the TEA algorithm, γ is NT_t.

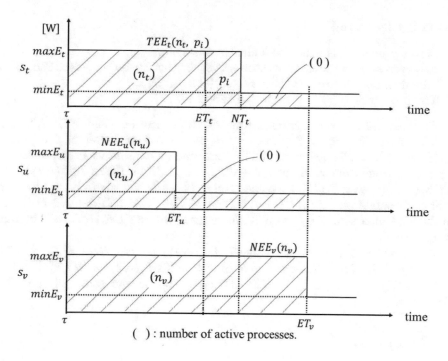

Fig. 2. Energy consumption of servers.

Suppose a server s_t is a host server of a new process p_i. Here, the server s_t consumes the energy $HTE_t(\gamma)$ from current time τ to time $\tau + \gamma$ for $\gamma \geq NT_t$:

$$HTE_t(\gamma) = TEE_t(n_t, \ p_i) + minE_t \cdot (\gamma - NT_t). \tag{7}$$

Another server $s_u \ (\neq \ s_t)$ consumes energy $OTE_u(\gamma)$ from time τ to time $\tau + \gamma$ as follows:

$$OTE_u(\gamma) = \begin{cases} NEE_u\,(n_u) + minE_u \cdot (\gamma - ET_u) & \text{if } \gamma \geq ET_u. \\ NEE_u(n_u) \cdot \gamma \, / \, ET_u & \text{otherwise.} \end{cases} \tag{8}$$

Thus, the total energy $TTE_t(\gamma)$ [W tu] to be consumed by all the servers from time τ to $\tau + \gamma$, where a server s_t is a host server of a new process p_i, is given as follows:

$$TTE_t\,(\gamma) = HTE_t\,(\gamma) + \sum_{s_u \in S(\neq s_t)} OTE_u\,(\gamma). \tag{9}$$

In this paper, we propose an MTES (Minimization of Total Energy consumption of Servers) algorithm to select a host server for a new process p_i as follows;

[MTES Algorithm]

1. A new process p_i is issued to a cluster of servers.
2. A server s_t is selected as a host server in the cluster, where $TTE_t(\gamma)$ is minimum and γ is $max(NT_t, \{ET_u | s_u \ (\neq s_t) \in S\})$.
3. The process p_i is performed on the server s_t.

Suppose ET_v is the maximum one of the execution time ET_t, ET_u and ET_v as shown in Fig. 2. Suppose a server s_t is a host server of the process p_i. Since $ET_v > NT_t$, γ is ET_v. The total energy consumption of the servers s_t, s_u, s_v from time τ to time $\tau + ET_v$ are $HTE_t(ET_v)$. $TTE_u(ET_v)$, and $TTE_v(ET_v)(= NEE_v(ET_v))$, respectively. The hatched area shows the total energy consumption $TTE_t(ET_v) = HTE_t(ET_v) + TTE_u(ET_v) + NEE_v(ET_v)$. In the TEA algorithm [11], $\gamma = NT_v$. A server s_t whose $TEE_t(NT_t)$ is minimum is selected.

Next, suppose NT_t is the longest, i.e. $NT_t > ET_u$ and $NT_t > ET_v$. Here, $\gamma = NT_t$. The total energy consumption $TTE_t(NT_t)$ is $HTE_t(NT_t)$ $(= TTE_t(NT_t)) + TTE_u(NT_t) + TTE_v(NT_t))$. This is the same as the TEA algorithm.

4 Evaluation

By using the EDS (Eco Distributed System) simulator [13], selection algorithms like the ESECS algorithm to select a server to perform a new process issued by a client are evaluated in terms of the total energy consumption of servers and the average execution time of processes.

In the evaluation, we consider a cluster S of sixteen servers $s_1, \cdots, s_{16}$ ($m = 16$). The performance and energy parameters of the servers are shown in Table 1. There are four types of homogeneous servers. The servers $s_1, \cdots, s_4$ are the fastest where the thread computation rates $TCR_1, \cdots, TCR_4$ are 1, i.e. $TCR_1 = \cdots = TCR_4 = 1$. The servers $s_5 \cdots s_8$ are the second type, where $TCR_5, \cdots, TCR_8$ are 0.8. The servers $s_9, \cdots, s_{12}$ are the third type where $TCR_9 = \cdots = TCR_{12} = 0.6$. The servers $s_1, \cdots s_{16}$ are the fourth type where $TCR_{13} = \cdots = TCR_{16} = 0.4$. For example, the server s_1 supports the maximum server computation rates $XSCR_1 = 16$ by sixteen threads, $nt_1 = 16$. The maximum power consumption $maxE_1, \cdots, maxE_4$ are 230 [W] and the minimum power consumption $minE_1, \cdots, minE_4$ are 150 [W]. The servers $s_{13}, \cdots, s_{16}$ support the maximum computation rates $XSCR_{13}, \cdots, XSCR_{16}$ which are 3.2 by eight threads, $nt_{13} = \cdots = nt_{16} = 8$, where $maxE_{13} = \cdots = maxE_{16} = 77$ [W] and $minE_{13} = \cdots = minE_{16} = 40$ [W]. The servers $s_5, \cdots, s_{12}$ support twelve threads, while $maxE_5, \cdots, maxE_8 > maxE_1, \cdots, maxE_4$ and $minE_5, \cdots, minE_8 > minE_1, \cdots, minE_4$. $XSCR_5 = \cdots = XSCR_8 = 9.6$ and $XSCR_9 = \cdots = XSCR_{12} = 7.2$. The servers $s_9, \cdots, s_{12}$ are more energy-efficient than the servers $s_5, \cdots, s_8$. The servers $s_{13}, \cdots, s_{16}$ show desktop PC servers. The servers $s_1, \cdots, s_4$ stand for server computers.

Let P be a set of processes $p_1, \cdots, p_n$ $(n \geq 1)$ to be issued to the cluster S. For each process p_i in the set P, the starting time $stime_i$ [tu] and the minimum execution time $minT_i$ [tu] are randomly given as $0 < stime_i \leq xtime$ and $5 \leq minT_i \leq 25$. Here, $xtime = 1,000$ [tu]. A variable P_t shows a set of active processes on each server s_t. Initially, $P_t = \phi$. At each time τ, if there is a process p_i whose start time $(stime_i)$ is τ, one server s_t is selected by a selection algorithm. The process p_i is added to the variable P_t, i.e. $P_t = P_t \cup \{p_i\}$. The computation residue RP_i of the process p_i is $minT_i$. For each server s_t, active processes in P_t are performed. The variable n_t denotes the number $|P_t|$ of active processes on each server s_t, i.e. $n_t = |P_t|$. The energy variable E_t is incremented by the power consumption $NE_t(|P_t|)$. If $|P_t| = \phi$, E_t is incremented by $minE_t$. If $|P_t| > 0$, the active time variable T_t [tu] is incremented by one [tu]. The variable T_t [tu] shows how long the server s_t is active, i.e. some active process is performed. For each process p_i in the set P_t, the computation residue RP_i of the process p_i is decremented by the process computation rate $NPR_t(n_t)$. If $RP_i \leq 0$, the process p_i terminates, i.e. $P_t = P_t - \{p_i\}$ and $P = P - \{p_i\}$. Here, the termination time $etime_i$ is τ and the execution time PT_i is $etime_i - stime_i + 1$. Until the set P gets empty, the steps are iterated. The variables E_t and T_t give the total energy consumption and execution time of each server s_t, respectively, and the variable PT_i shows the execution time of each process p_i.

Table 1. Parameters of servers.

s_t	np_t	nc_t	nt_t	TCR_t	$XSCR_t$	$minE_t$ [W]	$maxE_t$ [W]	pE_t [W]	cE_t [W]	tE_t [W]
$s_1, \cdots, s_4$	1	8	16	1.0	16.0	150.0	270.0	40.0	8.0	1.0
$s_5, \cdots, s_8$	1	6	12	0.8	9.6	128.0	200.0	30.0	5.0	1.0
$s_9, \cdots, s_{12}$	1	6	12	0.6	7.2	80.0	130.0	20.0	3.0	1.0
$s_{13}, \cdots, s_{16}$	1	4	8	0.4	3.2	40.0	67.0	15.0	2.0	0.5

Given the process set P and the server set S, the EDS simulator [12] obtains the total energy consumption E_t and total active time T_t of each server s_t $(t = 1, \cdots, 16)$ in the cluster S and the execution time PT_i of each process p_i $(i = 1, \cdots, n)$ in the set P. In the evaluation, the total energy consumption E [W tu] of the servers is $E_1 + \cdots + E_{16}$ and the total active time TAT [tu] is $T_1 + \cdots + T_{16}$. The average execution time APT [tu] of the n $(= |P|)$ processes is $\sum_{p_i \in P} PT_i / n$.

We consider the EA (Energy-aware) server selection [11], ESECS [12], and TEA [11] algorithms to compare with the MTES algorithm. In the EA and ESECS algorithm, only the energy consumption of a host server is considered. In the TEA and MTES algorithms, the total energy to be consumed by not only a host server but also the other servers is considered. The EA, TEA, and MTES algorithms take usage of the estimation model where only active processes are considered. The ESECS algorithm uses the ESECS estimation algorithm where possible processes to be issued after current time are taken into consideration in addition to active processes.

Figure 3 shows the total energy consumption E [W tu] of the servers of the MTES, TEA, ESECS, and EA algorithms obtained by the EDS simulator. The total energy consumption E of the servers in the MTES algorithm is smaller than the other algorithms as shown in Fig. 3. For example, the total energy consumption E of the servers of the MTES algorithm is 16.6% and 16.5% smaller than the EA algorithm for $n = 2,200$ and 3,000, respectively. For $n = 2,200$, the total energy consumption E of the MTES algorithm is 11.2% smaller than TEA algorithm.

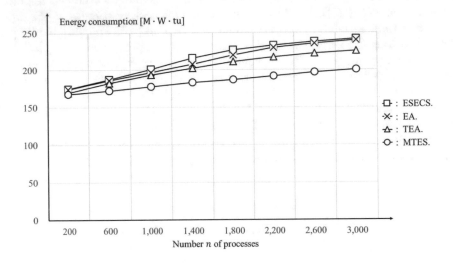

Fig. 3. Total energy consumption E of servers.

Figure 4 shows the total active time TAT [tu] of the servers for number n of processes. The total active time TAT of the MTES algorithm is shorter than the other algorithms. For $n = 2,200$ and 3,000, the TAT of the MTES algorithm is 82.3% and 78.8% smaller than the EA algorithm, respectively. For $n = 1,800$, the TAT of the MTES algorithm is 70.3% smaller than the TEA algorithm. This means, more number of processes can be performed on the servers in the MTES algorithm than the TEA algorithm.

Figure 5 shows the average execution time APT [tu] of n processes. The average execution time APT of the MTES algorithm is longer than the TEA algorithm. For $n = 2,200$ and 3,000, the APT of the servers is 4.1% and 3.8% longer than the TEA algorithm, respectively. In the TEA algorithm, the more number of processes are performed on the more powerful servers like the server s_1. On the other hand, processes are distributed to all the servers in the MTES algorithm. Some processes are performed on less powerful servers like the server s_4. This means, it takes longer time to perform the processes. Hence, the APT of the MTES algorithm is longer than the TEA algorithm.

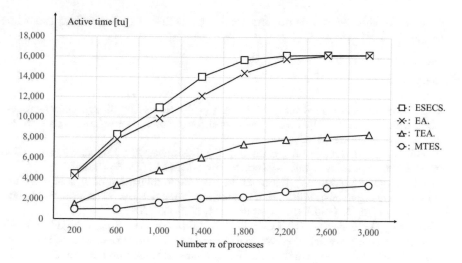

Fig. 4. Total active time TAT of servers.

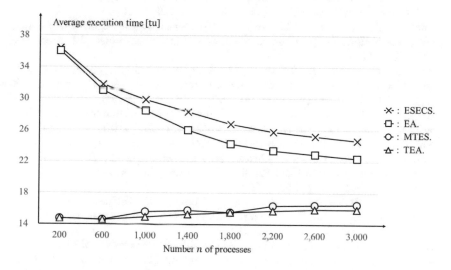

Fig. 5. Average execution time APT of processes.

5   Concluding Remarks

In this paper, we newly proposed the MTES algorithm to select a host server of a new application process. Here, the total energy consumption of not only the host server but also the other servers is tried to be more reduced. A host server of a new process is selected where the total energy to be consumed by all the servers to perform every current active process and the new process is the smallest. In the evaluation, we showed the total energy consumption and the average active time of servers are shorter in the MTES algorithm than the other algorithms.

References

1. Intel xeon processor 5600 series: The next generation of intelligent server processors, white paper (2010). http://www.intel.com/content/www/us/en/processors/xeon/xeon-5600-brief.html
2. Duolikun, D., Aikebaier, A., Enokido, T., Takizawa, M.: Energy-aware passive replication of processes. Int. J. Mob. Multimed. **9**(1,2), 53–65 (2013)
3. Duolikun, D., Enokido, T., Takizawa, M.: Dynamic migration of virtual machines to reduce energy consumption in a cluster international journal of grid and utility computing. Int. J. Grid Util. Comput. **9**(4), 357–366 (2018)
4. Duolikun, D., Kataoka, H., Enokido, T., Takizawa, M.: Simple algorithms for selecting an energy-efficient server in a cluster servers. In: Proceedings of the International Journal of Communication Networking and Distributed Systems, pp. 1–25 (2018)
5. Enokido, T., Ailixier, A., Takizawa, M.: A model for reducing power consumption in peer-to-peer systems. IEEE Syst. J. **4**(2), 221–229 (2010)
6. Enokido, T., Ailixier, A., Takizawa, M.: Process allocation algorithms for saving power consumption in peer-to-peer systems. IEEE Trans. Industr. Electron. **58**(6), 2097–2105 (2011)
7. Enokido, T., Ailixier, A., Takizawa, M.: An extended simple power consumption model for selecting a server to perform computation type processes in digital ecosystems. IEEE Trans. Industr. Electron. **10**(2), 1627–1636 (2014)
8. Enokido, T., Takizawa, M.: An integrated power consumption model for distributed system. IEEE Trans. Industr. Electron. **60**(2), 824–836 (2013)
9. Kataoka, H., Duolikun, D., Enokido, T., Takizawa, M.: Energy-efficient virtualisation of threads in a server cluster. In: Proceedings of the 10th International Conference on Broadband and Wireless Computing, Communication and Applications (BWCCA 2015), pp. 288–295 (2015)
10. Kataoka, H., Nakamura, S., Duolikun, D., Enokido, T., Takizawa, M.: Multi-level power consumption model and energy-aware server selection algorithm. Int. J. Grid Util. Comput. (IJGUC) **8**(3), 201–210 (2017)
11. Kataoka, H., Sawada, A., Duolikun, D., Enokido, T., Takizawa, M.: Energy-aware server selection algorithm in a scalable cluster. In: Proceedings of IEEE the 30th International Conference on Advanced Information Networking and Applications (AINA 2016), pp. 565–572 (2016)
12. Noguchi, K., Saito, T., Duolikun, D., Enokido, T., Takizawa, M.: An algorithm to select a server to minimize the total energy consumption of a cluster. In: Proceedings of the 15-th International Conference on P2P, Parallel, Grid, Cloud and Internet Computing (3PGCIC 2020), pp. 18–28 (2020)
13. Noguchi, K., Saito, T., Duolikun, D., Enokido, T., Takizawa, M.: An algorithm to select an energy-efficient server for an application process in a cluster of servers. In: Proceedings of the 12-th International Conference on Intelligent Networking and Collaborative Systems (INCoS 2020), pp. 101–111 (2020)

Performance Comparison of Constriction and Linearly Decreasing Inertia Weight Router Replacement Methods for WMNs by WMN-PSOSA-DGA Hybrid Simulation System Considering Chi-Square Distribution of Mesh Clients

Admir Barolli[1], Shinji Sakamoto[2], Phudit Ampririt[3], Seiji Ohara[3],
Leonard Barolli[4(✉)], and Makoto Takizawa[5]

[1] Department of Information Technology, Aleksander Moisiu University of Durres,
L.1, Rruga e Currilave, Durres, Albania
admir.barolli@gmail.com
[2] Department of Computer and Information Science, Seikei University,
3-3-1 Kichijoji-Kitamachi, Musashino-shi, Tokyo 180-8633, Japan
shinji.sakamoto@ieee.org
[3] Graduate School of Engineering, Fukuoka Institute of Technology,
3-30-1 Wajiro-Higashi, Higashi-Ku, Fukuoka 811-0295, Japan
iceattpon12@gmail.com, seiji.ohara.19@gmail.com
[4] Department of Information and Communication Engineering, Fukuoka Institute
of Technology, 3-30-1 Wajiro-Higashi, Higashi Ku, Fukuoka 811-0295, Japan
barolli@fit.ac.jp
[5] Department of Advanced Sciences, Faculty of Science and Engineering,
Hosei University, 3-7-2 Kajino-Cho, Koganei-Shi, Tokyo 184-8584, Japan
makoto.takizawa@computer.org

Abstract. Wireless Mesh Networks (WMNs) are gaining a lot of attention from researchers due to their advantages such as easy maintenance, low upfront cost, and high robustness. Connectivity and stability directly affect the performance of WMNs. However, WMNs have some problems such as node placement problem, hidden terminal problem and so on. In our previous work, we implemented a simulation system to solve the node placement problem in WMNs considering Particle Swarm Optimization (PSO), Simulated Annealing (SA) and Distributed Genetic Algorithm (DGA), called WMN-PSOSA-DGA. In this paper, we evaluate the performance of Constriction Method (CM) and Linearly Decreasing Inertia Weight Method (LDIWM) for WMNs using WMN-PSOSA-DGA hybrid simulation system considering Chi-square distribution of mesh clients. Simulation results show that a good performance is achieved for CM compared with the case of LDIWM.

1   Introduction

The wireless networks and devices are becoming increasingly popular and they provide users access to information and communication anytime and anywhere [1,3,12,13,15,19,22,23]. Wireless Mesh Networks (WMNs) are gaining a lot of attention because of their low cost nature that makes them attractive for providing wireless Internet connectivity. A WMN is dynamically self-organized and self-configured, with the nodes in the network automatically establishing and maintaining mesh connectivity among them-selves (creating, in effect, an ad hoc network). This feature brings many advantages to WMNs such as low up-front cost, easy network maintenance, robustness and reliable service coverage [2].

Mesh node placement in WMN can be seen as a family of problems, which are shown (through graph theoretic approaches or placement problems, e.g. [10,17]) to be computationally hard to solve for most of the formulations [35]. We consider the version of the mesh router nodes placement problem in which we are given a grid area where to deploy a number of mesh router nodes and a number of mesh client nodes of fixed positions (of an arbitrary distribution) in the grid area. The objective is to find a location assignment for the mesh routers to the cells of the grid area that maximizes the network connectivity and client coverage. Network connectivity is measured by Size of Giant Component (SGC) of the resulting WMN graph, while the user coverage is simply the number of mesh client nodes that fall within the radio coverage of at least one mesh router node and is measured by Number of Covered Mesh Clients (NCMC). Node placement problems are known to be computationally hard to solve [14,36]. In previous works, some intelligent algorithms have been investigated for node placement problem [5,11,18,20,21,27,28,37].

In [26], we implemented a Particle Swarm Optimization (PSO) and Simulated Annealing (SA) based simulation system, called WMN-PSOSA. Also, we implemented another simulation system based on Genetic Algorithm (GA), called WMN-GA [5,16], for solving node placement problem in WMNs. Then, we designed a hybrid intelligent system based on PSO, SA and DGA, called WMN-PSOSA-DGA [25].

In this paper, we evaluate the performance of Constriction Method (CM) and Linearly Decreasing Inertia Weight Method (LDIWM) for WMNs using WMN-PSOSA-DGA simulation system considering Chi-square distribution of mesh clients.

The rest of the paper is organized as follows. We present our designed and implemented hybrid simulation system in Sect. 2. The simulation results are given in Sect. 3. Finally, we give conclusions and future work in Sect. 4.

2   Proposed and Implemented Simulation System

Distributed Genetic Algorithm (DGA) has been focused from various fields of science. DGA has shown their usefulness for the resolution of many computationally hard combinatorial optimization problems. Also, Particle Swarm

Optimization (PSO) and Simulated Annealing (SA) are suitable for solving NP-hard problems.

2.1 Velocities and Positions of Particles

WMN-PSOSA-DGA decides the velocity of particles by a random process considering the area size. For instance, when the area size is $W \times H$, the velocity is decided randomly from $-\sqrt{W^2 + H^2}$ to $\sqrt{W^2 + H^2}$. Each particle's velocities are updated by simple rule [24].

For SA mechanism, next positions of each particle are used for neighbor solution s'. The fitness function f gives points to the current solution s. If $f(s')$ is larger than $f(s)$, the s' is better than s so the s is updated to s'. However, if $f(s')$ is not larger than $f(s)$, the s may be updated by using the probability of $\exp\left[\frac{f(s')-f(s)}{T}\right]$. Where T is called the "Temperature value" which is decreased with the computation so that the probability to update will be decreased. This mechanism of SA is called a cooling schedule and the next Temperature value of computation is calculated as $T_{n+1} = \alpha \times T_n$. In this paper, we set the starting temperature, ending temperature and number of iterations. We calculate α as

$$\alpha = \left(\frac{\text{SA ending temperature}}{\text{SA starting temperature}} \right)^{1.0/\text{number of iterations}}.$$

It should be noted that the positions are not updated but the velocities are updated in the case when the solusion s is not updated.

2.2 Routers Replacement Methods

A mesh router has x, y positions and velocity. Mesh routers are moved based on velocities. There are many router replacement methods. In this paper, we use CM and LDIWM.

Constriction Method (CM)
  CM is a method which PSO parameters are set to a week stable region ($\omega = 0.729$, $C_1 = C2 = 1.4955$) based on analysis of PSO by M. Clerc et al. [6,9,31].
Random Inertia Weight Method (RIWM)
  In RIWM, the ω parameter is changing randomly from 0.5 to 1.0. The C_1 and C_2 are kept 2.0. The ω can be estimated by the week stable region. The average of ω is 0.75 [8,29,33].
Linearly Decreasing Inertia Weight Method (LDIWM)
  In LDIWM, C_1 and C_2 are set to 2.0, constantly. On the other hand, the ω parameter is changed linearly from unstable region ($\omega = 0.9$) to stable region ($\omega = 0.4$) with increasing of iterations of computations [7,34].
Linearly Decreasing Vmax Method (LDVM)
  In LDVM, PSO parameters are set to unstable region ($\omega = 0.9$, $C_1 = C_2 = 2.0$). A value of V_{max} which is maximum velocity of particles is considered. With increasing of iteration of computations, the V_{max} is kept decreasing linearly [8,30,32].

Rational Decrement of Vmax Method (RDVM)

In RDVM, PSO parameters are set to unstable region ($\omega = 0.9$, $C_1 = C_2 = 2.0$). The V_{max} is kept decreasing with the increasing of iterations as

$$V_{max}(x) = \sqrt{W^2 + H^2} \times \frac{T - x}{x}.$$

Where, W and H are the width and the height of the considered area, respectively. Also, T and x are the total number of iterations and a current number of iteration, respectively [24].

2.3   DGA Operations

Population of individuals: Unlike local search techniques that construct a path in the solution space ,jumping from one solution to another one through local perturbations, DGA use a population of individuals giving thus the search a larger scope and chances to find better solutions. This feature is also known as "exploration" process in difference to "exploitation" process of local search methods.

Selection: The selection of individuals to be crossed is another important aspect in DGA as it impacts on the convergence of the algorithm. Several selection schemes have been proposed in the literature for selection operators trying to cope with premature convergence of DGA. There are many selection methods in GA. In our system, we implement 2 selection methods: Random method and Roulette wheel method.

Crossover operators: Use of crossover operators is one of the most important characteristics. Crossover operator is the means of DGA to transmit best genetic features of parents to offsprings during generations of the evolution process. Many methods for crossover operators have been proposed such as Blend Crossover (BLX-α), Unimodal Normal Distribution Crossover (UNDX), Simplex Crossover (SPX).

Mutation operators: These operators intend to improve the individuals of a population by small local perturbations. They aim to provide a component of randomness in the neighborhood of the individuals of the population. In our system, we implemented two mutation methods: uniformly random mutation and boundary mutation.

Escaping from local optimal: GA itself has the ability to avoid falling prematurely into local optimal and can eventually escape from them during the search process. DGA has one more mechanism to escape from local optimal by considering some islands. Each island computes GA for optimizing and they migrate its gene to provide the ability to avoid from local optimal.

Convergence: The convergence of the algorithm is the mechanism of DGA to reach to good solutions. A premature convergence of the algorithm would cause that all individuals of the population be similar in their genetic features and thus the search would result ineffective and the algorithm getting stuck into local optimal. Maintaining the diversity of the population is therefore very important to this family of evolutionary algorithms.

In following, we present fitness function, migration function, particle pattern and gene coding.

2.4   Fitness and Migration Functions

The determination of an appropriate fitness function, together with the chromosome encoding are crucial to the performance. Therefore, one of most important thing is to decide the determination of an appropriate objective function and its encoding. In our case, each particle-pattern and gene has an own fitness value which is comparable and compares it with other fitness value in order to share information of global solution. The fitness function follows a hierarchical approach in which the main objective is to maximize the SGC in WMN. Thus, the fitness function of this scenario is defined as

$$\text{Fitness} = 0.7 \times \text{SGC}(\boldsymbol{x}_{ij}, \boldsymbol{y}_{ij}) + 0.3 \times \text{NCMC}(\boldsymbol{x}_{ij}, \boldsymbol{y}_{ij}).$$

Our implemented simulation system uses Migration function as shown in Fig. 1. The Migration function swaps solutions between PSOSA part and DGA part.

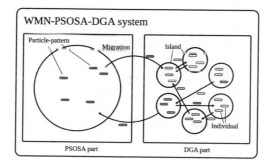

Fig. 1. Model of WMN-PSOSA-DGA migration.

2.5   Particle-Pattern and Gene Coding

In order to swap solutions, we design particle-patterns and gene coding carefully. A particle is a mesh router. Each particle has position in the considered area and velocities. A fitness value of a particle-pattern is computed by combination of mesh routers and mesh clients positions. In other words, each particle-pattern is a solution as shown is Fig. 2.

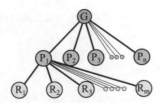

G: Global Solution
P: Particle-pattern
R: Mesh Router
n: Number of Particle-patterns
m: Number of Mesh Routers

Fig. 2. Relationship among global solution, particle-patterns and mesh routers in PSOSA part.

A gene describes a WMN. Each individual has its own combination of mesh nodes. In other words, each individual has a fitness value. Therefore, the combination of mesh nodes is a solution.

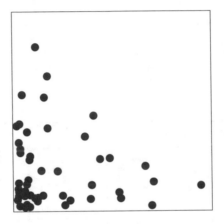

Fig. 3. Chi-square distribution.

3   Simulation Results

In this section, we show simulation results. In this work, we analyze the performance of CM and LDIWM for WMNs by WMN-PSOSA-DGA hybrid intelligent system considering Chi-square client distribution.

Our proposed system can generate many client distributions [4]. Here, we consider Chi-square distribution of mesh clients as shown in Fig. 3. The number of mesh routers is considered 16 and the number of mesh clients 48. We conducted simulations 10 times in order to avoid the effect of randomness and create a general view of results. We show the parameter setting for WMN-PSOSA-DGA in Table 1.

Table 1. WMN-PSOSA-DGA parameters.

Parameters	Values
Clients distribution	Chi-Square
Area size	32.0×32.0
Number of mesh routers	16
Number of mesh clients	48
Number of GA islands	16
Number of Particle-patterns	32
Number of migrations	200
Evolution steps	320
Radius of a mesh router	2.0–3.5
Selection method	Roulette wheel method
Crossover method	SPX
Mutation method	Boundary mutation
Crossover rate	0.8
Mutation rate	0.2
SA Starting value	10.0
SA Ending value	0.01
Total number of iterations	64000
Replacement method	CM, LDIWM

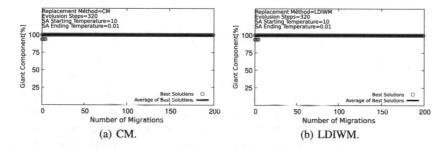

(a) CM.   (b) LDIWM.

Fig. 4. Simulation results of WMN-PSOSA-DGA for SGC.

We show simulation results in Fig. 4 and Fig. 5. We see that for SGC, the performance is almost the same for both replacement methods. However, for NCMC, CM performs better than LDIWM. The visualized simulation results are shown in Fig. 6. For both replacement methods, all mesh routers are connected, but some clients are not covered. We can see the number of covered mesh clients is larger for CM compared with the case of LDIWM.

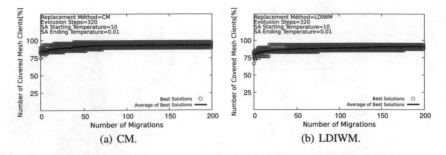

(a) CM.   (b) LDIWM.

Fig. 5. Simulation results of WMN-PSOSA-DGA for NCMC.

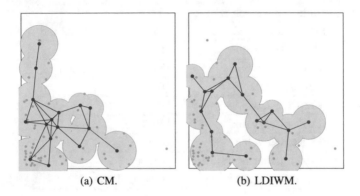

(a) CM.   (b) LDIWM.

Fig. 6. Visualized simulation results of WMN-PSOSA-DGA for different replacement methods.

4   Conclusions

In this work, we evaluated the performance of CM and LDIWM replacement methods for WMNs using a hybrid simulation system based on PSO, SA and DGA (called WMN-PSOSA-DGA) considering Chi-square distribution of mesh clients. Simulation results show that a good performance was achieved for CM compared with the case of LDIWM.

In our future work, we would like to evaluate the performance of the proposed system for different parameters and patterns.

References

1. Ahmed, S., Khan, M.A., Ishtiaq, A., Khan, Z.A., Ali, M.T.: Energy harvesting techniques for routing issues in wireless sensor networks. Int. J. Grid Util. Comput. **10**(1), 10–21 (2019)
2. Akyildiz, I.F., Wang, X., Wang, W.: Wireless mesh networks: a survey. Comput. Netw. **47**(4), 445–487 (2005)

3. Barolli, A., Sakamoto, S., Barolli, L., Takizawa, M.: A hybrid simulation system based on particle swarm optimization and distributed genetic algorithm for wmns: performance evaluation considering normal and uniform distribution of mesh clients. In: International Conference on Network-Based Information Systems, pp. 42–55. Springer (2018)
4. Barolli, A., Sakamoto, S., Barolli, L., Takizawa, M.: Performance analysis of simulation system based on particle swarm optimization and distributed genetic algorithm for WMNs considering different distributions of mesh clients. In: International Conference on Innovative Mobile and Internet Services in Ubiquitous Computing, pp. 32–45. Springer (2018)
5. Barolli, A., Sakamoto, S., Ozera, K., Barolli, L., Kulla, E., Takizawa, M.: Design and implementation of a hybrid intelligent system based on particle swarm optimization and distributed genetic algorithm. In: International Conference on Emerging Internetworking, Data & Web Technologies, pp. 79–93. Springer (2018)
6. Barolli, A., Sakamoto, S., Durresi, H., Ohara, S., Barolli, L., Takizawa, M.: A comparison study of constriction and linearly decreasing Vmax replacement methods for wireless mesh networks by WMN-PSOHC-DGA simulation system. In: International Conference on P2P, Parallel, Grid, Cloud and Internet Computing, pp. 26–34. Springer (2019)
7. Barolli, A., Sakamoto, S., Ohara, S., Barolli, L., Takizawa, M.: Performance analysis of WMNs by WMN-PSOHC-DGA simulation system considering linearly decreasing inertia weight and linearly decreasing Vmax replacement methods. In: International Conference on Intelligent Networking and Collaborative Systems, pp. 14–23. Springer (2019)
8 Barolli, A., Sakamoto, S., Ohara, S., Barolli, L., Takizawa, M.: Performance analysis of WMNs by WMN-PSOHC-DGA simulation system considering random inertia weight and linearly decreasing Vmax router replacement methods. In: Conference on Complex, Intelligent, and Software Intensive Systems, pp. 13–21. Springer (2019)
9. Clerc, M., Kennedy, J.: The particle swarm-explosion, stability, and convergence in a multidimensional complex space. IEEE Trans. Evol. Comput. **6**(1), 58–73 (2002)
10. Franklin, A.A., Murthy, C.S.R.: Node placement algorithm for deployment of two-tier wireless mesh networks. In: Proceedings of Global Telecommunications Conference, pp 4823–4827 (2007)
11. Girgis, M.R., Mahmoud, T.M., Abdullatif, B.A., Rabie, A.M.: Solving the wireless mesh network design problem using genetic algorithm and simulated annealing optimization methods. Int. J. Comput. Appl. **96**(11), 1–10 (2014)
12. Gorrepotu, R., Korivi, N.S., Chandu, K., Deb, S.: Sub-1GHz miniature wireless sensor node for IoT applications. Internet Things **1**, 27–39 (2018)
13. Islam, M.M., Funabiki, N., Sudibyo, R.W., Munene, K.I., Kao, W.C.: A dynamic access-point transmission power minimization method using PI feedback control in elastic WLAN system for IoT applications. Internet Things **8**(100), 089 (2019)
14. Maolin, T., et al.: Gateways placement in backbone wireless mesh networks. Int. J. Commun. Netw. Syst. Sci. **2**(1), 44 (2009)
15. Marques, B., Coelho, I.M., Sena, A.D.C., Castro, M.C.: A network coding protocol for wireless sensor fog computing. Int. J. Grid Util. Comput. **10**(3), 224–234 (2019)
16. Matsuo, K., Sakamoto, S., Oda, T., Barolli, A., Ikeda, M., Barolli, L.: Performance analysis of WMNs by WMN-GA simulation system for two WMN architectures and different TCP congestion-avoidance algorithms and client distributions. Int. J. Commun. Netw. Distrib. Syst. **20**(3), 335–351 (2018)

17. Muthaiah, S.N., Rosenberg, C.P.: Single gateway placement in wireless mesh networks. In: Proceedings of 8th International IEEE Symposium on Computer Networks, pp 4754–4759 (2008)
18. Naka, S., Genji, T., Yura, T., Fukuyama, Y.: A hybrid particle swarm optimization for distribution state estimation. IEEE Trans. Power Syst. **18**(1), 60–68 (2003)
19. Ohara, S., Barolli, A., Sakamoto, S., Barolli, L.: Performance analysis of WMNs by WMN-PSODGA simulation system considering load balancing and client uniform distribution. In: International Conference on Innovative Mobile and Internet Services in Ubiquitous Computing, pp. 25–38. Springer (2019)
20. Ozera, K., Sakamoto, S., Elmazi, D., Bylykbashi, K., Ikeda, M., Barolli, L.: A fuzzy approach for clustering in MANETs: performance evaluation for different parameters. Int. J. Space-Based Situated Comput. **7**(3), 166–176 (2017)
21. Ozera, K., Bylykbashi, K., Liu, Y., Barolli, L.: A fuzzy-based approach for cluster management in VANETs: performance evaluation for two fuzzy-based systems. Internet Things **3**, 120–133 (2018)
22. Ozera, K., Inaba, T., Bylykbashi, K., Sakamoto, S., Ikeda, M., Barolli, L.: A WLAN triage testbed based on fuzzy logic and its performance evaluation for different number of clients and throughput parameter. Int. J. Grid Util. Comput. **10**(2), 168–178 (2019)
23. Petrakis, E.G., Sotiriadis, S., Soultanopoulos, T., Renta, P.T., Buyya, R., Bessis, N.: Internet of Things as a Service (iTaaS): challenges and solutions for management of sensor data on the cloud and the fog. Internet Things **3**, 156–174 (2018)
24. Sakamoto, S., Oda, T., Ikeda, M., Barolli, L., Xhafa, F.: Implementation of a new replacement method in WMN-PSO simulation system and its performance evaluation. In: The 30th IEEE International Conference on Advanced Information Networking and Applications (AINA 2016), pp. 206–211 (2016). https://doi.org/10.1109/AINA.2016.42
25. Sakamoto, S., Barolli, A., Barolli, L., Takizawa, M.: Design and implementation of a hybrid intelligent system based on particle swarm optimization, hill climbing and distributed genetic algorithm for node placement problem in WMNs: a comparison study. In: The 32nd IEEE International Conference on Advanced Information Networking and Applications (AINA 2018), pp. 678–685. IEEE (2018)
26. Sakamoto, S., Ozera, K., Ikeda, M., Barolli, L.: Implementation of intelligent hybrid systems for node placement problem in WMNs considering particle swarm optimization, hill climbing and simulated annealing. Mobile Netw. Appl. **23**(1), 27–33 (2018)
27. Sakamoto, S., Barolli, A., Barolli, L., Okamoto, S.: Implementation of a web interface for hybrid intelligent systems. Int. J. Web Inf. Syst. **15**(4), 420–431 (2019)
28. Sakamoto, S., Barolli, L., Okamoto, S.: WMN-PSOSA: an intelligent hybrid simulation system for WMNs and its performance evaluations. Int. J. Web Grid Serv. **15**(4), 353–366 (2019)
29. Sakamoto, S., Ohara, S., Barolli, L., Okamoto, S.: Performance evaluation of WMNs by WMN-PSOHC system considering random inertia weight and linearly decreasing inertia weight replacement methods. In: International Conference on Innovative Mobile and Internet Services in Ubiquitous Computing, pp. 39–48. Springer (2019)
30. Sakamoto, S., Ohara, S., Barolli, L., Okamoto, S.: Performance evaluation of WMNs by WMN-PSOHC system considering random inertia weight and linearly decreasing Vmax replacement methods. In: International Conference on Network-Based Information Systems, pp. 27–36. Springer (2019)

31. Sakamoto, S., Ohara, S., Barolli, L., Okamoto, S.: Performance evaluation of WMNs WMN-PSOHC system considering constriction and linearly decreasing inertia weight replacement methods. In: International Conference on Broadband and Wireless Computing, Communication and Applications, pp. 22–31. Springer (2019)
32. Schutte, J.F., Groenwold, A.A.: A Study of Global Optimization using Particle Swarms. J. Glob. Optim. **31**(1), 93–108 (2005)
33. Shi, Y.: Particle swarm optimization. IEEE Connect. **2**(1), 8–13 (2004)
34. Shi, Y., Eberhart, R.C.: Parameter selection in particle swarm optimization. In: Evolutionary Programming VII, pp. 591–600 (1998)
35. Vanhatupa, T., Hannikainen, M., Hamalainen, T.: Genetic algorithm to optimize node placement and configuration for WLAN planning. In: The 4th IEEE International Symposium on Wireless Communication Systems, pp. 612–616 (2007)
36. Wang, J., Xie, B., Cai, K., Agrawal, D.P.: Efficient mesh router placement in wireless mesh networks. In: Proceedings of IEEE International Conference on Mobile Adhoc and Sensor Systems (MASS 2007), pp. 1–9 (2007)
37. Yaghoobirafi, K., Nazemi, E.: An autonomic mechanism based on ant colony pattern for detecting the source of incidents in complex enterprise systems. Int. J. Grid Util. Comput. **10**(5), 497–511 (2019)

A Comparison Study of Linearly Decreasing Inertia Weight Method and Rational Decrement of Vmax Method for WMNs Using WMN-PSOHC Intelligent System Considering Normal Distribution of Mesh Clients

Shinji Sakamoto[1(✉)], Leonard Barolli[2], and Shusuke Okamoto[1]

[1] Department of Computer and Information Science, Seikei University,
3-3-1 Kichijoji-Kitamachi, Musashino-shi, Tokyo 180-8633, Japan
shinji.sakamoto@ieee.org, okam@st.seikei.ac.jp
[2] Department of Information and Communication Engineering, Fukuoka Institute
of Technology, 3-30-1 Wajiro-Higashi, Higashi-Ku, Fukuoka 811-0295, Japan
barolli@fit.ac.jp

Abstract. Wireless Mesh Networks (WMNs) are becoming an important networking infrastructure. However, they have some problems such as node placement, security, transmission power and so on. To solve node placement problem in WMNs, we have implemented a hybrid simulation system based on PSO and HC called WMN-PSOHC. In this paper, we present the performance evaluation of WMNs by using WMN-PSOHC intelligent system considering Linearly Decreasing Inertia Weight Method (LDIWM) and Rational Decrement of Vmax Method (RDVM). The simulation results show that a better performance is achieved for LDIWM compared with the RDVM.

1 Introduction

The wireless networks and devices are becoming increasingly popular and they provide users access to information and communication anytime and anywhere [1, 3–5, 9–12, 14, 15, 17, 19, 20, 25, 29]. Wireless Mesh Networks (WMNs) are gaining a lot of attention because of their low cost nature that makes them attractive for providing wireless Internet connectivity. A WMN is dynamically self-organized and self-configured, with the nodes in the network automatically establishing and maintaining mesh connectivity among them-selves (creating, in effect, an ad hoc network). This feature brings many advantages to WMNs such as low up-front cost, easy network maintenance, robustness and reliable service coverage [2]. Moreover, such infrastructure can be used to deploy community networks, metropolitan area networks, municipal and corporative networks, and to support applications for urban areas, medical, transport and surveillance systems.

In this work, we deal with node placement problem in WMNs. We consider the version of the mesh router nodes placement problem in which we are given a grid area where to deploy a number of mesh router nodes and a number of mesh client nodes of fixed positions (of an arbitrary distribution) in the grid area. The objective is to find a location assignment for the mesh routers to the cells of the grid area that maximizes the network connectivity and client coverage. Network connectivity is measured by Size of Giant Component (SGC) of the resulting WMN graph, while the user coverage is simply the number of mesh client nodes that fall within the radio coverage of at least one mesh router node and is measured by Number of Covered Mesh Clients (NCMC). Node placement problems are known to be computationally hard to solve [13,33]. In some previous works, intelligent algorithms have been recently investigated [8,16,18,27,28,35]. We already implemented a Particle Swarm Optimization (PSO) based simulation system, called WMN-PSO [23]. Also, we implemented a simulation system based on Hill Climbing (HC) for solving node placement problem in WMNs, called WMN-HC [22].

In our previous work [23,26], we presented a hybrid intelligent simulation system based on PSO and HC. We called this system WMN-PSOHC. In this paper, we analyze the performance of Linearly Decreasing Inertia Weight Method (LDIWM) and Rational Decrement of Vmax Method (RDVM) by WMN-PSOHC simulation system considering Normal distribution of mesh clients.

The rest of the paper is organized as follows. We present our designed and implemented hybrid simulation system in Sect. 2. In Sect. 3, we introduce WMN-PSOHC Web GUI tool. The simulation results are given in Sect. 4. Finally, we give conclusions and future work in Sect. 5.

2 Proposed and Implemented Simulation System

2.1 Particle Swarm Optimization

In Particle Swarm Optimization (PSO) algorithm, a number of simple entities (the particles) are placed in the search space of some problem or function and each evaluates the objective function at its current location. The objective function is often minimized and the exploration of the search space is not through evolution [21]. However, following a widespread practice of borrowing from the evolutionary computation field, in this work, we consider the bi-objective function and fitness function interchangeably. Each particle then determines its movement through the search space by combining some aspect of the history of its own current and best (best-fitness) locations with those of one or more members of the swarm, with some random perturbations. The next iteration takes place after all particles have been moved. Eventually the swarm as a whole, like a flock of birds collectively foraging for food, is likely to move close to an optimum of the fitness function.

Each individual in the particle swarm is composed of three $\mathcal{D}$-dimensional vectors, where $\mathcal{D}$ is the dimensionality of the search space. These are the current position $\vec{x}_i$, the previous best position $\vec{p}_i$ and the velocity $\vec{v}_i$.

The particle swarm is more than just a collection of particles. A particle by itself has almost no power to solve any problem; progress occurs only when the particles interact. Problem solving is a population-wide phenomenon, emerging from the individual behaviors of the particles through their interactions. In any case, populations are organized according to some sort of communication structure or topology, often thought of as a social network. The topology typically consists of bidirectional edges connecting pairs of particles, so that if j is in i's neighborhood, i is also in j's. Each particle communicates with some other particles and is affected by the best point found by any member of its topological neighborhood. This is just the vector $\vec{p_i}$ for that best neighbor, which we will denote with $\vec{p_g}$. The potential kinds of population "social networks" are hugely varied, but in practice certain types have been used more frequently.

In the PSO process, the velocity of each particle is iteratively adjusted so that the particle stochastically oscillates around $\vec{p_i}$ and $\vec{p_g}$ locations.

2.2 Hill Climbing

Hill Climbing (HC) algorithm is a heuristic algorithm. The idea of HC is simple. In HC, the solution s' is accepted as the new current solution if $\delta \leq 0$ holds, where $\delta = f(s') - f(s)$. Here, the function f is called the fitness function. The fitness function gives points to a solution so that the system can evaluate the next solution s' and the current solution s.

The most important factor in HC is to define the neighbor solution, effectively. The definition of the neighbor solution affects HC performance directly. In our WMN-PSOHC system, we use the next step of particle-pattern positions as the neighbor solutions for the HC part.

2.3 WMN-PSOHC System Description

In following, we present the initialization, client distributions, particle-pattern, fitness function and router replacement methods.

Initialization
Our proposed system starts by generating an initial solution randomly, by *ad hoc* methods [34]. We decide the velocity of particles by a random process considering the area size. For instance, when the area size is $W \times H$, the velocity is decided randomly from $-\sqrt{W^2 + H^2}$ to $\sqrt{W^2 + H^2}$.

Client Distributions
Our system can generate many client distributions. In this paper, we consider Normal distribution of mesh clients. In Normal distribution, clients are located around the center of considered area as shown in Fig. 1.

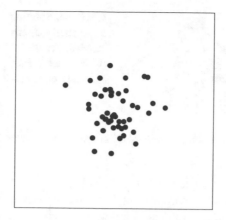

Fig. 1. Normal distribution of mesh clients.

Particle-Pattern

A particle is a mesh router. A fitness value of a particle-pattern is computed by combination of mesh routers and mesh clients positions. In other words, each particle-pattern is a solution as shown is Fig. 2. Therefore, the number of particle-patterns is a number of solutions.

Fitness Function

One of most important thing is to decide the determination of an appropriate objective function and its encoding. In our case, each particle-pattern has an own fitness value and compares other particle-patterns fitness value in order to share information of global solution. The fitness function follows a hierarchical approach in which the main objective is to maximize the SGC in WMN. Thus, we use α and β weight-coefficients for the fitness function and the fitness function of this scenario is defined as:

$$\text{Fitness} = \alpha \times \text{SGC}(\boldsymbol{x}_{ij}, \boldsymbol{y}_{ij}) + \beta \times \text{NCMC}(\boldsymbol{x}_{ij}, \boldsymbol{y}_{ij}).$$

Router Replacement Methods

A mesh router has x, y positions and velocity. Mesh routers are moved based on velocities. There are many router replacement methods in PSO field [7,30–32]. In this paper, we consider LDIWM and RDVM.

Linearly Decreasing Inertia Weight Method (LDIWM)

In LDIWM, C_1 and C_2 are set to 2.0, constantly. On the other hand, the ω parameter is changed linearly from unstable region ($\omega = 0.9$) to stable region ($\omega = 0.4$) with increasing of iterations of computations [6,32].

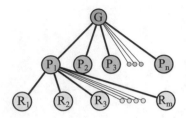

G: Global Solution
P: Particle-pattern
R: Mesh Router
n: Number of Particle-patterns
m: Number of Mesh Routers

Fig. 2. Relationship among global solution, particle-patterns and mesh routers.

Rational Decrement of Vmax Method (RDVM)

In RDVM, PSO parameters are set to unstable region ($\omega = 0.9$, $C_1 = C_2 = 2.0$). A value of V_{max} which is maximum velocity of particles is considered. The V_{max} is kept decreasing with the increasing of iterations as

$$V_{max}(x) = \sqrt{W^2 + H^2} \times \frac{T - x}{x}.$$

Where, W and H are the width and the height of the considered area, respectively. Also, T and x are the total number of iterations and a current number of iteration, respectively [24].

3   WMN-PSOHC Web GUI Tool

The Web application follows a standard Client-Server architecture and is implemented using LAMP (Linux + Apache + MySQL + PHP) technology (see Fig. 3). We show the WMN-PSOHC Web GUI tool in Fig. 4. Remote users (clients) submit their requests by completing first the parameter setting. The parameter values to be provided by the user are classified into three groups, as follows.

- Parameters related to the problem instance: These include parameter values that determine a problem instance to be solved and consist of number of router nodes, number of mesh client nodes, client mesh distribution, radio coverage interval and size of the deployment area.
- Parameters of the resolution method: Each method has its own parameters.
- Execution parameters: These parameters are used for stopping condition of the resolution methods and include number of iterations and number of independent runs. The former is provided as a total number of iterations and depending on the method is also divided per phase (e.g., number of iterations in a exploration). The later is used to run the same configuration for the same problem instance and parameter configuration a certain number of times.

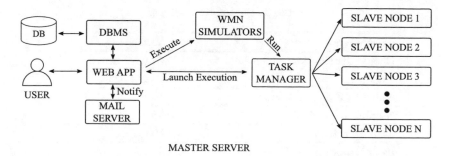

Fig. 3. System structure for web interface.

Simulator parameters, Particle Swarm Optimization and Hill Climbing			
Distribution	Normal ⌄		
Number of clients	48	(integer)(min:48 max:128)	
Number of routers	16	(integer) (min:16 max:48)	
Area size (WxH)	32	(positive real number)	32	(positive real number)
Radius (Min & Max)	2	(positive real number)	2	(positive real number)
Independent runs	10	(integer) (min:1 max:100)	
Replacement method	Constriction Method ⌄		
Number of Particle-patterns	9	(integer) (min:1 max:64)	
Max iterations	800	(integer) (min:1 max:6400)	
Iteration per Phase	4	(integer) (min:1 max:Max iterations)	
Send by mail	⬚		

Run

Fig. 4. WMN-PSOHC Web GUI Tool.

4   Simulation Results

In this section, we show simulation results using WMN-PSOHC system. In this work, we consider Normal distribution of mesh clients. The number of mesh routers is considered 16 and the number of mesh clients 48. We consider the number of particle-patterns 9. We conducted simulations 100 times, in order to avoid the effect of randomness and create a general view of results. The total number of iterations is considered 800 and the iterations per phase is considered 4. We show the parameter setting for WMN-PSOHC in Table 1.

We show the simulation results in Fig. 5 and Fig. 6. For SGC, both replacement methods reach the maximum (100%). This means that all mesh routers are connected to each other. We see that LDIWM converges faster than RDVM for SGC. Also, for the NCMC, LDIWM performs better than RDVM. Therefore, we conclude that the performance of LDIWM is better compared with RDVM.

Table 1. Parameter settings.

Parameters	Values
Clients distribution	Normal distribution
Area size	32.0×32.0
Number of mesh routers	16
Number of mesh clients	48
Total iterations	800
Iteration per phase	4
Number of particle-patterns	9
Radius of a mesh router	2.0
Fitness function weight-coefficients (α, β)	0.7, 0.3
Replacement method	LDIWM, RDVM

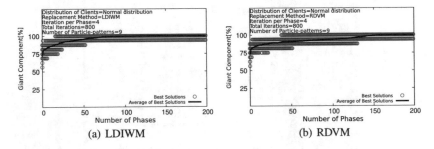

(a) LDIWM   (b) RDVM

Fig. 5. Simulation results of WMN-PSOHC for SGC.

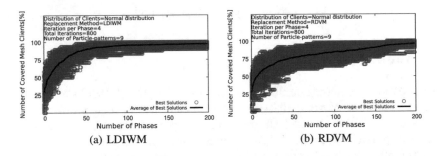

(a) LDIWM   (b) RDVM

Fig. 6. Simulation results of WMN-PSOHC for NCMC.

5   Conclusions

In this work, we evaluated the performance of LDIWM and RDVM router replacement methods for WMNs by WMN-PSOHC hybrid intelligent simulation system. Simulation results show that the performance of LDIWM is better compared with RDVM.

In our future work, we would like to evaluate the performance of the proposed system for different parameters and scenarios.

References

1. Ahmed, S., Khan, M.A., Ishtiaq, A., Khan, Z.A., Ali, M.T.: Energy harvesting techniques for routing issues in wireless sensor networks. Int. J. Grid Utility Comput. **10**(1), 10–21 (2019)
2. Akyildiz, I.F., Wang, X., Wang, W.: Wireless mesh networks: a survey. Comput. Networks **47**(4), 445–487 (2005)
3. Barolli, A., Sakamoto, S., Barolli, L., Takizawa, M.: A hybrid simulation system based on particle swarm optimization and distributed genetic algorithm for WMNs: performance evaluation considering normal and uniform distribution of mesh clients. In: International Conference on Network-Based Information Systems, pp. 42–55, Springer (2018)
4. Barolli, A., Sakamoto, S., Barolli, L., Takizawa, M.: Performance analysis of simulation system based on particle swarm optimization and distributed genetic algorithm for WMNs considering different distributions of mesh clients. In: International Conference on Innovative Mobile and Internet Services in Ubiquitous Computing, pp. 32–45, Springer (2018)
5. Barolli, A., Sakamoto, S., Barolli, L., Takizawa, M.: Performance evaluation of WMN-PSODGA system for node placement problem in WMNs considering four different crossover methods. In: The 32nd IEEE International Conference on Advanced Information Networking and Applications (AINA-2018), pp 850–857. IEEE (2018)
6. Barolli, A., Sakamoto, S., Ohara, S., Barolli, L., Takizawa, M.: Performance analysis of WMNs by WMN-PSOHC-DGA simulation system considering linearly decreasing inertia weight and linearly decreasing Vmax replacement methods. In: International Conference on Intelligent Networking and Collaborative Systems, pp. 14–23, Springer (2019)
7. Clerc, M., Kennedy, J.: The particle swarm-explosion, stability, and convergence in a multidimensional complex space. IEEE Trans. Evol. Comput. **6**(1), 58–73 (2002)
8. Girgis, M.R., Mahmoud, T.M., Abdullatif, B.A., Rabie, A.M.: Solving the wireless mesh network design problem using genetic algorithm and simulated annealing optimization methods. Int. J. Comput. Appl. **96**(11), 1–10 (2014)
9. Gorrepotu, R., Korivi, N.S., Chandu, K., Deb, S.: Sub-1 GHz miniature wireless sensor node for IoT applications. Internet of Things **1**, 27–39 (2018)
10. Inaba, T., Obukata, R., Sakamoto, S., Oda, T., Ikeda, M., Barolli, L.: Performance evaluation of a QoS-aware fuzzy-based CAC for LAN access. Int. J. Space-Based Situated Comput. **6**(4), 228–238 (2016)
11. Inaba, T., Sakamoto, S., Oda, T., Ikeda, M., Barolli, L.: A testbed for admission control in WLAN: a fuzzy approach and its performance evaluation. In: International Conference on Broadband and Wireless Computing, pp. 559–571, Springer, Communication and Applications (2016)
12. Islam, M.M., Funabiki, N., Sudibyo, R.W., Munene, K.I., Kao, W.C.: A dynamic access-point transmission power minimization method using pi feedback control in elastic WLAN system for IoT applications. Internet of Things **8**(100), 089 (2019)
13. Maolin, T., et al.: Gateways placement in backbone wireless mesh networks. Int. J. Commun. Network Syst. Sci. **2**(1), 44 (2009)

14. Marques, B., Coelho, I.M., Sena, A.D.C., Castro, M.C.: A network coding protocol for wireless sensor fog computing. Int. J. Grid Utility Comput. **10**(3), 224–234 (2019)

15. Matsuo, K., Sakamoto, S., Oda, T., Barolli, A., Ikeda, M., Barolli, L.: Performance analysis of WMNs by WMN-GA simulation system for two WMN architectures and different TCP congestion-avoidance algorithms and client distributions. Int. J. Commun. Networks Distrib. Syst. **20**(3), 335–351 (2018)

16. Naka, S., Genji, T., Yura, T., Fukuyama, Y.: A hybrid particle swarm optimization for distribution state estimation. IEEE Trans. Power Syst. **18**(1), 60–68 (2003)

17. Ohara, S., Barolli, A., Sakamoto, S., Barolli, L.: Performance analysis of WMNs by WMN-PSODGA simulation system considering load balancing and client uniform distribution. In: International Conference on Innovative Mobile and Internet Services in Ubiquitous Computing, pp. 25–38, Springer (2019)

18. Ozera, K., Bylykbashi, K., Liu, Y., Barolli, L.: A fuzzy-based approach for cluster management in VANETs: performance evaluation for two fuzzy-based systems. Internet of Things **3**, 120–133 (2018)

19. Ozera, K., Inaba, T., Bylykbashi, K., Sakamoto, S., Ikeda, M., Barolli, L.: A WLAN triage testbed based on fuzzy logic and its performance evaluation for different number of clients and throughput parameter. Int. J. Grid Utility Comput. **10**(2), 168–178 (2019)

20. Petrakis, E.G., Sotiriadis, S., Soultanopoulos, T., Renta, P.T., Buyya, R., Bessis, N.: Internet of Things as a service (iTaaS): challenges and solutions for management of sensor data on the cloud and the fog. Internet of Things **3**, 156–174 (2018)

21. Poli, R., Kennedy, J., Blackwell, T.: Particle swarm optimization. Swarm Intell. **1**(1), 33–57 (2007)

22. Sakamoto, S., Lala, A., Oda, T., Kolici, V., Barolli, L., Xhafa, F.: Analysis of WMN-HC simulation system data using friedman test. In: The Ninth International Conference on Complex, Intelligent, and Software Intensive Systems (CISIS-2015), pp. 254–259. IEEE (2015)

23. Sakamoto, S., Oda, T., Ikeda, M., Barolli, L., Xhafa, F.: Implementation and evaluation of a simulation system based on particle swarm optimisation for node placement problem in wireless mesh networks. Int. J. Commun. Networks Distrib. Syst. **17**(1), 1–13 (2016)

24. Sakamoto, S., Oda, T., Ikeda, M., Barolli, L., Xhafa, F.: Implementation of a new replacement method in WMN-PSO simulation system and its performance evaluation. The 30th IEEE International Conference on Advanced Information Networking and Applications (AINA-2016), pp. 206–211 (2016). https://doi.org/10.1109/AINA.2016.42

25. Sakamoto, S., Obukata, R., Oda, T., Barolli, L., Ikeda, M., Barolli, A.: Performance analysis of two wireless mesh network architectures by WMN-SA and WMN-TS simulation systems. J. High Speed Networks **23**(4), 311–322 (2017)

26. Sakamoto, S., Ozera, K., Ikeda, M., Barolli, L.: Implementation of intelligent hybrid systems for node placement problem in WMNs considering particle swarm optimization, hill climbing and simulated annealing. Mobile Networks Appl. **23**(1), 27–33 (2018)

27. Sakamoto, S., Barolli, A., Barolli, L., Okamoto, S.: Implementation of a web interface for hybrid intelligent systems. Int. J. Web Inf. Syst. **15**(4), 420–431 (2019)

28. Sakamoto, S., Barolli, L., Okamoto, S.: WMN-PSOSA: an intelligent hybrid simulation system for WMNs and its performance evaluations. Int. J. Web Grid Serv. **15**(4), 353–366 (2019)

29. Sakamoto, S., Ozera, K., Barolli, A., Ikeda, M., Barolli, L., Takizawa, M.: Implementation of an intelligent hybrid simulation systems for WMNs based on particle swarm optimization and simulated annealing: performance evaluation for different replacement methods. Soft Comput. **23**(9), 3029–3035 (2019)
30. Schutte, J.F., Groenwold, A.A.: A study of global optimization using particle swarms. J. Global Optim. **31**(1), 93–108 (2005)
31. Shi, Y.: Particle swarm optimization. IEEE Connections **2**(1), 8–13 (2004)
32. Shi, Y., Eberhart, R.C.: Parameter selection in particle swarm optimization. Evol. Program. VII 591–600 (1998)
33. Wang, J., Xie, B., Cai, K., Agrawal, D.P.: Efficient mesh router placement in wireless mesh networks. In: Proceedings of IEEE International Conference on Mobile Ad hoc and Sensor Systems (MASS-2007), pp. 1–9 (2007)
34. Xhafa, F., Sanchez, C., Barolli, L.: Ad hoc and neighborhood search methods for placement of mesh routers in wireless mesh networks. In: Proceedings of 29th IEEE International Conference on Distributed Computing Systems Workshops (ICDCS-2009), pp. 400–405 (2009)
35. Yaghoobirafi, K., Nazemi, E.: An autonomic mechanism based on ant colony pattern for detecting the source of incidents in complex enterprise systems. Int. J. Grid Utility Comput. **10**(5), 497–511 (2019)

Implementation of Disaster Information Acquisition Method Using CCN Moving Router

Taisuke Ono[✉] and Tomoyuki Ishida

Fukuoka Institute of Technology, Fukuoka, Fukuoka 811-0295, Japan
s17b2011@bene.fit.ac.jp, t-ishida@fit.ac.jp

Abstract. Japan is prone to natural disasters such as earthquakes and typhoons. A large amount of traffic is generated during the disaster, and there are concerns about congestion and deterioration of communication quality. Therefore, we focused on a new network architecture called Content Centric Network (CCN). In this study, we conducted field experiments to acquire disaster information in event of disasters by utilizing drones and bicycles as CCN moving routers, and verified the usefulness of CCN moving routers.

1 Introduction

With the development of information system technology and information communication network technology, services using these technologies have become indispensable in everyday life. In particular, tablet terminals, such as smartphones, have become widespread and can even be considered as daily necessities. In the event of a disaster, tablet terminals are used as a means of confirming the safety of family and friends and collecting and sharing information.

However, when a large-scale natural disaster occurs, communication is often restricted due to congestion, which is a problem, and access is generally limited to Internet and telephone lines. During the Great East Japan Earthquake that occurred in March 2011, communication was restricted to prevent congestion on telephone lines but data communication was barely restricted. However, as Internet usage becomes increasingly content- and service-oriented, the total traffic in the network increases yearly. Therefore, in the event of an earthquake directly beneath the Tokyo metropolitan area or a huge Nankai Trough earthquake, restrictions on data communication are inevitable.

Information-centric network (ICN) technology has drawn attention amid concerns about the aforementioned communication problems during natural disasters. This technology requests content by specifying the content name instead of the Internet Protocol (IP) address of the server. Then, if the neighboring router or node that receives the content request has the corresponding content, the content is directly transferred to the recipient. This enables users to quickly acquire content and effectively utilize servers and network resources [1].

In other words, it is reused by leaving a cache of the passed content in the router or node. When the same content is requested, the information is acquired from the router without requesting it from the target server. Currently, the usefulness of ICN technology

L. Barolli et al. (Eds.): EIDWT 2021, LNDECT 65, pp. 114–121, 2021.
https://doi.org/10.1007/978-3-030-70639-5_11

during various natural disasters is being studied. Among various ICN technologies, the content-centric network (CCN) is attracting a great deal of attention as a means of confirming safety and collecting information in the event of a large-scale natural disaster.

2 Previous Work

Yamazaki et al. [2] studied the effectiveness of using a cache in the event of a large-scale natural disaster by examining the usefulness of a cache on virtual networks and conducting comparative experiments between IP networks and a CCN using a network simulator. They constructed a pseudo-network using an Ubuntu virtual machine and a VyOS virtual machine, which is a software router. In addition, they created a pseudo-content router and a small CCN network to store cache in the router. In this virtual network environment, they acquired a web page using a cache on an IP network and verified the usefulness of the cache. Furthermore, they constructed CCN simulations using ccnSim and IP network simulations using the INET Framework and compared the usefulness of the CCN in the event of a disaster with IP networks. They determined the usefulness of the CCN using the number of hops, cache hit rate, and data acquisition time from the simulation results. However, their experiments were limited to experiments on simulators, and the effectiveness of CCNs using actual machines was not verified.

3 Related Work

Kitagawa et al. [3] proposed an information-oriented network architecture in which content can be acquired simply by specifying the name of the content in a remote location that cannot be acquired by incorporating route control into the information-oriented network. Kitagawa et al. realized information sharing between independent networks that were not interconnected using a flying router (FR) that combined a CCN router and an unmanned air vehicle.

Furthermore, Gao et al. [4] proposed a FR incorporating a moving control system that utilized the ICN naming scheme and strategy layer to develop a system that enabled the active control of moving routers. The study demonstrated that the ICN naming scheme can be used as a movement control application programming interface for a FR and that active control of a FR is possible.

However, to the best of our knowledge, in the field experiments of all studies, only a drone was used as the moving router and verification using multiple moving routers, as in the case of an actual disaster, was not performed.

4 Research Objective

In this study, we constructed a real environment using moving routers via a drone and a bicycle to examine the usefulness of information acquisition using cache information in the event of a natural disaster. In the field experiment using moving routers (drone and bicycle), a Raspberry Pi was attached to the drone and bicycle to make them function as a CCN moving router. The CCN moving router serves to connect nodes at the same time, and using the CCN moving router and cache, it is possible to acquire content even if the server goes down in the event of a disaster.

5 System Configuration

Figure 1 presents the network configuration and data flow in the field experiment (assuming a local natural disaster) in this study. The network consists of one Publisher, two Consumers, and two CCN moving routers (drone and bicycle). The Publisher in a CCN refers to the content holder, and Consumer refers to the content requester.

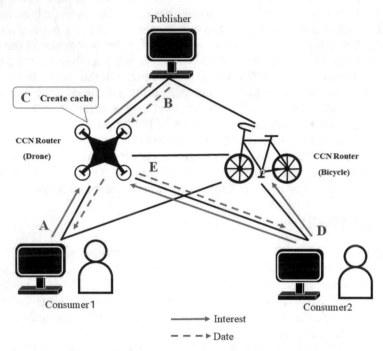

Fig. 1. Network configuration in field experiment.

The following is an example of requesting and acquiring information in the field experiment assuming a local natural disaster.

1) Consumer1 transmits "Interest" (content request message), which is a packet used when requesting content by a CCN to acquire disaster information from the Publisher.
2) The Publisher, which receives "Interest," transmits disaster information to Consumer1 if it has the requested disaster information.
3) When the disaster information transmitted from the Publisher to Consumer1 passes through the CCN router (drone), a disaster information cache is created in the CCN router (drone).
4) Consumer2 transmits "Interest" regarding the same disaster information requested by Consumer1 to the Publisher in A to the Publisher.
5) In C, since the cache of the disaster information requested by Consumer1 from the Publisher is created in the CCN router (drone), Consumer2 acquires the disaster information from the CCN router (drone) instead of from the Publisher.

6   Field Experiment Assuming Local Natural Disasters

In this study, it was assumed that the server that provides content goes down in the event of a disaster; thus, the network topology was changed during the experimental process (Fig. 2). Figure 3 illustrates the drone equipped with a Raspberry Pi, and Fig. 4 illustrates the bicycle equipped with a Raspberry Pi.

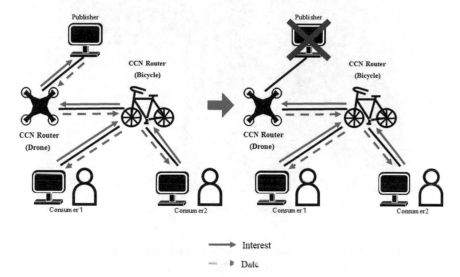

    ⟶   Interest

    — ▶ Date

Fig. 2. Changes to network topology.

Fig. 3. Drone with Raspberry Pi.

Fig. 4. Bicycle with Raspberry Pi.

In the field experiment, we used disaster information content such as the disaster prevention information top page (web page compressed file) on the Fukuoka City website; the disaster/disaster prevention information top page (web page compressed file) on the Ministry of Land, Infrastructure, Transport, and Tourism website; and the hazard map on the Fukuoka City website, as presented in Table 1.

Table 1. List of acquired content.

Website name	Capacity	Content name
Disaster prevention information top page on the Fukuoka City website	410 KB	ccn_disaster/www.city.fukuoka.lg.jp.tar.gz
Disaster/disaster prevention information top page on the Ministry of Land, Infrastructure, Transport and Tourism website	60 KB	ccn_disaster/www.mlit.go.jp.tar.gz
Hazard map on the Fukuoka City website	8 MB	ccn_disaster/bm-higasiku_light.pdf

Figure 5 presents the data flow when disaster information is acquired from the Publisher and when it is acquired from the cache when the Publisher goes down. In the figure, A-D display the data flow when acquiring disaster information from the Publisher and

E and F display the data flow when data are acquired from the cache when the Publisher goes down.

The time taken to acquire the data from the Publisher and from the cache for each type of disaster information is summarized in Tables 2, 3 and 4. Table 2 pertains to the disaster prevention information top page on the Fukuoka City website; Table 3 pertains to the disaster/disaster prevention information top page on the Ministry of Land, Infrastructure, Transport, and Tourism website; Table 4 pertains to the hazard map on the Fukuoka City website.

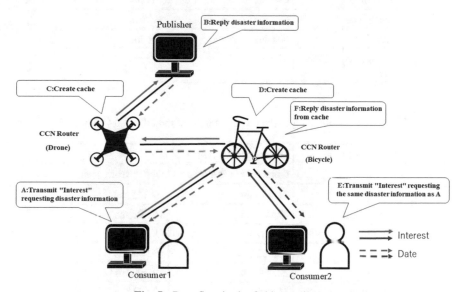

Fig. 5. Data flow in the field experiment.

Table 2. Acquisition time of disaster prevention information from the top page of the Fukuoka City website.

Number of times	Data acquisition time	
	Obtain from Publisher	Obtain from cache
First time	1,730 ms	918 ms
Second time	1,810 ms	909 ms
Third time	1,803 ms	955 ms
Fourth time	1,750 ms	925 ms
Fifth time	1,808 ms	913 ms
Average acquisition time	1,780 ms	924 ms

Table 3. Acquisition time of disaster/disaster prevention information from the top page of the Ministry of Land, Infrastructure, Transport, and Tourism website.

Number of times	Data acquisition time	
	Obtain from Publisher	Obtain from cache
First time	735 ms	507 ms
Second time	622 ms	494 ms
Third time	686 ms	503 ms
Fourth time	618 ms	493 ms
Fifth time	607 ms	504 ms
Average acquisition time	654 ms	500 ms

Table 4. Acquisition time of the hazard map from the Fukuoka City website.

Number of times	Data acquisition time	
	Obtain from Publisher	Obtain from cache
First time	22,450 ms	9,305 ms
Second time	22,967 ms	10,393 ms
Third time	17,237 ms	9,661 ms
Fourth time	15,981 ms	9,697 ms
Fifth time	22,305 ms	9,100 ms
Average acquisition time	20,188 ms	9,631 ms

Figure 6 presents a comparison of the average acquisition time of each type of disaster information. This figure indicates that the data acquisition time for the disaster prevention information top page and hazard map on the Fukuoka City website was reduced by approximately half by acquiring the data from the cache. The disaster/disaster prevention information top page on the Ministry of Land, Infrastructure, Transport, and Tourism website also displayed the effectiveness of data acquisition from the cache, although the difference in the average acquisition time was small.

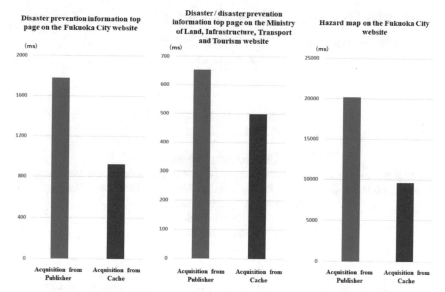

Fig. 6. Comparison of average acquisition time.

7 Conclusion

In this study, we conducted an experiment on a disaster information acquisition method using a content-oriented network and moving routers. Through a field experiment, we verified the usefulness of information acquisition using moving routers and caches during local natural disasters. As mobile routers, we used a drone, which is currently popular in various fields, and a bicycle with high mobility; we equipped the mobile routers with a Raspberry Pi that functioned as a router. The results of the field experiment demonstrated the effectiveness of cache utilization using moving routers in the event of a local natural disaster.

Acknowledgments. This study was supported by JSPS KAKENHI Grant Number JP19K04972 and the research grand of The Telecommunications Advancement Foundation (TAF) of Japan.

References

1. Asaeda, H., Matsuzono, K.: Prototype implementations and evaluation techniques for information-centric networking. Comput. Softw. **33**(3), 3–15 (2016)
2. Yamazaki, M., Takahagi, K., Ishida, T., Sugita, K., Uchida, N., Shibata, Y.: Proposal of information acquisition method utilizing CCN in a time of large scale natural disaster. In: Proceeding of the 10th International Conference on P2P, Parallel, Grid, Cloud and Internet Computing, pp. 645–650 (2015)
3. Kitagawa, T., Ata, S., Murata, M.: Retrieving information with autonomously moving router in information-centric network guide to the technical report and template. In: Proceeding of the 1st IEICE Technical Committee on ICN, pp. 1–6 (2015)
4. Gao, Y., Kitagawa, T., Ata, S., Eum, S., Murata, M.: Design of flying router control system based on name of packed in information-centric networking. In: Proceeding of the 8th IEICE Technical Committee on ICN, pp. 1–6 (2017)

Realistic Topic-Based Data Transmission Protocol in a Mobile Fog Computing Model

Takumi Saito[1]([⊠]), Shigenari Nakamura[2], Tomoya Enokido[3],
and Makoto Takizawa[1]

[1] Hosei University, Tokyo, Japan
takumi.saito.3j@stu.hosei.ac.jp, makoto.takizawa@computer.org
[2] Tokyo Metropolitan Industrial Technology Research Institute, Tokyo, Japan
nakamura.shigenari@gmail.com
[3] Rissho University, Tokyo, Japan
eno@ris.ac.jp

Abstract. In the fog computing (FC) models, a fog node supports not only routing functions but also application processes. By the application processes, output data is calculated on input data from sensors and other fog nodes and sent to target fog nodes which can calculate on the output data. In this paper, we consider the MPSFC (Mobile topic-based PS (publish/subscribe) FC) model where mobile fog nodes communicate with one another by publishing and subscribing messages of data in wireless networks. Subscription topics of a fog node denote input data on which the fog node can calculate. Publication topics of a message show data carried by the message. In the TBDT (Topic-Based Data Transmission) and ETBDT (Epidemic and TBDT) protocols proposed in our previous studies, a fog node only publishes a message of the output data to a target fog node in the communication range. Here, while a fewer number of messages are transmitted, the delivery ratio of messages is smaller than the epidemic routing protocol. In this paper, we newly propose PTBDT (Probability and TBDT) and TTLBDT (Time-To-Live Based Data Transmission) protocols in order to increase the delivery ratio. If another node is found in the communication range, a fog node forwards messages to the node. Even if the node is not a target node, the node receives the message with some probability. In the evaluation, we show the delivery ratio in the PTBDT protocol is larger than the TTLBDT protocol.

Keywords: IoT · Mobile fog computing (MFC) model · Probability and topic-based data transmission (PTBDT) protocol · Time-to-live based data transmission (TTLBDT) protocol · Mobile topic-based publish/subscribe fog computing (MPSFC) model

© The Author(s), under exclusive license to Springer Nature Switzerland AG 2021
L. Barolli et al. (Eds.): EIDWT 2021, LNDECT 65, pp. 122–133, 2021.
https://doi.org/10.1007/978-3-030-70639-5_12

1 Introduction

Fog computing (FC) models [8,14] to efficiently realize the IoT are composed of fog nodes in addition to server clouds and sensor and actuator devices. A fog node supports application processes to calculate on sensor data. Output data is calculated on sensor data by an application process supported by a fog node and then is sent to other fog nodes to do further calculation on the output data. In the TBFC (Tree-based FC) model [5,6,9,12], fog nodes are hierarchically structured in a tree to reduce the energy consumption of the fog nodes. In order to make the TBFC model tolerant of faults, the FTBFC (Fault-tolerant TBFC) model is also proposed [10–12]. The DTBFC (Dynamic TBFC) model [7,13] is also proposed where the tree structure of fog nodes is dynamically changed.

The MFC (Mobile FC) model [3,4] is composed of mobile fog nodes which communicate with other fog nodes in wireless communication links. Mobile fog nodes communicate with other fog nodes only in the communication range of wireless ad-hoc networks. Thus, mobile fog nodes communicate with other fog nodes by taking advantage of opportunistic routing protocols [2,19].

A PS (Publish/Subscribe) model is a contents-aware, event-driven model of a distributed system [20,21]. In topic-based PS models [18,22], data carried by messages is denoted by topics. We consider the P2PPS (P2P (peer-to-peer) type of topic-based PS) model [20,21] to realize the FC model. Each fog node f_i is a peer which can publish a message m with publication topics $m.P$ and subscribe messages by specifying subscription topics $f_i.S$. The subscription topics $f_i.S$ denote input data on which the fog node f_i can calculate. The publication topics $m.P$ denote output data od_i of the source fog node f_i. A fog node f_i only receives a message m published by a fog node f_j whose publication topics $m.P$ and the subscription topics $f_i.S$ include some common topic, i.e. $m.P \cap f_i.S \neq \phi$. Here, the fog node f_i is a target fog node of the source node f_j. Thus, topics denote data on which a fog node calculates and which a message carries to a target fog node. In the TBDT (Topic-Based Data Transmission) protocol [16,17], each fog node f_i sends a message of the output data to only a target node f_j in the communication range. Here, while a fewer number of messages are transmitted, the delivery ratio of each message is smaller than the epidemic routing protocol [1]. In the ETBDT (Epidemic and TBDT) protocol [15], a fog node f_i to deliver output data to the root node sends the output data to a fog node f_j in the communication range even if the fog node f_j is not a target node. In this paper, we propose a TTLBDT (Time-To-Live-Based Data Transmission) protocol and a PTBDT (Probability and Topic-Based Data Transmission) protocol. Every fog node f_i sends the output data to not only a target but also non-target node f_j in the communication range. In the PTBDT protocol, the non-target node f_j receives the output data with some probability. In the TTLBDT protocol, each data has a TTL (Time-To-Live) field. The TTL of output data is decremented by one as one time unit passes. If the TTL gets zero, the output data in a fog node is removed. In the evaluation, we show the number of messages transmitted in the PTBDT protocol is fewer than the epidemic routing protocol [1] and the delivery ratio of messages is larger than the TBDT protocol.

In Sect. 2, we present the MPSFC model. In Sect. 3, we propose the PTBDT and TTLBDT protocols. In Sect. 4, we evaluate the PTBDT and TTLBDT protocols.

2   MPSFC (Mobile Publish/Subscribe Fog Computing) Model

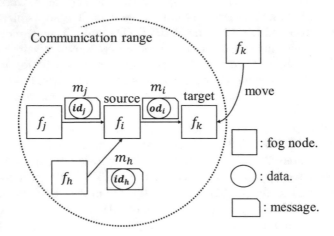

Fig. 1. MFC model.

In this paper, we consider the MPSFC (Mobile topic-based Publish/Subscribe Fog Computing) model [17] to realize the MFC (mobile fog computing) model [3,4] of the IoT by taking advantage of the PS (publish/subscribe) model [20,21]. The MPSFC model is composed of mobile fog nodes which communicate with one another by publishing and subscribing messages in wireless communication networks. Mobile fog nodes communicate with one another only in the communication range [1]. A mobile fog node is also equipped with sensors and actuators. A fog node collects data from sensors and activates actuators. In addition, a fog node f_i supports an application process $p(f_i)$ to calculate output data on the input data from sensors and other fog nodes and sends the output data to other fog nodes. A fog node f_i receives input data id_j from each source fog node f_j. Then, the fog node f_i calculates the output data od_i on the input data and then forwards the output data od_i to a target fog node f_k [Fig. 1]. The data od_i is the input data id_i of the fog node f_k.

A fog node f_i specifies subscription topics $f_i.S$ which denote input data on which the fog node f_i can calculate. A message m is characterized by publication topics $m.P$ which denote output data od_j of the source fog node f_j. A fog node f_i only receives a message m whose $m.P \cap f_i.S \neq \phi$. This means, the fog node f_i supports a process $p(f_i)$ to calculate on data od_j carried by the message m.

Here, the fog node f_i is a *target* node of the fog node f_j and f_j is a *source* fog node of f_i ($f_j \rightarrow f_i$). A fog node f_i receives a collection ID_i of input data, i.e. $ID_i = \{id_{ij} \mid f_j \rightarrow f_i\}$. Then, on receipt of messages from every source node, the fog node f_i calculates output data od_i on input data ID_i carried by the messages. The output data od_i is characterized by publication topics $od_i.P$ ($\subseteq P$). The publication topics $f_i.P$ of a fog node f_i show types of output data of a process $p(f_i)$.

Let D be a collection of data in a system. A process $p(f_i)$ supported by a fog node f_i is considered to be a function which uses inputs data $id_{i1}, ..., id_{i,l_i}$ ($l_i \geq 1$) and returns outputs data od_i, i.e. $od_i = p(f_i)$ ($id_{i1}, ..., id_{i,l_i}$), where $id_{ij} \in D$ ($j = 1, ..., l_i$) and $od_i \in D$. Let T_{ij} be the publication topics $id_{ij}.P$ and T_i be $od_i.P$. Here, the process $p(f_i)$ is specified as $p(f_i)$: $T_{i1} \times \cdots \times T_{i,l_i} \rightarrow T_i$. Here, topics T_{ij} and T_i are referred to as *sorts* of the process $p(f_i)$. Here, $T_{i1} \cup \cdots \cup T_{i,l_i} \subseteq f_i.S$ and $T_i \subseteq f_i.P$. The publication topics $m.P$ of a message m carrying the output data of a fog node f_i are $f_i.P$.

A fog node f_i *precedes* a fog node f_j ($f_i \rightarrow f_j$) if the publication topics $f_i.P$ of the fog node f_i and the subscription topics $f_j.S$ include at least one common topic, i.e. $f_i.P \cap f_j.S \neq \phi$. Then, the process $p(f_i)$ *precedes* the process $p(f_j)$ ($p(f_i) \Rightarrow p(f_j)$) if and only if (iff) the publication topics $f_i.P$ are one of the input sorts of the subscription topics $f_j.S$.

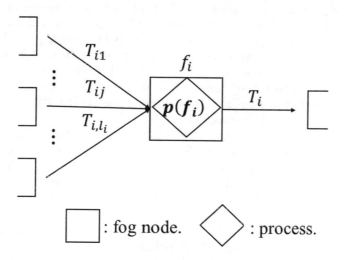

Fig. 2. Fog node.

A fog node f_i can communicate with another fog node f_j ($f_i \leftrightarrow f_j$) only if the fog node f_j is in the communication range of the fog node f_i. Let $FN(f_i)$ be a set of fog nodes which are in the communication range of a node f_i, i.e. $\{f_j \mid f_i \leftrightarrow f_j\}$. Each fog node f_i can only communicate with another fog node f_j in the set $FN(f_i)$. A message m published by a source fog node f_i is only received by a target fog node f_j in the communication range, where $f_i \rightarrow f_j$, i.e. $m.P \cap f_j.S \neq \phi$.

Let $TN(f_i)$ ($\subseteq F$) be a set $\{f_j \mid f_i \rightarrow f_j\}$ of target fog nodes of a fog node f_i. Let $TFN(f_i)$ ($\subseteq F$) be a set of target fog nodes of a fog node f_i with which the fog node f_i can communicate, i.e. $\{f_j \mid f_i \rightarrow f_j$ and $f_j \leftrightarrow f_i\}$, i.e. $TFN(f_i)$ $= TN(f_i) \cap FN(f_i)$. A fog node f_i can only deliver a message m to a target fog node f_j in the set $TFN(f_i)$.

3   TTLBDT (Time-to-live Based Data Transmission) and PTBDT (Probability and TBDT) Protocols

In the TBDT (Topic-based Data Transmission) protocol proposed in our previous papers [16,17], each fog node f_i only delivers a message m of output data od_i to a target fog node f_j in the communication range, i.e. $f_i \rightarrow f_j$ and $f_i \leftrightarrow f_j$. A fog node which forwards the output data to a root fog node is referred to as *last* node. In the ETBDT (Epidemic and TBDT) protocol [15], a last fog node forwards the output data to not only target fog nodes but also non-target fog nodes. A fog node which forwards the output data to non-target fog nodes is referred to as *epidemic*. The more number of fog nodes, the more number of messages are transmitted and stored in fog nodes, while the larger delivery ratio. In this paper, we propose a TTLBDT (Time-to-live based data transmission) protocol and a PTBDT (Probability and TBDT) protocol to reduce the number of messages and buffer size. In the TTLBDT protocol, each data has a TTL (Time-to-live) field. Then, each fog node f_i determines an initial value of the TTL field ttl_{f_i} of the data. The TTL field ttl_i of the data d_i is set up by each fog node f_i when the fog node f_i receives the data d_i, i.e. $ttl_i = ttl_{f_i}$. The TTL is decremented by one as one time unit passes. If the TTL gets zero, the output data in a fog node is removed.

First, suppose a fog node f_i obtains input data from sensors and calculates output data od_i on the sensor data. Then, a fog node f_i sends the output data od_i to a target fog node f_j if the target fog node f_j is in the communication range of the source fog node f_i ($f_i \leftrightarrow f_j$). Suppose a target fog node f_j is in the communication range of a fog node f_i, i.e. $f_i \leftrightarrow f_j$. Here, the fog node f_i includes the output data od_i to a message m_i. In the TTLBDT protocol, a message carries output data in the buffer $f_i.M$. A message m_i carries a collection of data $d_{i1}, \ldots, d_{i,l_i}$. The fog node f_i publishes the message m_i to the target fog node f_j. $d_{ij}.P$ shows topics denoting the data d_{ij}. The publication $m_i.P$ of the message m_i is a set of the publication topics $d_{i1}.P, \ldots, d_{i,l_i}.P$ of the data $d_{i1}, \ldots, d_{i,l_i}$ carried by the message m. Then, the message m_i arrives at the fog node f_j. Here, only at most one third of the maximum size of the buffer $f_j.M$ can be used to store the data of the message m_i.

In the PTBDT protocol, the fog node f_j checks if the publication $m_i.P$ of the message m_i and the subscription $f_j.S$ of the fog node f_j include at least one common topic, i.e. $m_i.P \cap f_j.S \neq \phi$. Here, if $m_i.P \cap f_j.S \neq \phi$, the message m_i is received by a fog node f_j since f_j is a target node of f_i. Otherwise, the fog node f_j receives the message m_i at probability λ. At probability $1 - \lambda$, the fog node f_i neglects the message m_i Then, if the fog node f_j receives the message m_i, each data d_{ij} in the message m_i is stored in the buffer $f_j.M$.

Each fog node f_i publishes and receives a message m and calculates output data on input data carried by the message m in each protocol as follows:

We present the procedure of the TTLBDT protocol;

[**Fog node** f_i **sends a message** m_i **to a fog node** f_j]

1. A fog node f_i finds a fog node f_j in the communication range of the fog node f_i;
2. The fog node f_i adds output data to a message m_i;
3. The fog node f_i **publishes** the message m_i to the target fog node f_j;

[**Fog node** f_i **receives a message** m_j **from a fog node** f_j]

1. A message m_j from a node f_j arrives at a fog node f_i;
2. If a buffer $f_i.M$ of the fog node f_i is full, the fog node f_i **deletes** the data d_k whose TTL is smallest in the buffer $f_i.M$;
3. The fog node f_i stores the data od_j of the message m_i as the input data id_i in a buffer $f_i.M$;
4. Here, only at most one third of the maximum size of the buffer $f_i.M$ can be used to store the data of the message m_j
5. The TTL field ttl_i of the input data id_i is reset to be ttl_{f_i};

[**Fog node** f_i **calculates on input data** id_i]

1. The fog node f_i calculates output data od_i on the input data id_i in the message m_j using the process $p(f_i)$;
2. The output data od_i is stored in the buffer $f_i.M$;

We present the procedure of the PTBDT protocol
[**Fog node** f_i **publishes a message** m_i **to a fog node** f_j]

1. A fog node f_i finds a fog node f_j in the communication range of the fog node f_i, i.e. $f_i \leftrightarrow f_j$;
2. The fog node f_i adds the output data od_i in the buffer $f_i.M$ to a message m_i;
3. The publication $m_i.P$ is a set of topics $od_i.P$ ($\subseteq T$), i.e. $m_i.P = od_i.P$;
4. The fog node f_i **publishes** the message m_i to the target fog node f_j;

[**Fog node** f_i **receives a message** m_j **from a fog node** f_j]

1. A message m_j from a node f_j arrives at a fog node f_i;
2. If $m_j.P \cap f_i.S \neq \phi$, the fog node f_i **receives** the message m_j, else if $m_j.P \cap f_i.S = \phi$, f_i randomly receives m_j with the probability λ. Otherwise, the fog node f_i **neglects** the message m_j;
3. If the buffer $f_i.M$ of the fog node f_i is full, the fog node f_i **deletes** the data d_k whose TTL is smallest and where $f_i.P \cap d_k.T = \phi$ in the buffer $f_i.M$;
4. The fog node f_i stores the data od_j as the input data id_i in the buffer $f_i.M$;

[Fog node f_i calculates on input data id_i]

1. The fog node f_i calculates output data od_i on the input data id_i in the message m_j using the process $p(f_i)$;
2. The output data od_i is stored in the buffer $f_i.M$;

4   Evaluation

We evaluate the TTLBDT and PTBDT protocols of the MPSFC model in terms of the number of messages exchanged among fog nodes and delivery ratio of messages compared with the TBDT [16,17] and ETBDT [15] protocols and the epidemic routing protocol [1]. There are mobile fog nodes f_1, ..., f_n ($n \geq 1$) and sensor nodes n_1,..., n_5 on an $m \times m$ mesh M. We assume the distance d between a pair of neighboring points is one in the mesh M. Each fog node f_i moves on the mesh M in a random walk way. Let cr_i be the communication range of a fog node f_i and a sensor node n_i. Each fog node f_i moves with velocity s_i in a random walk.

Let T be a set of all topics t_1, ..., t_l ($l \geq 1$) in a system. Let D be a collection of data in a system. First, each sensor node n_i has one data d_i and a topic $T(d_i)$ of the data d_i is randomly taken from the set T. Then, each fog node f_i has a process $p(f_i)$. Each fog node f_i supports a process $p(f_i) : T_{i1} \times \cdots \times T_{i,l_i} \rightarrow T_i$. The subscription $f_i.S$ and the publication $f_i.P$ of a fog node f_i are a collection $\{T_{i1}, \ldots T_{i,l_i}\}$ of input sorts and the output sort T_i, respectively. Then, data has a TTL (time-to-live) field. Initially, the TTL field ttl_{f_i} of the fog node f_i is 70. Value of the TTL field ttl_i of the data d_i is set up by the fog node f_i when f_i stores the data d_i in the buffer $f_i.M$, i.e. $ttl_i = 70$. The TTL field is decremented by one at one simulation step. If TTL field of the data is 0, the data is deleted from the buffer.

Next, every fog node f_i randomly moves in the mesh M with the velocity s_i for each simulation step. Let d_{ij} show the distance between a pair of fog nodes f_i and f_j or a pair of a sensor node s_i and a fog node f_j. If each sensor node n_i finds a fog node f_j in the communication range cr_i, i.e. $d_{ij} \leq cr_i$, the sensor node n_i sends sensor data sd_i to the fog node f_j. If each fog node f_i also finds a fog node f_j in the communication range cr_i, i.e. $d_{ij} \leq cr_i$, the fog node f_i publishes a message m_i to the fog node f_j. Here, the message m carries the output data od_i of the fog node f_i. The publication $m_i.P$ of the message m_i is the publication topics $od_i.P$ of the output data od_i. Then, the message m_i arrives at the fog node f_j. The fog node f_j receives a message m_i and stores output data od_i as the input data id_i in the buffer $f_j.M$. The fog node f_j calculates the output data od_j on the input data id_i by using the process $p(f_j)$. The output data od_j is stored in the buffer $f_j.M$. Finally, the delivery ratios of messages in the PTBDT protocol, the TTLBDT protocol, and the epidemic routing protocol [1] are calculated. In the epidemic routing protocol [1], if a message m arrives at the fog node f_i, the fog node f_i receives the message m and stores data in the message m in the buffer $f_i.M$.

In the evaluation, an application process is composed of four stages. In the first stage, the process calculates the average value on input data. In the second stage, the process merges input data into output data. In the third stage, the process joins input data into output data. The processes $p_1, \ldots, p_5$ do the first stage. The processes $p_6, \ldots, p_{10}$ do the second stage. The processes $p_{11}, \ldots,$ p_{15} do the third stage. The processes $p_{16}, \ldots, p_{20}$ receive the final data. The processes $p_1, \ldots, p_5$ have five sorts of sensor data $sd_1, \ldots, sd_5$ and five sorts of output data $od_{1,1}, \ldots, od_{1,5}$. The processes $p_6, \ldots, p_{10}$ also have five sorts of input data $id_{1,1}, \ldots, id_{1,5}$ and five sorts of output data $od_{1,6}, \ldots, od_{1,10}$. The processes $p_{11}, \ldots, p_{15}$ also have five sorts of input data $id_{1,6}, \ldots, id_{1,10}$ and five sorts of output data $od_{1,11}, \ldots, od_{1,15}$. Then, each fog node f_i is equipped with a buffer $f_i.M$. In the simulator, the size of each buffer $f_i.M$ is 10 and size of sensor data sd_i is 1. Then, the output ratio of each process is 0.5. There are a 500 $\times$ 500 mesh ($m = 500$), and twenty topics ($l = 20$). In the evaluation, we change the number fn of fog nodes which support each process. For example, if each process is supported by two fog nodes ($fn = 2$), there are forty fog nodes in the simulator. Then, fog nodes are divided into two types: vehicle fog node and human fog node. The velocity s_i of each vehicle fog node and human fog node f_i is 1 and 2 at one simulation step, respectively. Here, the communication range cr_i of each fog node f_i is 3.

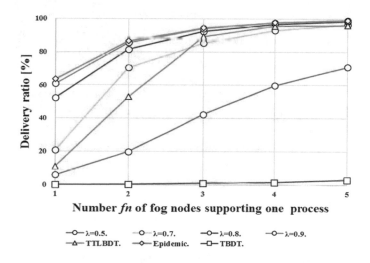

Fig. 3. Delivery ratio.

Figures 3 and 4 show the delivery ratios of data to the root node in the each protocol. In the PTBDT protocol, probability λ is 0.5, 0.7, 0.8, 0.9. Then, the PTBDT protocol with the probability $\lambda = 1.0$ means the epidemic routing protocol and the probability $\lambda = 0$ stands for the TBDT protocol. The delivery ratios of the PTBDT protocol with probability $\lambda = 0.9$ and the epidemic routing

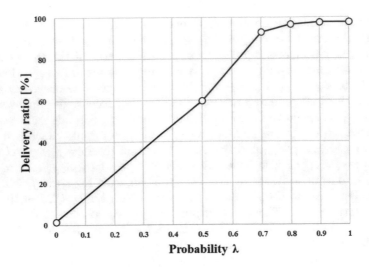

Fig. 4. Delivery ratio for probability λ.

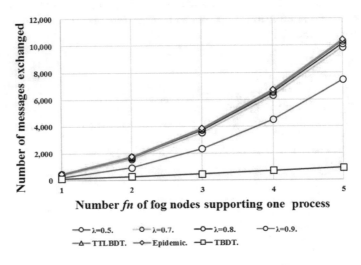

Fig. 5. Number of messages exchanged.

protocol are almost the same. The delivery ratios of the TTLBDT protocol and the epidemic routing protocol are almost the same at $fn \geq 3$. The delivery ratios of the PTBDT protocol with probability $\lambda = 0.7$ and the epidemic routing protocol are almost the same when the number fn of each fog node which supports one process is four ($fn = 4$).

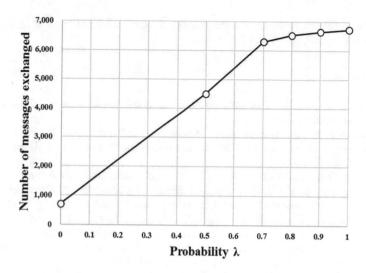

Fig. 6. Number of messages exchanged for probability λ.

Figures 5 and 6 show the number of messages exchanged among fog nodes in each protocol. The numbers of messages exchanged among fog nodes in the TTLBDT protocol, the PTBDT protocol with probability $\lambda = 0.9$, and the epidemic routing protocol are almost the same. The number of messages exchanged among fog nodes in the epidemic routing protocol is a little bit larger than the PTBDT protocol with probability $\lambda = 0.8$, 0.9. Thus, in the PTBDT protocol with probability $\lambda = 0.8$, 0.9, the number of messages received and the number of messages exchanged among fog nodes can be reduced in a same delivery ratio compared with the epidemic routing protocol.

5   Concluding Remarks

In this paper, we considered the MPSFC (Mobile Publish/Subscribe Fog Computing) model to efficiently realize the IoT, where mobile fog nodes like vehicles communicate with other nodes in wireless networks. Here, each fog node calculates the output data on the input data received from other fog nodes and forwards the output data to target fog nodes in the epidemic routing way. In the TBDT protocol [16,17], each fog node delivers a message with output data to only a target fog node. The number of messages can be reduced in the TBDT protocol but the delivery ratio of messages is smaller in the TBDT protocol than the epidemic routing protocol. In this paper, we newly proposed the TTLBDT (Time-To-Live Based Data Transmission) and PTBDT (Probability and TBDT) protocols where each fog node f_i is delivered not only messages of data on which the fog node f_i can calculate by taking advantage of the topic-based PS model but also data on which the fog node f_i cannot calculate. In the evaluation, we showed the delivery ratio can be increased in the TTLBDT and PTBDT protocols compared with the TBDT protocol.

References

1. Amin, V., David, B.: Epidemic routing for partially-connected adhoc networks. Technical report (2000)
2. Dhurandher, S.K., Sharma, D.K., Woungang, I., Saini, A.: An energy-efficient history-based routing scheme for opportunistic networks. Int. J. Commun. Syst. **30**(7), e2989 (2015)
3. Gima, K., Oma, R., Nakamura, S., Enokido, T., Takizawa, M.: A model for mobile fog computing in the IoT (accepted). In: Proceedings of the 22nd International Conference on Network-Based Information Systems (NBiS-2019) (2019)
4. Gima, K., Oma, R., Nakamura, S., Enokido, T., Takizawa, M.: Parallel data transmission protocols in the mobile fog computing model. In: Proceedings of the 14th International Conference on Broad-Band Wireless Computing, Communication and Applications (BWCCA-2019), pp. 494–503 (2019)
5. Guo, Y., Oma, R., Nakamura, S., Duolikun, D., Enokido, T., Takizawa, M.: Data and subprocess transmission on the edge node of TWTBFC model. In: Proceedings of the 11th International Conference on Intelligent Networking and Collaborative Systems (INCoS-2019), pp. 80–90 (2019)
6. Guo, Y., Oma, R., Nakamura, S., Duolikun, D., Enokido, T., Takizawa, M.: Evaluation of a two-way tree-based fog computing (TWTBFC) model. In: Proceedings of the 13th International Conference on Innovative Mobile and Internet Services in Ubiquitous Computing (IMIS-2019), pp. 72–81 (2019)
7. Mukae, K., Saito, T., Nakamura, S., Enokido, T., Takizawa, M.: A dynamic tree-based fog computing (DTBFC) model for the energy-efficient IoT. In: Proceedings of the 15th International Conference on Broad-Band and Wireless Computing, Communication and Applications (BWCCA-2020), pp. 330–340 (2020)
8. Oma, R., Nakamura, S., Duolikun, D., Enokido, T., Takizawa, M.: An energy-efficient model for fog computing in the Internet of Things (IoT). Internet Things **1–2**, 14–26 (2018)
9. Oma, R., Nakamura, S., Duolikun, D., Enokido, T., Takizawa, M.: Evaluation of an energy-efficient tree-based model of fog computing. In: Proceedings of the 21st International Conference on Network-Based Information Systems (NBiS-2018), pp. 99–109 (2018)
10. Oma, R., Nakamura, S., Duolikun, D., Enokido, T., Takizawa, M.: Energy-efficient recovery algorithm in the fault-tolerant tree-based fog computing (FTBFC) model. In: Proceedings of the 33rd International Conference on Advanced Information Networking and Applications (AINA-2019), pp. 132–143 (2019)
11. Oma, R., Nakamura, S., Duolikun, D., Enokido, T., Takizawa, M.: A fault-tolerant tree-based fog computing model (accepted). Int. J. Web Grid Serv. (IJWGS) (2019)
12. Oma, R., Nakamura, S., Enokido, T., Takizawa, M.: A tree-based model of energy-efficient fog computing systems in IoT. In: Proceedings of the 12th International Conference on Complex, Intelligent, and Software Intensive Systems (CISIS-2018), pp. 991–1001 (2018)
13. Oma, R., Nakamura, S., Enokido, T., Takizawa, M.: A dynamic tree-based fog computing (DTBFC) model for the energy-efficient IoT. In: Proceedings of the 8th International Conference on Emerging Internet, Data and Web Technologies (EIDWT-2020), pp. 24–34 (2020)

14. Rahmani, A.M., Liljeberg, P., Preden, J.S., Jantsch, A.: Fog Computing in the Internet of Things. Springer, Heidelberg (2018)
15. Saito, T., Nakamura, S., Enokido, T., Takizawa, M.: Epidemic and topic-based data transmission protocol in a mobile fog computing model. In: Proceedings of the 15th International Conference on Broad-Band and Wireless Computing, Communication and Applications (BWCCA-2020), pp. 34–43 (2020)
16. Saito, T., Nakamura, S., Enokido, T., Takizawa, M.: Topic-based processing protocol in a mobile fog computing model. In: Proceedings of the 23nd International Conference on Network-Based Information Systems (NBiS-2020) Accepted (2020)
17. Saito, T., Nakamura, S., Enokido, T., Takizawa, M.: A topic-based publish/subscribe system in a fog computing model for the IoT. In: Proceedings of the 14th International Conference on Complex, Intelligent and Software Intensive Systems (CISIS-2020), pp. 12–21 (2020)
18. Setty, V., van Steen, M., Vintenberg, R., Voulgais, S.: Poldercast: Fast, robust, and scalable architecture for p2p topic-based pub/sub. In: Proceedings of ACM/IFIP/USENIX 13th International Conference on Middleware (Middleware-2012), pp. 271–291 (2012)
19. Spaho, E., Barolli, L., Kolici, V., Lala, A.: Evaluation of single-copy and multiple-copy routing protocols in a realistic VDTN scenario. In: Proceedings of the 10th International Conference on Complex, Intelligent, and Software Intensive Systems (CISIS-2016), pp. 285–289 (2016)
20. Tarkoma, S.: Publish/Subscribe System: Design and Principles, 1st edn. Wiley, Hoboken (2012)
21. Tarkoma, S., Rin, M., Visala, K.: The publish/subscribe internet routing paradigm (PSIRP): Designing the future internet architecture. In: Future Internet Assembly, pp. 102–111 (2009)
22. Yamamoto, Y., Hayashibara, N.: Merging topic groups of a publish/subscribe system in causal order. In: Proceedings of the 31st International Conference on Advanced Information Networking and Applications Workshops (WAINA-2017), pp. 172–177 (2017)

The Improved Active Time-Based (IATB) Algorithm with Multi-threads Allocation

Tomoya Enokido[1][✉] and Makoto Takizawa[2]

[1] Faculty of Business Administration, Rissho University, 4-2-16, Osaki,
Shinagawa-ku, Tokyo 141-8602, Japan
eno@ris.ac.jp
[2] Department of Advanced Sciences, Faculty of Science and Engineering,
Hosei University, 3-7-2, Kajino-cho, Koganei-shi, Tokyo 184-8584, Japan
makoto.takizawa@computer.org

Abstract. In order to provide scalable and high performance distributed application services, an energy-efficient server cluster system equipped with virtual machines like cloud computing systems are required. Here, processing load of virtual machines has to balance with one another to not only achieve performance objectives but also reduce the total electric energy consumption of a server cluster. In this paper, an improved active time-based (IATB) algorithm is newly proposed to reduce the total electric energy consumption of a server cluster and the response time of each application process.

Keywords: Energy-efficient server cluster · Virtual machine · Active time-based (ATB) algorithm · Improved ATB (IATB) algorithm

1 Introduction

In current information systems, virtual machines [1] supported by server cluster systems [2–5] like cloud computing systems [6] are widely used to provide various types of scalable and high performance distributed application services. Application services are supported by taking usage of high performance and scalable computing environments provided by a server cluster. On the other hand, a server cluster system consumes a large amount of electric energy to provide application services since a server cluster system is composed of a large number of physical servers and a huge number of application processes are performed on virtual machines installed in each physical server. Hence, energy-efficient server cluster systems [2–5] are required to realize distributed application services as discussed in Green computing [6].

In our previous studies, the *active time-based* (*ATB*) algorithm [7] is proposed to allocate computation type application processes (computation processes) to virtual machines in a server cluster system so that the total electric energy consumption of a server cluster and the average response time of each computation process can be reduced. However, in the ATB algorithm, only one

L. Barolli et al. (Eds.): EIDWT 2021, LNDECT 65, pp. 134–142, 2021.
https://doi.org/10.1007/978-3-030-70639-5_13

thread on a CPU is allocated to each virtual machine in each physical server and computation processes are performed on each virtual machine by using only one thread even if some threads are not used in a physical server. Hence, each physical server consumes more electric energy to perform computation processes on virtual machines since it takes longer time to perform computation processes if more number of processes are performed on each virtual machine.

In this paper, the *Improved ATB (IATB)* algorithm is newly proposed to furthermore reduce the total electric energy consumption of a server cluster equipped with virtual machines. In the IATB algorithm, the total electric energy consumption of a server cluster and the average response time of each computation process can be reduced by allocating idle threads to virtual machines which are performing computation processes in each physical server if the total electric energy of the physical server does not increase. Then, the computation rate of each active virtual machine to perform computation processes increases and the average response time of each computation process can be more reduced in the IATB algorithm than the ATB algorithm. As a result, the total electric energy consumption of a server cluster can be more reduced in the IATB algorithm than the ATB algorithm. The evaluation results show the total electric energy consumption of a server cluster and the response time of each computation process can be more reduced in the IATB algorithm than the ATB algorithm.

In Sect. 2, we discuss the computation model of a virtual machine and power consumption model of a physical server. In Sect. 3, we discuss the IATB algorithm. In Sect. 4, we evaluate the IATB algorithm compared with the ATB algorithm.

2 Computation and Power Consumption Model

2.1 Computation Model of a Virtual Machine

A server cluster system is composed of multiple physical servers s_1, ..., s_n ($n \geq 1$). In this paper, we assume a server s_t is equipped with one CPU composed of multiple homogeneous cores. Let nc_t be the total number of cores in a server s_t ($nc_t \geq 1$) and C_t be a set of cores c_{1t}, ..., $c_{nc_t t}$ in the server s_t. We assume the Hyper-Threading Technology [8] is enabled on a CPU. Let ct_t be the number of threads on each core c_{ht} in a server s_t. Let nt_t be the total number of threads in a server s_t, i.e. $nt_t = nc_t \cdot ct_t$. Let TH_t be a set of threads th_{1t}, ..., $th_{nt_t t}$ ($nt_t \geq 1$) in a server s_t. Threads $th_{(h-1)\cdot ct_t+1}$, ..., $th_{h\cdot ct_t}$ ($1 \leq h \leq nc_t$) are bounded to a core c_{ht}. Let V_t be a set of virtual machines VM_{1t}, ..., VM_{nt} supported by a server s_t. A virtual machine VM_{vt} is performed on threads in a server s_t. Let $VT_{vt}(\tau)$ be a set of threads allocated to a virtual machine VM_{vt} at time τ. $nVT_{vt}(\tau)$ is $|VT_{vt}(\tau)|$ and $1 \leq nVT_{vt}(\tau) \leq nt_t$. A virtual machine VM_{vt} is *active* iff (if and only if) at least one process is performed on the virtual machine VM_{vt}. Otherwise, the virtual machine VM_{vt} is *idle*. A thread th_{kt} is *active* iff at least one virtual machine VM_{vt} is active on the thread th_{kt}. Otherwise, the thread th_{kt} is *idle*. A core c_{ht} is *active* iff at least one thread th_{kt} is active on the core c_{ht}. Otherwise, the core c_{ht} is *idle*. In this paper, we assume each thread

th_{kt} in a server s_t is not allocated to multiple virtual machines at time τ, i.e. there is no overcommitment of a thread in a server s_t.

In this paper, we consider *computation processes* which mainly consume CPU resources of a virtual machine. A term *process* stands for a computation process in this paper. A notation pt^i_{kt} stands for a process p^i performed on a thread th_{kt} in a server s_t. Let Th^i_{kt} be the total computation time [msec] of a process pt^i_{kt} which the process p^i is performed on a thread th_{kt}. Let $minTh^i_{kt}$ be the minimum computation time of a process pt^i_{kt} where the process pt^i_{kt} is exclusively performed on one thread th_{kt} on a core c_{ht} and the other threads on the core c_{ht} are *idle* in a server s_t. We assume $minTh^i_{1t} = minTh^i_{2t} = \cdots = minTh^i_{rt}$ in a server s_t. $minTh^i = minTh^i_{kt}$ on the fastest server s_t. We assume one virtual computation step [vs] is performed for one time unit [tu] on one thread th_{kt} in the fastest server s_t. That is, the maximum computation rate $Maxf_{kt}$ of a thread th_{kt} on a core c_{ht} in the fastest server s_t where only the thread th_{kt} is active on the core c_{ht} is 1 [vs/tu]. $Maxf_{ku} \leq Maxf_{kt}$ on the slower server s_u. We assume $Maxf_{1t} = Maxf_{2t} = \cdots = Maxf_{rt}$ in a server s_t. $Maxf = max(Maxf_{k1}, ..., Maxf_{kn})$. A process p^i is considered to be composed of VS^i virtual computation steps. $VS^i = minTh^i \cdot Maxf = minTh^i$ [vs]. The maximum computation rate $maxf^i_{kt}$ of a process pt^i_{kt} is $VS^i/minTh^i_{kt}$ ($0 \leq maxf^i_{kt} \leq 1$) where the process p^i_{kt} is exclusively performed on a thread th_{kt} on a core c_{ht} and only the thread th_{kt} is active the a core c_{ht} in a server s_t.

The computation rate $FT_{kt}(\tau)$ of a thread th_{kt} on a core c_{ht} in a server s_t at time τ is given as follows:

$$FT_{kt}(\tau) = Maxf_{kt} \cdot \beta_{kt}(at_{kt}(\tau)). \tag{1}$$

Here, $at_{kt}(\tau)$ is the number of active threads on the core c_{ht} at time τ where the thread th_{kt} is bounded to the core c_{ht}. Let $\beta_{kt}(at_{kt}(\tau))$ be the *performance degradation ratio* of a thread th_{kt} on a core c_{ht} at time τ ($0 \leq \beta_{kt}(at_{kt}(\tau)) \leq 1$) where multiple threads are active on the same core c_{ht}. $\beta_{kt}(at_{kt}(\tau)) = 1$ if $at_{kt}(\tau) = 1$. $\beta_{kt}(at_{kt}(\tau_1)) \leq \beta_{kt}(at_{kt}(\tau_2)) \leq 1$ if $at_{kt}(\tau_1) \geq at_{kt}(\tau_2)$.

Suppose a virtual machine VM_{vt} is performed on a set $VT_{vt}(\tau)$ of threads in a server s_t at time τ. The computation rate $FV_{vt}(\tau)$ of the virtual machine VM_{vt} at time τ is given as follows:

$$FV_{vt}(\tau) = \sum_{th_{kt} \in VT_{vt}(\tau)} FT_{kt}(\tau). \tag{2}$$

In this paper, processes are performed on virtual machines installed in each server s_t. A notation p^i_{vt} stands for an instance of a process p^i performed on a virtual machine VM_{vt} in a server s_t. Processes which are being performed on a virtual machine VM_{vt} at time τ are *current*. Let $CP_{vt}(\tau)$ be a set of current processes on a virtual machine VM_{vt} at time τ and $NC_{vt}(\tau)$ be $|CP_{vt}(\tau)|$. In this paper, we assume the computation rate $FV_{vt}(\tau)$ of a virtual machine VM_{vt} at time τ is uniformly allocated to every current process on the virtual machine VM_{vt}.

The computation rate $f_{vt}^i(\tau)$ of a process p_{vt}^i performed on a virtual machine VM_{vt} at time τ is given as follows:

$$f_{vt}^i(\tau) = \begin{cases} \alpha_{vt}(\tau) \cdot FV_{vt}(\tau) \, / \, NC_{vt}(\tau), & \text{if } NC_{vt}(\tau) > nVT_{vt}(\tau), \\ FT_{kt}(\tau), & \text{otherwise.} \end{cases} \quad (3)$$

Here, $\alpha_{vt}(\tau)$ is the *computation degradation ratio* of a virtual machine VM_{vt} at time τ $(0 \leq \alpha_{vt}(\tau) \leq 1)$. $\alpha_{vt}(\tau_1) \leq \alpha_{vt}(\tau_2) \leq 1$ if $NC_{vt}(\tau_1) \geq NC_{vt}(\tau_2)$. $\alpha_{vt}(\tau) = 1$ if $NC_{vt}(\tau) \leq 1$. Here, $\alpha_{vt}(\tau)$ is assumed to be $\varepsilon_{vt}^{NC_{vt}(\tau)-1}$ where $0 \leq \varepsilon_{vt} \leq 1$. The formula (3) means the computation rate $f_{vt}^i(\tau)$ of each process p_{vt}^i performed on a virtual machine VM_{vt} at time τ decreases as the number of current processes increases on the virtual machine VM_{vt} if $NC_{vt}(\tau) > nVT_{vt}(\tau)$. If $NC_{vt}(\tau) \leq nVT_{vt}(\tau)$, each process p_{vt}^i in a set $CP_{vt}(\tau)$ is exclusively performed on one thread th_{kt} in a set $VT_{vt}(\tau)$ of threads bounded to the virtual machine VM_{vt}.

Suppose that a process p_{vt}^i starts and terminates on a virtual machine VM_{vt} at time st_{vt}^i and et_{vt}^i, respectively. Let T_{vt}^i be the total computation time of a process p_{vt}^i performed on a virtual machine VM_{vt}. Here, $T_{vt}^i = et_{vt}^i - st_{kt}^i$ and $\sum_{\tau=st_{vt}^i}^{et_{vt}^i} f_{vt}^i(\tau) = VS^i$. At time st_{vt}^i a process p_{vt}^i starts, the computation laxity $lc_{vt}^i(\tau)$ is VS^i [vs]. The computation laxity $lc_{vt}^i(\tau)$ [vs] of a process p_{vt}^i at time τ is given as follows:

$$lc_{vt}^i(\tau) - VS^i - \sum_{x=st_{vt}^i}^{\tau} f_{vt}^i(x), \quad (4)$$

2.2 Power Consumption Model of a Server

A notation $E_t(\tau)$ shows the electric power [W] of a server s_t at time τ. Let $maxE_t$ and $minE_t$ be the maximum and minimum electric power [W] of a server s_t, respectively. Let $ac_t(\tau)$ be the number of active cores in a server s_t at time τ. A notation $minC_t$ shows the electric power [W] where at least one core c_{ht} is active on a server s_t. Let cE_t be the electric power [W] consumed by a server s_t to make one core active.

The *power consumption model of a server with virtual machines* (*PCSV model*) [9] to perform computation processes on virtual machines is proposed. According to the PCSV model, the electric power $E_t(\tau)$ [W] of a server s_t to perform processes on virtual machines at time τ is given as follows [9]:

$$E_t(\tau) = minE_t + \sigma_t(\tau) \cdot (minC_t + ac_t(\tau) \cdot cE_t). \quad (5)$$

Here, $\sigma_t(\tau) = 1$ if at least one core c_{ht} is active on a server s_t at time τ. Otherwise, $\sigma_t(\tau) = 0$. The electric power $E_t(\tau)$ of a server s_t depends on the number of active cores at time τ.

The processing power $PE_t(\tau)$ [W] is $E_t(\tau) - minE_t$ at time τ in a server s_t. The total processing electric energy $TPE_t(\tau_1, \tau_2)$ [J] of a server s_t from time τ_1 to τ_2 is $\sum_{\tau=\tau_1}^{\tau_2} PE_t(\tau)$.

3   Virtual Machine Selection Algorithm

3.1   System Model

Suppose a client cl^i first issues a request process p^i to a load balancer K. The load balancer K selects one virtual machine VM_{vt} in a server cluster S for the request process p^i and forwards the process p^i to the virtual machine VM_{vt}. On receipt of the request process p^i, a process p^i_{vt} is created and performed on a virtual machine VM_{vt}. On termination of the process p^i_{vt}, the virtual machine VM_{vt} sends a reply r^i_{vt} to the load balancer K. A load balancer K has a variable $CP_{vt}(\tau)$. If the request process p^i is forwarded to a virtual machine VM_{vt} at time τ, $CP_{vt}(\tau) = CP_{vt}(\tau) \cup \{p^i_{vt}\}$. If the load balancer K receives a reply r^i_{vt} of a process p^i_{vt}, $CP_{vt}(\tau) = CP_{vt}(\tau) - \{p^i_{vt}\}$. Hence, the load balancer K can check how many number of current processes are performed on each virtual machine VM_{vt} at time τ.

3.2   The Active Time-Based (ATB) Algorithm

In our previous studies, the *active time-based* (*ATB*) algorithm [7] is proposed to select a virtual machine VM_{vt} for each request process p^i so that the total electric energy consumption of a server cluster S and the average response time of each process can be reduced. The ATB algorithm assumes only one thread th_{kt} is allocated to each virtual machine VM_{vt} in a server s_t and the minimum computation time $minT^i$ of each process p^i is shorter than the one time unit [msec], i.e. $minT^i \leq 1$ [msec]. The minimum computation time $minT^i$ of each process p^i is set to one millisecond [msec] to estimate the minimum response time $minRT^i_{vt}$ of a process p^i_{vt} performed on a virtual machine VM_{vt}. Let d_{Kt} be the delay time [msec] between a load balancer K and a server s_t. The minimum response time $minRT^i_{vt}$ of a process p^i_{vt} where the process p^i_{vt} is exclusively performed on a virtual machine VM_{vt} and only VM_{vt} is active on a core c_{ht} in a server s_t is calculated as $minRT^i_{vt} = 2d_{Kt} + minT^i \cdot Maxf/Maxf_{vt} = 2d_{Kt} + 1 \cdot 1/Maxf_{vt} = 2d_{Kt} + 1/Maxf_{vt}$ [msec]. The ATB algorithm estimates the increased active time $iACT_{ht}(\tau)$ of each core c_{ht} in a server s_t at time τ based on the response time RT^i_{vt} of each process p^i_{vt} performed on each virtual machine VM_{vt} in the server s_t. Suppose a new request process p^{new} is allocated to a virtual machine VM_{vt} performed on a server s_t at time τ. In the PCSV model, the electric power $E_t(\tau)$ [W] of a server s_t at time τ depends on the number $ac_t(\tau)$ of active cores in the server s_t at time τ as shown in Eq. (5). The total processing electric energy laxity $tpel_t(\tau)$ [J] shows how much electric energy a server s_t has to consume to perform every current process on every active virtual machine in the server s_t at time τ. In the ATB algorithm, the total processing electric energy laxity $tpel_t(\tau)$ of a server s_t at time τ is estimated by the following Eq. (6):

$$tpel_t(\tau) = minC_t + \sum_{h=1}^{nc_t}(iATC_{ht}(\tau) \cdot cE_t)). \tag{6}$$

Let $TPEL_{vt}(\tau)$ be the total processing electric energy laxity of a server cluster S where a new request process p^{new} is allocated to a virtual machine VM_{vt} performed on a server s_t at time τ. In the ATB algorithm, a virtual machine VM_{vt} where the total processing electric energy laxity $TPEL_{vt}(\tau)$ of a server cluster S is the minimum at time τ is selected for a new request p^{new}.

3.3   The Improved ATB (IATB) Algorithm

In the ATB algorithm [7], only one thread is allocated to a virtual machine in a server. Hence, some threads may be idle while the other threads are active to perform many replicas. As a result, a server consumes more electric energy since it takes longer time to perform processes if more number of processes are performed on each virtual machine. In this paper, we newly propose an *Improved ATB (IATB)* algorithm to furthermore reduce the total processing electric energy consumption of a server cluster S. In the IATB algorithm, the total processing electric energy consumption of a server cluster S and the average response time of each process can be reduced by allocating idle threads to active virtual machines in each server s_t. In the *power consumption model of a server with virtual machines (PCSV model)* [9], the electric power consumption $E_t(\tau)$ of a server s_t depends on the number $ac_t(\tau)$ of active cores in the server s_t at time τ. At time τ, the electric power consumption $E_t(\tau)$ of the server s_t where current processes are performed on a virtual machine VM_{vt} with one thread th_{kt} on a core c_{ht} is the same as the electric power where current processes are performed on the virtual machine VM_{vt} with multiple threads on the same core c_{ht} since $ac_t(\tau) = 1$.

In the IATB algorithm, if a thread $th_{k't}$ on a core c_{ht} allocated to a virtual machine $VM_{v't}$ is idle, i.e. the virtual machine $VM_{v't}$ is idle, the thread $th_{k't}$ is used for another active virtual machine VM_{vt} performed on the same core c_{ht} in each server s_t at time τ. Then, the computation time of each process performed on the virtual machine VM_{vt} can be reduced in the IATB algorithm than the ATB algorithm since the computation rate $FV_{vt}(\tau)$ of the virtual machine VM_{vt} increases. As a result, the total processing electric energy consumption of each server s_t can be reduced in the IATB algorithm than the ATB algorithm since the active time of each core can be reduced. Let $idle_{ht}(\tau)$ be a set of idle threads on a core c_{ht} in a server s_t at time τ. Let $avm_{ht}(\tau)$ be a set of active virtual machines on a core c_{ht} at time τ. At time τ, idle threads on a core c_{ht} are allocated to each active virtual machine VM_{vt} performed on the core c_{ht} in a server s_t by the following **Thread_Alloc()** procedure:

Thread_Alloc(τ) {
    for each core c_{ht} in a server s_t, {
        $avm_{ht}(\tau)$ = a set of active VMs on a core c_{ht} at time τ;
        if $|idle_{ht}(\tau)| \geq 1$ and $|avm_{ht}(\tau)| \geq 1$, {
            while $|idle_{ht}(\tau)| > 0$, {
                $th = th_{kt} \in idle_{ht}(\tau)$;
                VM_{vt} = a virtual machine where $CP_{vt}(\tau)$ is the maximum in $avm_{ht}(\tau)$;

$VT_{vt}(\tau) = VT_{vt}(\tau) \cup \{th\}$; /* a thread th is bounded to VM_{vt}. */
$idle_{ht}(\tau) = idle_{ht}(\tau) - \{th\}$;
$avm_{ht}(\tau) = avm_{ht}(\tau) - \{VM_{vt}\}$;
if $avm_{ht}(\tau) = \phi$, $avm_{ht}(\tau)$ = a set of active VMs on c_{ht} at time τ;
        } /* while end. */
    } /* if end. */
} /* for end. */
}

4   Evaluation

The IATB algorithm is evaluated in terms of the total processing electric energy consumption [KJ] of a homogeneous server cluster S and the response time of each process p^i compared with the ATB algorithm [7]. We consider a homogeneous cluster S composed of five servers s_1, ..., s_5 ($t = 5$). Every server s_t ($1 \leq t \leq 5$) follows the same power consumption model as shown in Table 1. The parameters of each server s_t are obtained from the experiment [9]. Every server s_t is equipped with a dual-core CPU ($nc_t = 2$). Two threads are bounded to each core in a server s_t, i.e. $ct_t = 2$. Hence, the number nt_t of threads in each server s_t is four. The total number of threads in the server cluster S is twenty. Initially, each thread th_{kt} is allocated to one virtual machine VM_{vt} in a server s_t (k, $v = 1$, ..., 4 and $t = 1$, ..., 5). Hence, there are twenty virtual machines in the server cluster S. Every virtual machine VM_{vt} follows the same computation model as shown in Table 2. The parameters of each virtual machine VM_{vt} are obtained from the experiment [9].

Table 1. Homogeneous cluster S.

Server	nc_t	ct_t	nt_t	$minE_t$	$minC_t$	cE_t	$maxE_t$
s_t	2	2	4	14.8 [W]	6.3 [W]	3.9 [W]	33.8 [W]

Table 2. Parameters of virtual machine.

Virtual machine	$Max f_{vt}$	ε_{vt}	$\beta_{vt}(1)$	$\beta_{vt}(2)$
VM_{vt}	1 [vs/msec]	1	1	0.6

The number m of processes p^1, ..., p^m ($0 \leq m \leq 10{,}000$) are issued. The starting st^i time of each process p^i is randomly selected in a unit of one millisecond between 1 and 360 [msec]. The minimum computation time $minT^i$ of every process p^i is assumed to be 1 [msec]. The delay time d_{Kt} of every pair of a load balancer K and every server s_t is 1 [msec] in the server cluster S ($1 \leq t \leq 5$). The minimum response time $minRT_{vt}^i$ of every process p_{vt}^i is $2d_{Kt} + minT_{vt}^i = 2 \cdot 1 + 1 = 3$ [msec].

4.1   Response Time

The response time RT_{vt}^i for each process p^i in the server cluster S is measured five times for the total number m of processes. Let $RT_{vt}^{i,tm}$ be the average response time RT_{vt}^i obtained in the tm-th simulation for total number m of processes. The average response time ART is $\sum_{tm=1}^{5}\sum_{i=1}^{m} RT_{vt}^{i,tm}/(m \cdot 5)$. Figure 1 shows the average response time ART [msec] in the server cluster S for the total number m of processes in the IATB and ATB algorithms. From the evaluation, the average response time ART in the IATB algorithm can be more reduced than the ATB algorithm. In the IATB algorithm, idle threads in each server s_t are allocated to active virtual machines in the server s_t if the total processing electric energy consumption of the server s_t does not increase. Then, the computation rate of each active virtual machine to perform processes increases. As a result, the average response time of each process can be more reduced in the IATB algorithm than the ATB algorithm since the computation resources in the server cluster S can be more efficiently utilized in the IATB algorithm than the ATB algorithm.

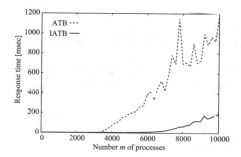

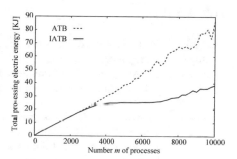

Fig. 1. Average response time ART.   **Fig. 2.** Forwarding a request process.

4.2   Total Energy Consumption of a Server Cluster

Let TEC_{tm} be the total processing electric energy consumption [KJ] to perform the number m of processes in the server cluster S obtained in the tm-th simulation. The total processing electric energy consumption TEC_{tm} is measured five times for each number m of processes. Then, the average total processing electric energy $ATEC$ [KJ] is calculated as $\sum_{tm=1}^{5} TEC_{tm}/5$ for each number m of processes. Figure 2 shows the average total processing electric energy consumption $ATEC$ of the server cluster S to perform the number m of processes in the IATB and ATB algorithms. In the IATB algorithm, the response time of each process can be more reduced than the ATB algorithm as shown in Fig. 1 since idle threads in each server s_t are allocated to active virtual machines in the server s_t. Then, the active time of each virtual machine can be more reduced in the IATB algorithm than the ATB algorithm. As a result, the average total processing electric energy consumption of the server cluster S can be more reduced in the IATB algorithm than the ATB algorithm. Following the evaluation, the IATB algorithm is more useful than the ATB algorithm.

5   Concluding Remarks

In this paper, the IATB algorithm is newly proposed to reduce the total electric energy consumption of a server cluster and the response time of each process by allocating idle threads in each server s_t to active virtual machines. We showed the total electric energy consumption of a server cluster and the response time of each process can be more reduced in the IATB algorithm than the ATB algorithm. The evaluation results show the IATB algorithm is more useful than the ATB algorithm.

Acknowledgements. This work was supported by the Japan Society for the Promotion of Science (JSPS) KAKENHI Grant Number 19K11951.

References

1. KVM: Main Page - KVM (Kernel Based Virtual Machine) (2015). http://www.linux-kvm.org/page/Mainx_Page
2. Enokido, T., Aikebaier, A., Takizawa, M.: Process allocation algorithms for saving power consumption in peer-to-peer systems. IEEE Trans. Ind. Electron. **58**(6), 2097–2105 (2011)
3. Enokido, T., Aikebaier, A., Takizawa, M.: A model for reducing power consumption in peer-to-peer systems. IEEE Syst. J. **4**(2), 221–229 (2010)
4. Enokido, T., Aikebaier, A., Takizawa, M.: An extended simple power consumption model for selecting a server to perform computation type processes in digital ecosystems. IEEE Trans. Ind. Inf. **10**(2), 1627–1636 (2014)
5. Enokido, T., Takizawa, M.: Integrated power consumption model for distributed systems. IEEE Trans. Ind. Electron. **60**(2), 824–836 (2013)
6. Natural Resources Defense Council (NRDS): Data center efficiency assessment - scaling up energy efficiency across the data center industry: Evaluating key drivers and barriers (2014). http://www.nrdc.org/energy/files/data-center-efficiency-assessment-IP.pdf
7. Enokido, T., Duolikun, D., Takizawa, M.: An energy efficient load balancing algorithm based on the active time of cores. In: Proceedings of the 12th International Conference on Broad-Band Wireless Computing, Communication and Applications (BWCCA-2017), pp. 185–196 (2017)
8. Intel: Intel Xeon Processor 5600 Series : The Next Generation of Intelligent Server Processors (2010). http://www.intel.com/content/www/us/en/processors/xeon/xeon-5600-brief.html
9. Enokido, T., Takizawa, M.: Power consumption and computation models of virtual machines to perform computation type application processes. In: Proceedings of the 9th International Conference on Complex, Intelligent and Software Intensive Systems (CISIS-2015), pp. 126–133 (2015)

Effect of Vehicle Technical Condition on Real-Time Driving Risk Management in VANETs

Kevin Bylykbashi[1(✉)], Ermioni Qafzezi[1], Makoto Ikeda[2], Keita Matsuo[2], Leonard Barolli[2], and Makoto Takizawa[3]

[1] Graduate School of Engineering, Fukuoka Institute of Technology (FIT), 3-30-1 Wajiro-Higashi, Higashi-Ku, Fukuoka 811–0295, Japan
bylykbashi.kevin@gmail.com, eqafzezi@gmail.com
[2] Department of Information and Communication Engineering, Fukuoka Institute of Technology (FIT), 3-30-1 Wajiro-Higashi, Higashi-Ku, Fukuoka 811-0295, Japan
makoto.ikd@acm.org, {kt-matsuo,barolli}@fit.ac.jp
[3] Department of Advanced Sciences, Faculty of Science and Engineering, Hosei University, 3-7-2, Kajino-machi, Koganei-shi, Tokyo 184-8584, Japan
makoto.takizawa@computer.org

Abstract. In this paper, we present a fuzzy-based driving-support system for real-time risk management in Vehicular Ad hoc Networks (VANETs) considering vehicle technical condition as a new parameter. The proposed system, called Fuzzy-based Simulation System for Driving Risk Management (FSSDRM), considers the current condition of different parameters which have an impact on the driver and vehicle performance to assess the risk level. The parameters include the vehicle speed, the weather and road condition, and factors that affect the driver's ability to drive, such as his/her current health condition and the inside environment in which he/she is driving in addition to the vehicle technical condition. The data for input parameters can come from different sources, such as on-board and on-road sensors and cameras, sensors and cameras in the infrastructure and from the communications between the vehicles. Based on the driving risk level, the system can invoke a certain action, which when performed, it reduces the driving risk and provides a better driving support. We show through simulations the effect of the considered parameters on the determination of the driving risk and demonstrate a few actions that can be performed accordingly.

1 Introduction

Road traffic accidents claim approximately 1.35 million lives each year and cause up to 50 million non-fatal injuries, with many of those injured people incurring a disability as a result of those injuries. The fact is that each of those deaths and injuries is totally preventable [9]. In this regard, industry, governmental institutions and academia researchers are conducting substantial research to provide proper systems and infrastructure for car accident prevention. The initiatives of many governments for a collaboration of such researchers has concluded to the

L. Barolli et al. (Eds.): EIDWT 2021, LNDECT 65, pp. 143–154, 2021.
https://doi.org/10.1007/978-3-030-70639-5_14

establishment of Intelligent Transport Systems (ITSs). ITSs focus on the deployment of intelligent transportation technologies by combining cutting-edge information, communication, and control technologies to design sustainable information networks based on people, vehicles and roads.

As a main component of ITS, Vehicular Ad hoc Networks (VANETs) aim not only to save lives but also to improve the traffic mobility, increase efficiency, and promote travel convenience of drivers and passengers. In VANETs, network nodes (vehicles) are equipped with networking functions to exchange essential information such as safety messages and traffic/road information with one another via vehicle-to-vehicle (V2V), and with roadside units (RSUs) through vehicle-to-infrastructure (V2I) communications. Although VANETs are already implemented in reality introducing several applications, the current architectures face numerous challenges, thus are far from achieving full marks in safety.

To overcome the encountered challenges, the integration of many emerging technologies—Software Defined Networking (SDN), Cloud Computing, Edge/Fog Computing, 5G, Information-Centric Networking (ICN), Blockchain and so forth—within current VANETs is actively being proposed. Although each one of these emerging technologies promises to solve several issues, other approaches and technologies—Wireless Sensor Networks (WSNs), Internet of Things (IoT)—that have been around us for years, can be used as an effective complement to alleviate various limitations.

Alongside these technologies, various artificial intelligence approaches including fuzzy logic and machine learning, are paving the way not only for a complete deployment of VANETs but also for reaching a bigger goal, that of putting the fully autonomous vehicles on the roads. Nevertheless, fully driverless cars still have a long way to go and the current advances fall only between the Level 2 and 3 of the Society of Automotive Engineers (SAE) levels [8]. Until we have self-driving vehicles, many automotive companies and academia researchers will continue working on Driver-Assistance Systems (DASs) as a principal safety feature to enhance the driving safety in non-automated vehicles. These intelligent systems reside inside the vehicle and rely on the measurement and perception of the surrounding environment, and based on the acquired information, a variety of actions is taken to ease the driving operation.

Apart from the external factors, there are other factors that have an effect on the driving operation as well. The first and the foremost is the driver itself, as the driver's ability to drive is rigorously related with the driving operation. By considering these important factors, we aim to realize a non-complex and non-intrusive intelligent driving-support system to detect a danger or a risky situation and warn the driver in real-time about the danger. Our system is based on fuzzy logic and makes use of the information acquired from various in-car sensors as well as from communications with other vehicles and infrastructure to evaluate the condition of the considered parameters. The parameters include factors that affect the driver's ability to drive, such as his/her current health condition and the inside environment in which he/she is driving, the vehicle speed and factors related to the outside environment such as the weather and road condition.

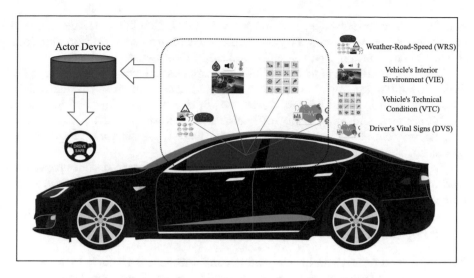

Fig. 1. A visualization of the proposed system model.

In addition to these parameters, the vehicle technical condition is included in this work as a new parameter for its crucial impact on the driving risk, especially when this factor is combined with the other considered factors. A model of the proposed system is presented in Fig. 1. We evaluate the proposed system by computer simulation. Based on the proposed system's output value, it can be decided if an action is needed, and if so, which is the appropriate task to be performed in order to reduce the driving risk and to provide a better driving support.

The structure of the paper is as follows. Section 2 presents a brief overview of VANETs. Section 3 describes the proposed fuzzy-based simulation system and its implementation. Section 4 discusses the simulation results. Finally, conclusions and future work are given in Sect. 5.

2   Overview of VANETs

VANETs are a special case of Mobile Ad hoc Networks (MANETs) in which the mobile nodes are vehicles. In VANETs, nodes (vehicles) have high mobility and tend to follow organized routes instead of moving randomly. Moreover, vehicles offer attractive features such as higher computational capability and localization through GPS.

VANETs have huge potential to enable applications ranging from road safety, traffic optimization, infotainment, commercial to rural and disaster scenario connectivity. Among these, the road safety and traffic optimization are considered the most important ones as they have the goal to reduce drastically the high number of accidents, guarantee road safety, make traffic management and create new forms of inter-vehicle communications in ITSs. The ITSs manage the vehicle traffic, support drivers with safety and other information, and provide some services such as automated toll collection and driver assist systems [3].

Despite the attractive features, VANETs are characterized by very large and dynamic topologies, variable capacity wireless links, bandwidth and hard delay constrains, and by short contact durations which are caused by the high mobility, high speed and low density of vehicles. In addition, limited transmission ranges, physical obstacles and interferences, make these networks characterized by disruptive and intermittent connectivity.

To make VANETs applications possible, it is necessary to design proper networking mechanisms that can overcome relevant problems that arise from vehicular environments.

3   Proposed Fuzzy-Based Simulation System

The highly competitive and rapidly advancing autonomous vehicle race has been on for several years now and it is a matter of time until we have these vehicles on the roads. However, even if the automotive companies do all what it takes on their end to create the fully automated cars, there will still be one big obstacle, the infrastructure. In addition, this could take decades even in the most developed countries. Moreover, 93% of the world's fatalities on the roads occur in low- and middle-income countries [9] and considering all these facts, DASs should remain the focus of interest for the foreseeable future.

DASs can be very helpful in many situations as they do not depend on the infrastructure as much as the driverless vehicles do. Furthermore, DASs can provide driving-support with very little cost, thus help the low- and middle-income countries in the long battle against car accidents. Most of these systems focus on driver's situation and intervene when he/she seems incapable of driving safely.

Our research work focuses on developing an intelligent non-complex driving support system which determines the driving risk level in real-time by considering different types of parameters. In the previous works, we have considered different parameters including in-car environment parameters such as the ambient temperature and noise, and driver's vital signs, i.e., heart and respiratory rate for which we implemented a testbed and conducted experiments in a real scenario [2]. The considered parameters include environmental factors and driver's health condition which can affect the driver capability and vehicle performance. In [1], we presented an integrated fuzzy-based system, which in addition to those parameters, it considers also the following inputs: the vehicle speed, the weather and road condition, the driver's body temperature and vehicle interior relative humidity. The inputs were categorized based on their characteristics of how they effect the driving operation. In this work, we include the vehicle technical condition, as sometimes it is a vehicle failure which becomes a factor of many car crashes. Although crashes related to vehicle techincal conditions alone account for less than 10% of all car crashes, this factor can combine with other factors, and as a result, it increases the extent of caused injuries.

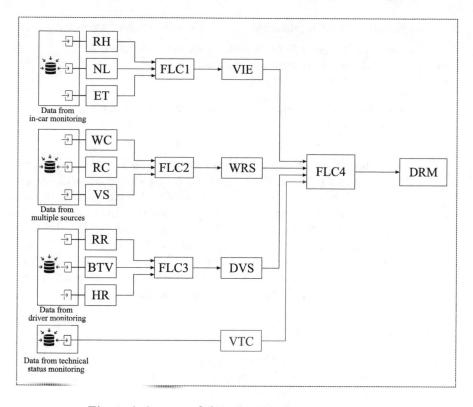

Fig. 2. A diagram of the proposed fuzzy-based system.

We use fuzzy logic to implement the proposed system as it can make a real-time decision based on the uncertainty and vagueness of the provided information [4–7, 10, 11]. The proposed system called Fuzzy-based Simulation System for Driving Risk Management (FSSDRM) is shown in Fig. 2. It consists of four FLCs, three of which are used to lower the overall complexity of the system. While it seems more complex by the way our system is built, it is far better than having ten input parameters in a single FLC because this would result in a very complex Fuzzy Rule Base (FRB) composed of thousands of rules, which, in turn, would increase the overall complexity of the system. FSSDRM has the following inputs: vehicle's Environment Temperature (ET), Relative Humidity (RH), Noise Level (NL), driver's Body Temperature Variation (BTV), Heart Rate (HR) and Respiratory Rate (RR), Weather Condition (WC), Road Condition (RC), Vehicle Speed (VS) and Vehicle Technical Condition (VTC), and its output is the Driving Risk Management (DRM). FLC1 makes use of RH, NL and ET, FLC2 has WC, RC and VS as inputs, and RR, BTV and RR are the inputs of FLC3. Vehicle's Inside Environment (VIE), Weather-Road-Speed (WRS) and Driver's Vital Signs are the outputs of these three FLCs (FLC1, FLC2 and FLC3, respectively) and together with VTC, serve as input parameters for FLC4. Since FLC1, FLC2, FLC3 are presented in a previous work [1], in this work we describe in detail only FLC4 which includes VTC as a new parameter.

The term sets of system parameters of FLC4 are defined respectively as:

$T(WRS) = \{No/Minor\ Danger\ (N/MD),\ Moderate\ Danger(MD),\ Danger(D)\};$

$T(VIE) = \{Uncomfortable\ (UC),\ Moderate\ (Mo),\ Comfortable\ (C)\};$

$T(VTC) = \{Bad\ (Ba),\ Fair\ (Fa),\ Good\ (Go)\};$

$T(DVS) = \{Bad\ (B),\ Fair\ (F),\ Good\ (G)\};$

$T(DRM) = \{Safe\ (Sf),\ Low\ (Lw),\ Moderate\ (Md),\ High\ (Hg),\ Very\ High\ (VH),\ Severe\ (Sv),$
$\qquad\qquad Danger\ (Dg)\}.$

Based on the linguistic description of input and output parameters, we make the Fuzzy Rule Base (FRB) of the FLC. The FRB forms a fuzzy set of dimensions $|\ T(x_1)\ |\ \times\ |\ T(x_2)\ |\ \times\ \cdots\ \times\ |\ T(x_n)\ |$, where $|\ T(x_i)\ |$ is the number of terms on $T(x_i)$ and n is the number of FLC input parameters. FLC4 has four input parameters with three linguistic terms each, therefore there are 81 rules in the FRB, which are shown in Table 1. The control rules of FRB have the form: IF "conditions" THEN "control action". The membership functions are shown in Fig. 3. We use triangular and trapezoidal membership functions because these types of functions are more suitable for real-time operation.

4   Simulation Results

In this section, we present the simulation results for our proposed system. The simulation results are presented in Figs. 4, 5 and 6. We consider the WRS and VIE as constant parameters. We show the relation between DRM and DVS for different VTC values. The VTC values considered for simulations are 0.1, 0.5 and 0.9, which simulate a vehicle with bad, fair and good technical condition, respectively.

In Fig. 4 we consider the WRS value 0.1 and change VIE from 0.1 to 0.9. From Fig. 4(a), we can see that there is not any situation considered as safe. The lowest DRM value is decided when a driver with good vital signs drives a vehicle in good technical condition and it is decided as a situation with low risk. This is due to the bad condition of the vehicle interior environment which can make the driver feel uncomfortable and it can cause the driver not to focus totally on the driving operation. When the inside environment is comfortable (see Fig. 4(b)), we can see that the risk is reduced and safe situations are now present. However, this happens only when vehicle is in optimal technical condition and when the conditions deteriorate, the DRM is increased.

Figure 5 shows the simulation results for WRS = 0.5. When the inside environment is uncomfortable, all the scenarios include situations with risk present to certain degrees based on the driver's vital signs and vehicle technical condition. For example, when the VTC is not good, many DRM values are decided as very high, severe and danger. From Fig. 5(b), it can be seen that the risk is present (DRM > 0.2) in all situations even when the inside environment is comfortable. This is due to WRS, because a WRS value of 0.5 is decided when the road or weather condition is not good and the driver is driving at a high speed [1].

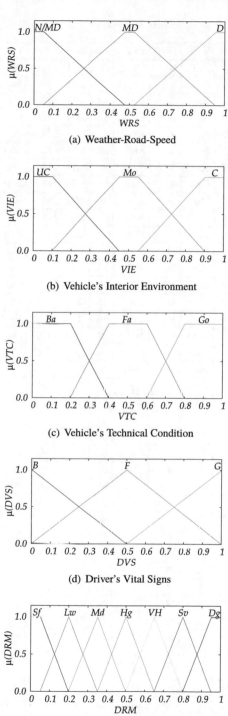

(a) Weather-Road-Speed

(b) Vehicle's Interior Environment

(c) Vehicle's Technical Condition

(d) Driver's Vital Signs

(e) Driving Risk Management

Fig. 3. Membership functions.

Table 1. FRB of FLC4.

No	WRS	VIE	VTC	DVS	DRM	No	WRS	VIE	VTC	DVS	DRM
1	N/MD	UC	Ba	B	Dg	41	MD	Mo	Fa	F	VH
2	N/MD	UC	Ba	F	Sv	42	MD	Mo	Fa	G	Hg
3	N/MD	UC	Ba	G	Hg	43	MD	Mo	Go	B	VH
4	N/MD	UC	Fa	B	Sv	44	MD	Mo	Go	F	Hg
5	N/MD	UC	Fa	F	VH	45	MD	Mo	Go	G	Md
6	N/MD	UC	Fa	G	Md	46	MD	C	Ba	B	Dg
7	N/MD	UC	Go	B	VH	47	MD	C	Ba	F	VH
8	N/MD	UC	Go	F	Hg	48	MD	C	Ba	G	Hg
9	N/MD	UC	Go	G	Lw	49	MD	C	Fa	B	Sv
10	N/MD	Mo	Ba	B	Sv	50	MD	C	Fa	F	Hg
11	N/MD	Mo	Ba	F	VH	51	MD	C	Fa	G	Md
12	N/MD	Mo	Ba	G	Md	52	MD	C	Go	B	VH
13	N/MD	Mo	Fa	B	VH	53	MD	C	Go	F	Md
14	N/MD	Mo	Fa	F	Hg	54	MD	C	Go	G	Lw
15	N/MD	Mo	Fa	G	Lw	55	D	UC	Ba	B	Dg
16	N/MD	Mo	Go	B	Hg	56	D	UC	Ba	F	Dg
17	N/MD	Mo	Go	F	Md	57	D	UC	Ba	G	Dg
18	N/MD	Mo	Go	G	Sf	58	D	UC	Fa	B	Dg
19	N/MD	C	Ba	B	VH	59	D	UC	Fa	F	Dg
20	N/MD	C	Ba	F	Hg	60	D	UC	Fa	G	Sv
21	N/MD	C	Ba	G	Md	61	D	UC	Go	B	Dg
22	N/MD	C	Fa	B	Hg	62	D	UC	Go	F	Dg
23	N/MD	C	Fa	F	Md	63	D	UC	Go	G	VH
24	N/MD	C	Fa	G	Lw	64	D	Mo	Ba	B	Dg
25	N/MD	C	Go	B	Md	65	D	Mo	Ba	F	Dg
26	N/MD	C	Go	F	Lw	66	D	Mo	Ba	G	Sv
27	N/MD	C	Go	G	Sf	67	D	Mo	Fa	B	Dg
28	MD	UC	Ba	B	Dg	68	D	Mo	Fa	F	Dg
29	MD	UC	Ba	F	Dg	69	D	Mo	Fa	G	VH
30	MD	UC	Ba	G	VH	70	D	Mo	Go	B	Dg
31	MD	UC	Fa	B	Dg	71	D	Mo	Go	F	Sv
32	MD	UC	Fa	F	Sv	72	D	Mo	Go	G	Hg
33	MD	UC	Fa	G	Hg	73	D	C	Ba	B	Dg
34	MD	UC	Go	B	Sv	74	D	C	Ba	F	Dg
35	MD	UC	Go	F	VH	75	D	C	Ba	G	VH
36	MD	UC	Go	G	Md	76	D	C	Fa	B	Dg
37	MD	Mo	Ba	B	Dg	77	D	C	Fa	F	Sv
38	MD	Mo	Ba	F	Sv	78	D	C	Fa	G	Hg
39	MD	Mo	Ba	G	VH	79	D	C	Go	B	Dg
40	MD	Mo	Fa	B	Sv	80	D	C	Go	F	VH
						81	D	C	Go	G	Md

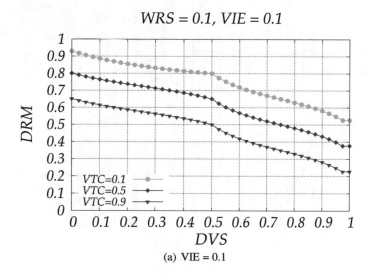

(a) VIE = 0.1

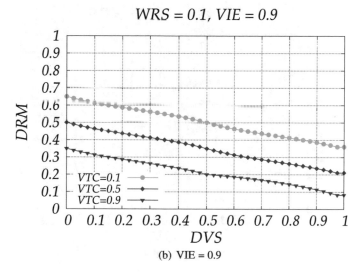

(b) VIE = 0.9

Fig. 4. Simulation results for WRS = 0.1.

When WRS is increased even more (Fig. 6), we see that the DRM values are increased drastically. These scenarios include driving at a high speed in bad weather/road conditions which is dangerous even when it is not combined with other factors. Therefore, all situations must be taken seriously and an appropriate action should be taken immediately.

WRS = 0.5, VIE = 0.1

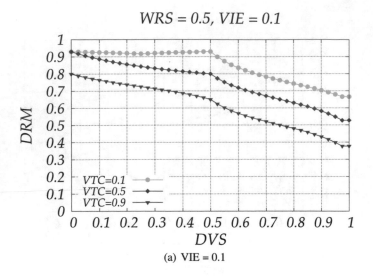

(a) VIE = 0.1

WRS = 0.5, VIE = 0.9

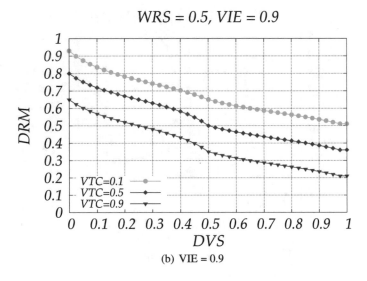

(b) VIE = 0.9

Fig. 5. Simulation results for WRS = 0.5.

When a number of consecutive decided DRM values are slightly above 0.2, the system can perform a certain action. For instance, if VIE is the cause of the risk, the system could take action to lift the driver's mood or improve the vehicle inside environment. On the other hand, if the high DRM values are caused by DVS or WRS, the system suggests or urges the driver to pull over while limiting the vehicle's operating speed to a speed that the risk level is decreased significantly.

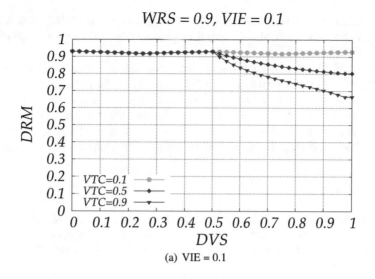

(a) VIE = 0.1

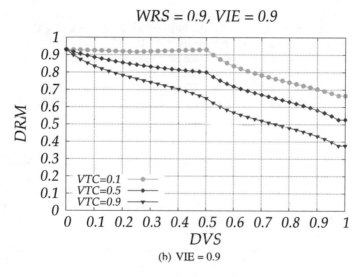

(b) VIE = 0.9

Fig. 6. Simulation results for WRS = 0.9.

5   Conclusions

In this paper, we presented an integrated fuzzy-based driving-support system which determines the driving risk in real time and provides an output that can be used to take different actions when the risk exceeds certain limits. By performing the appropriate actions, the risk can be reduced, thus increasing driving safety. For the implementation of the proposed system we considered four parameters: WRS, VIE, VTC and DVS. We showed through simulations the effect of the considered parameters on the determination of the driving risk level.

The only situation determined as safe is the scenario that includes a healthy driver who feels comfortable driving a vehicle in optimal technical condition in a good outside environment.

However, it may occur that the system provides an output which determines a low risk, when actually the chances for an accident to happen are high, or the opposite scenario, which is the case when the system's output implies a false alarm. Therefore, we intend to implement the system in a testbed, estimate the system accuracy and find which parameters can mostly lead to a false positive output and what should be improved to reduce the false negative outputs.

References

1. Bylykbashi, K., Qafzezi, E., Ampririt, P., Ikeda, M., Matsuo, K., Barolli, L.: Performance evaluation of an integrated fuzzy-based driving-support system for real-time risk management in VANETs. Sensors **20**(22), 6537 (2020). https://doi.org/10.3390/s20226537
2. Bylykbashi, K., Qafzezi, E., Ikeda, M., Matsuo, K., Barolli, L.: Fuzzy-based driver monitoring system (FDMS): implementation of two intelligent FDMSs and a testbed for safe driving in VANETs. Future Gener. Comput. Syst. **105**, 665–674 (2020). https://doi.org/10.1016/j.future.2019.12.030
3. Hartenstein, H., Laberteaux, L.: A tutorial survey on vehicular ad hoc networks. IEEE Commun. Mag. **46**(6), 164–171 (2008)
4. Kandel, A.: Fuzzy Expert Systems. CRC Press, Boca Raton (1991)
5. Klir, G.J., Folger, T.A.: Fuzzy Sets, Uncertainty, and Information. Prentice Hall Inc., Upper Saddle River (1987)
6. McNeill, F.M., Thro, E.: Fuzzy Logic: A Practical Approach. Academic Press, Cambridge (1994)
7. Munakata, T., Jani, Y.: Fuzzy systems: an overview. Commun. ACM **37**(3), 69–77 (1994). https://doi.org/10.1145/175247.175254
8. SAE On-Road Automated Driving (ORAD) committee: Taxonomy and definitions for terms related to driving automation systems for on-road motor vehicles. Technical report, Society of Automotive Engineers (SAE) (2018). https://doi.org/10.4271/J3016 201806
9. World Health Organization: Global status report on road safety 2018: summary. World Health Organization, Geneva, Switzerland (2018). (WHO/NMH/NVI/18.20). Licence: CC BY-NC-SA 3.0 IGO
10. Zadeh, L.A., Kacprzyk, J.: Fuzzy Logic for the Management of Uncertainty. John Wiley & Sons Inc, New York (1992)
11. Zimmermann, H.J.: Fuzzy Set Theory and Its Applications. Springer Science & Business Media. New York (1996). https://doi.org/10.1007/978-94-015-8702-0

Resource Management in SDN-VANETs Using Fuzzy Logic: Effect of Average Processing Capability per Neighbor Vehicle on Management of Cloud-Fog-Edge Resources

Ermioni Qafzezi[1]([⊠]), Kevin Bylykbashi[1], Phudit Ampririt[1], Makoto Ikeda[2], Leonard Barolli[2], and Makoto Takizawa[3]

[1] Graduate School of Engineering, Fukuoka Institute of Technology (FIT), 3-30-1 Wajiro-Higashi, Higashi-Ku, Fukuoka 811–0295, Japan
eqafzezi@gmail.com, bylykbashi.kevin@gmail.com, iceattpon12@gmail.com
[2] Department of Information and Communication Engineering, Fukuoka Institute of Technology (FIT), 3-30-1 Wajiro-Higashi, Higashi-Ku, Fukuoka 811-0295, Japan
makoto.ikd@acm.org, barolli@fit.ac.jp
[3] Department of Advanced Sciences, Faculty of Science and Engineering, Hosei University, 3-7-2 Kajino-machi, Koganei-shi, Tokyo 184-8584, Japan
makoto.takizawa@computer.org

Abstract. Traditional Vehicular Ad hoc Networks (VANETs) experience several challenges in deployment and management due to poor scalability, low flexibility, bad connectivity and lack of intelligence. The integration of Cloud, Fog and Edge Computing in VANETs together with the use of Software Defined Networking (SDN) are seen as a way to deal with these communication challenges. In this work, we propose a new fuzzy-based system for coordination and management of resources in a layered cloud-fog-edge SDN-VANETs architecture. The proposed system decides the appropriate resources to be used by a particular vehicle when the tasks it has to accomplish require computing resources that go beyond those of the vehicle. Based on the output of the system, it is decided whether the vehicle could use the resources of its neighbors, fog or cloud servers. The decision is made by considering the average processing capability of the neighboring vehicles, density of neighbors, latency constrains and the complexity of the task. We demonstrate by simulations the feasibility of the proposed system to improve the management of the network resources.

1 Introduction

The long distances separating homes and workplaces/facilities/schools as well as the traffic present in these distances make people spend a significant amount of time in vehicles. Thus, it is important to offer drivers and passengers ease of driving, convenience, efficiency and safety. This has led to the emerging of

L. Barolli et al. (Eds.): EIDWT 2021, LNDECT 65, pp. 155–167, 2021.
https://doi.org/10.1007/978-3-030-70639-5_15

Vehicular Ad hoc Networks (VANETs), where vehicles are able to communicate and share important information among them. VANETs are a relevant component of Intelligent Transportation Systems (ITS) which offer more safety and better transportation.

VANETs are capable to offer numerous services such as road safety, enhanced traffic management, as well as travel convenience and comfort. To achieve road safety, emergency messages must be transmitted in real-time, which stands also for the actions that should be taken accordingly in order to avoid potential accidents. Thus, it is important for the vehicles to always have available connections to infrastructure and to other vehicles on the road. On the other hand, traffic efficiency is achieved by managing traffic dynamically according to the situation and by avoiding congested roads, whereas comfort is attained by providing in-car infotainment services.

The advances in vehicle technology have made it possible for the vehicles to be equipped with various forms of smart cameras and sensors, wireless communication modules, storage and computational resources. While more and more of these smart cameras and sensors are incorporated in vehicles, massive amounts of data are generated from monitoring the on-road and in-board status. This exponential growth of generated vehicular data, together with the boost of the number of vehicles and the increasing data demands from in-vehicle users, has led to a tremendous amount of data in VANETs [9]. Moreover, applications like autonomous driving require even more storage capacity and complex computational capability. As a result, traditional VANETs face huge challenges in meeting such essential demands of the ever-increasing advancement of VANETs.

The integration of Cloud-Fog-Edge Computing in VANETs is the solution to handle complex computation, provide mobility support, low latency and high bandwidth. Each of them serves different functions, but also complements each-other in order to enhance the performance of VANETs. Even though the integration of Cloud, Fog and Edge Computing in VANETs solves significant challenges, this architecture lacks the needed mechanisms for resource and connectivity management because the network is controlled in a decentralized manner. The prospective solution to solve these problems is the augmentation of Software Defined Networking (SDN) in this architecture.

The SDN is a promising choice in managing complex networks with minimal cost and providing optimal resource utilization. SDN offers a global knowledge of the network with a programmable architecture which simplifies network management in such extremely complicated and dynamic environments like VANETs [8]. In addition, it will increase flexibility and programmability in the network by simplifying the development and deployment of new protocols and by bringing awareness into the system, so that it can adapt to changing conditions and requirements, i.e., emergency services [3]. This awareness allows SDN-VANET to make better decisions based on the combined information from multiple sources, not just individual perception from each node.

We have previously proposed an intelligent approach which can be used to manage the cloud, fog and edge resources in SDN-VANETs through a fuzzy-based system implemented in the SDN Controllers (SDNCs). This fuzzy system decided the appropriate computing resources to be used by a particular vehicle which needs additional resources, based on the vehicle relative speed with neighboring vehicles, number of neighboring vehicles, time-sensitivity and data complexity [7]. In another work [6], we implemented an intelligent system that considers the available resources of each neighboring vehicle and the predicted contact duration between them and the vehicle to calculate each neighbor's capability to process data. The average processing capability of the neighboring vehicles is a parameter that gives an accurate assessment of the edge resources, thus we include it in this work. Based on the final output value, our system decides which is the best layer among the cloud, fog and edge layers to process the data while satisfying the application requirements.

The remainder of the paper is as follows. In Sect. 2, we present an overview of Cloud-Fog-Edge SDN-VANETs. In Sect. 3, we describe the proposed fuzzy-based system. In Sect. 4, we discuss the simulation results. Finally, conclusions and future work are given in Sect. 5.

2   Cloud-Fog-Edge SDN-VANETs

While cloud, fog and edge computing in VANETs offer scalable access to storage, networking and computing resources, SDN provides higher flexibility, programmability, scalability and global knowledge. In Fig. 1, we give a detailed structure of this VANET architecture. It includes the topology structure, its logical structure and the content distribution on the network. As it is shown, it consists of Cloud Computing data centers, fog servers with SDNCs, roadside units (RSUs), RSU Controllers (RSUCs), Base Stations and vehicles. We also illustrate the infrastructure-to-infrastructure (I2I), vehicle-to-infrastructure (V2I), and vehicle-to-vehicle (V2V) communication links. The fog devices (such as fog servers and RSUs) are located between vehicles and the data centers of the main cloud environments. The safety applications data generated through on board and on road sensors are processed first in the vehicles as they require real-time processing. If more storing and computing resources are needed, the vehicle can request to use those of the other adjacent vehicles, assuming a connection can be established and maintained between them for a while. With the vehicles having created multiple virtual machines on other vehicles, the virtual machine migration must be achievable in order to provide continuity as one/some vehicle may move out of the range. However, to set-up virtual machines on the nearby vehicles, multiple requirements must be met and when these demands are not satisfied, the fog servers are used.

Cloud servers are used as a repository for software updates, control policies and for the data that need long-term analytics and are not delay-sensitive. On the other side, SDN modules which offer flexibility and programmability, are used to simplify the network management by offering mechanisms that improve the network traffic control and coordination of resources. The implementation of this architecture promises to enable and improve VANET applications such as road and vehicle safety services, traffic optimization, video surveillance, commercial and entertainment applications.

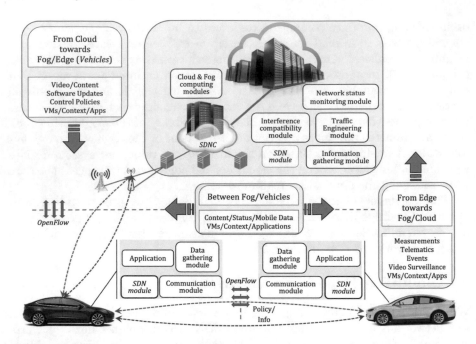

Fig. 1. Logical architecture of cloud-fog-edge SDN-VANET with content distribution.

3   Proposed Fuzzy-Based System

In this section, we present the layered cloud-fog-edge SDN-VANETs architecture which is coordinated by the global intelligence of SDNC. In this architecture, SDNC manages not only the computing and storage resources of fog, but also those of edge and cloud. An illustration of this layered architecture is given in Fig. 2.

Our proposed system, named Fuzzy-based System for Resource Management (FSRM), operates in all three modes of the SDN-VANET i.e., central control mode, distributed control mode and hybrid control mode [3]. A vehicle that needs storage and computing resources for a particular application can use those of neighboring vehicles, fog servers or cloud data centers based on the application requirements and available connections. For instance, for a temporary application that needs real-time processing, the vehicle can use the resources of adjacent vehicles if the requirements to realize such operations are fulfilled. Otherwise, it will use the resources of fog servers, which offer low latency as well. Whereas real-time applications require the usage of edge and fog layer resources, for delay tolerant applications vehicles can use the cloud resources as these applications do not require low latency.

FSRM is implemented in the SDNC and in the vehicles which are equipped with SDN modules. If a vehicle does not have an SDN module, it sends the information to SDNC which sends back its decision. The FSRM uses the beacon messages received from the adjacent vehicles to extract information such as their

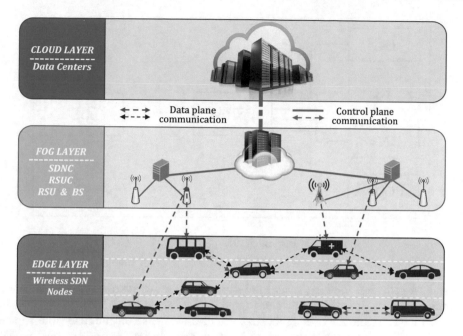

Fig. 2. Layered architecture of cloud-fog-edge SDN-VANETs.

current position, velocity, direction, and based on the application requirements, it decides the appropriate layer to run and process the application data.

The structure of the proposed FSRM is shown in Fig. 3. For the implementation of our system, we consider four input parameters: Data Complexity (DC), Time Sensitivity (TS), Number of Neighboring Vehicles (NNV) and Average Processing Capability per Neighbor Vehicle (APCpNV) to determine the Layer Selection Decision (LSD) value.

DC: Vehicles not only consume but also generate a huge amount of data, which include data from other sources (internet, cloud servers, fog servers, smart traffic light nodes, traffic information center infrastructures, other vehicles etc.) as well as data generated by in-vehicle smart cameras and sensors. On the other hand, given their enormous computing resources, vehicles can be used in the Big Data era also for computing big data non-related to VANETs. Regarding the complexity, there are many factors which determine the data' complexity and the volume is only one of them. Even only one application uses data which differs in the matter of type and structure, not to mention that they may come from many disparate sources. Besides, there are different kinds of applications which include not only VANET applications but also other applications which will be using vehicles' computational resources to process their data.

TS: Different applications have different requirements in terms of latency. For instance, safety applications require a strict latency to be guaranteed, ideally <1 ms, whereas comfort and entertainment applications can tolerate the latency up to some seconds and are considered as delay-tolerant. System updates and

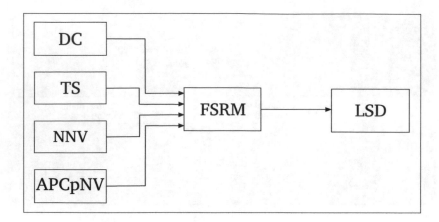

Fig. 3. FSRM structure.

the data collected for long-term analytics can tolerate even more and for such applications the latency is not considered a requirement at all.

NNV: The number of neighboring vehicles changes continuously due to the vehicles which move out of the *vehicle's* communication range and the ones which appear. Vehicles moving at the opposite direction make the changes even more frequent. Given that the bigger the angle between the directions, the bigger the distance created between the vehicles, we include in the calculations only the neighboring vehicles whose directions create an angle less than 90° with the direction of the *vehicle*. Vehicles moving at directions which create bigger angles move out of the communication range very quickly; therefore, it is impossible for the vehicle to use their resources.

APCpNV: This parameter is an important indicator of edge-layer's capability to process data since it considers the amount of resources that the neighboring vehicles are willing to share. We calculate the processing capability for each neighbor individually (we give the way to determine the processing capability of a vehicle in [6]) and the sum of processing capability of all neighboring vehicles divided by the number of neighboring vehicles gives the value of this parameter.

LSD: The output parameter values consist of values between 0 and 1, denoting three decisions on the resources to be used by the vehicle. Values in the intervals [0, 0.3], (0.3, 0.7) and [0.7, 1] determine the edge, fog and cloud layer, respectively.

We consider fuzzy logic to implement the proposed system because our system parameters are not correlated with each other. Having three or more parameters which are not correlated with each other results in a non-deterministic polynomial-time hard (NP-hard) problem and fuzzy logic can deal with these problems. Moreover, we want our system to make decisions in real time and fuzzy logic can give very good results in decision making and control problems [1, 2, 4, 5, 10, 11].

The input and output parameters are fuzzified using the membership functions showed in Fig. 4. We use triangular and trapezoidal membership functions because they are suitable for real-time operation. The term sets for each linguistic parameter are shown in Table 1. We decided the number of term sets by carrying out many simulations. In Table 2, we show the Fuzzy Rule Base (FRB) of FSRM, which consists of 81 rules. The control rules have the form: IF "conditions" THEN "control action". For instance, for Rule 1: "IF DC is Lo, TS is Lw, NNV is Sp and APCpNV is L, THEN LSD is DL6" or for Rule 48: "IF DC is Mo, TS is Hg, NNV is Sp and APCpNV is H, THEN LSD is DL3".

Table 1. Parameters and their term sets for FSRM.

Parameters	Term sets
Data Complexity (DC)	Low (Lo), Moderate (Mo), High (Hi)
Time Sensitivity (TS)	Low (Lw), Middle (Md), High (Hg)
Number of Neighboring Vehicles (NNV)	Sparse (Sp), Medium Density (Me), Dense (De)
Average Processing Capability per Neighbor Vehicle (APCpNV)	Low (L), Moderate (M), High (H)
Layer Selection Decision (LSD)	Decision Level 1 (DL1), DL2, DL3, DL4, DL5, DL6, DL7

4   Simulation Results

In this section, we present the simulation results for our proposed system. The simulations are conducted with FuzzyC, which is a simulation tool based on Fuzzy Logic written in C language. The simulation results are presented in Fig. 5, Fig. 6 and Fig. 7. We consider the DC and TS as constant parameters. We change the NNV value from 0.1 to 0.9 units which represent a sparse and dense environment of vehicles, respectively. The NNV = 0.5 simulates a network of vehicles which is neither sparse nor dense.

In Fig. 5, we consider a scenario where the data to be processed are not complex. When a low latency is not a strong requirement, we can see that most the data are processed in fog and edge layer. Even though the data are not time sensitive, there is no need to send them in the cloud as it can be processed in the fog or edge given its low complexity. In the case of time-sensitive data (TS = 0.9) we can see that all data are processed in edge and fog depending on the edge-layer's processing capability. When the number of neighboring vehicle is high and the APCpNV is not zero, these data are processed in the edge layer.

As shown in Fig. 6, when DC = 0.5, most of the delay tolerant data will be processed in cloud or fog layer. On the other hand, time-critical data will

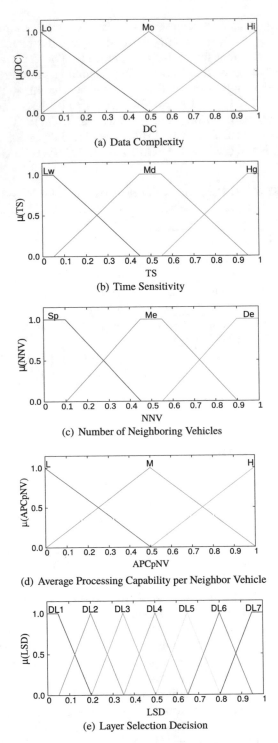

Fig. 4. Membership functions.

Table 2. The fuzzy rule base of FSRM.

Rule	DC	TS	NNV	APCpNV	LSD	Rule	DC	TS	NNV	APCpNV	LSD
1	Lo	Lw	Sp	L	DL6	41	Mo	Md	Me	M	DL4
2	Lo	Lw	Sp	M	DL4	42	Mo	Md	Me	H	DL3
3	Lo	Lw	Sp	H	DL3	43	Mo	Md	De	L	DL5
4	Lo	Lw	Me	L	DL6	44	Mo	Md	De	M	DL3
5	Lo	Lw	Me	M	DL3	45	Mo	Md	De	H	DL2
6	Lo	Lw	Me	H	DL2	46	Mo	Hg	Sp	L	DL5
7	Lo	Lw	De	L	DL6	47	Mo	Hg	Sp	M	DL4
8	Lo	Lw	De	M	DL2	48	Mo	Hg	Sp	H	DL3
9	Lo	Lw	De	H	DL1	49	Mo	Hg	Me	L	DL4
10	Lo	Md	Sp	L	DL5	50	Mo	Hg	Me	M	DL3
11	Lo	Md	Sp	M	DL3	51	Mo	Hg	Me	H	DL2
12	Lo	Md	Sp	H	DL2	52	Mo	Hg	De	L	DL3
13	Lo	Md	Me	L	DL4	53	Mo	Hg	De	M	DL2
14	Lo	Md	Me	M	DL2	54	Mo	Hg	De	H	DL1
15	Lo	Md	Me	H	DL1	55	Hi	Lw	Sp	L	DL7
16	Lo	Md	De	L	DL3	56	Hi	Lw	Sp	M	DL7
17	Lo	Md	De	M	DL1	57	Hi	Lw	Sp	H	DL6
18	Lo	Md	De	H	DL1	58	Hi	Lw	Me	L	DL7
19	Lo	Hg	Sp	L	DL4	59	Hi	Lw	Me	M	DL6
20	Lo	Hg	Sp	M	DL3	60	Hi	Lw	Me	H	DL5
21	Lo	Hg	Sp	H	DL2	61	Hi	Lw	De	L	DL7
22	Lo	Hg	Me	L	DL3	62	Hi	Lw	De	M	DL5
23	Lo	Hg	Me	M	DL2	63	Hi	Lw	De	H	DL4
24	Lo	Hg	Me	H	DL1	64	Hi	Md	Sp	L	DL7
25	Lo	Hg	De	L	DL2	65	Hi	Md	Sp	M	DL6
26	Lo	Hg	De	M	DL1	66	Hi	Md	Sp	H	DL5
27	Lo	Hg	De	H	DL1	67	Hi	Md	Me	L	DL7
28	Mo	Lw	Sp	L	DL7	68	Hi	Md	Me	M	DL5
29	Mo	Lw	Sp	M	DL6	69	Hi	Md	Me	H	DL4
30	Mo	Lw	Sp	H	DL4	70	Hi	Md	De	L	DL7
31	Mo	Lw	Me	L	DL7	71	Hi	Md	De	M	DL4
32	Mo	Lw	Me	M	DL5	72	Hi	Md	De	H	DL3
33	Mo	Lw	Me	H	DL3	73	Hi	Hg	Sp	L	DL5
34	Mo	Lw	De	L	DL6	74	Hi	Hg	Sp	M	DL5
35	Mo	Lw	De	M	DL4	75	Hi	Hg	Sp	H	DL4
36	Mo	Lw	De	H	DL2	76	Hi	Hg	Me	L	DL5
37	Mo	Md	Sp	L	DL7	77	Hi	Hg	Me	M	DL4
38	Mo	Md	Sp	M	DL5	78	Hi	Hg	Me	H	DL3
39	Mo	Md	Sp	H	DL4	79	Hi	Hg	De	L	DL4
40	Mo	Md	Me	L	DL6	80	Hi	Hg	De	M	DL3
						81	Hi	Hg	De	H	DL2

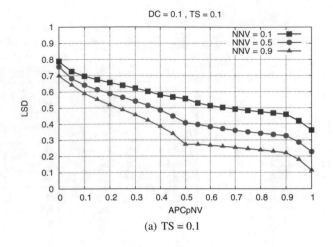

(a) TS = 0.1

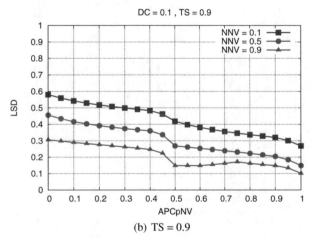

(b) TS = 0.9

Fig. 5. Simulation results for DC = 0.1.

be processed either in fog or edge layer, depending on the average processing capability for neighbor vehicle (see Fig. 6(b)). Both edge and fog layer offer low latency, thus the application requirements related to the time sensitivity are satisfied.

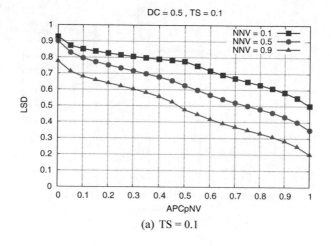

(a) TS = 0.1

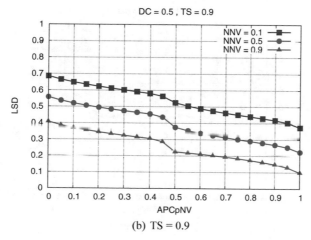

(b) TS = 0.9

Fig. 6. Simulation results for DC = 0.5.

With the increase of data complexity (DC = 0.9), as we can see from Fig. 7, none of the data will be processed in edge. High data complexity requires great computing capability, therefore the majority of these data will be computed in cloud layer. On the other side, for time-critical applications, the data will be processed in edge only when APCpNV is high and there is a dense environment of neighboring vehicles.

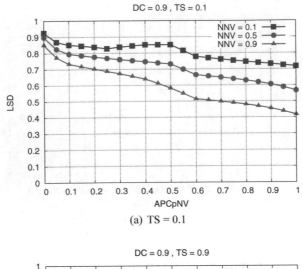

(a) TS = 0.1

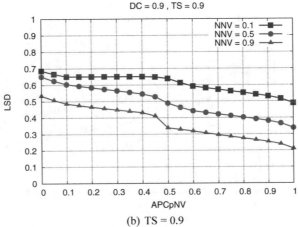

(b) TS = 0.9

Fig. 7. Simulation results for DC = 0.9.

5   Conclusions

In this paper, we implemented a fuzzy-based system for resource management in a cloud-fog-edge layered architecture for SDN-VANET considering Average Processing Capability per Neighbor Vehicle as a new parameter. In addition to APCpNV, FSRM takes into consideration the following parameters: DC, TS and NNV. We evaluated the performance of FSRM by computer simulations. From the simulations results, we conclude as follows.

- With the increase of data complexity less data is processed in the edge. No data will be processed in the edge layer if data complexity is high and time sensitivity is low.

- In a dense environment, moderate complex data can be processed in the edge only if the APCpNV is high.
- Time-critical data are processed either in Edge or Fog Layer.

In the future, we would like to evaluate the performance of the proposed system and compare it with other systems.

References

1. Kandel, A.: Fuzzy Expert Systems. CRC Press Inc., Boca Raton (1992)
2. Klir, G.J., Folger, T.A.: Fuzzy Sets, Uncertainty, and Information. Prentice Hall, Upper Saddle River (1988)
3. Ku, I., Lu, Y., Gerla, M., Gomes, R.L., Ongaro, F., Cerqueira, E.: Towards software-defined vanet: architecture and services. In: 13th Annual Mediterranean Ad Hoc Networking Workshop (MED-HOC-NET), pp. 103–110 (2014)
4. McNeill, F.M., Thro, E.: Fuzzy Logic: A Practical Approach. Academic Press Professional Inc, San Diego (1994)
5. Munakata, T., Jani, Y.: Fuzzy systems: an overview. Commun. ACM **37**(3), 69–77 (1994)
6. Qafzezi, E., Bylykbashi, K., Ampririt, P., Ikeda, M., Barolli, L., Takizawa, M.: Assessment of available edge computing resources in SDN-VANETs by a fuzzy-based system considering trustworthiness as a new parameter. In: International Conference on P2P, Parallel, Grid, Cloud and Internet Computing, pp. 102–112. Springer (2020)
7. Qafzezi, E., Bylykbashi, K., Ikeda, M., Matsuo, K., Barolli, L., Takizawa, M.: Resource management in SDN-VANETs using fuzzy logic: effect of data complexity on coordination of cloud-fog-edge resources. In: International Conference on Complex, Intelligent and Software Intensive Systems, pp. 498–509. Springer (2020)
8. Truong, N.B., Lee, G.M., Ghamri-Doudane, Y.: Software defined networking-based vehicular adhoc network with fog computing. In: 2015 IFIP/IEEE International Symposium on Integrated Network Management (IM), pp. 1202–1207 (2015)
9. Xu, W., Zhou, H., Cheng, N., Lyu, F., Shi, W., Chen, J., Shen, X.: Internet of vehicles in big data era. IEEE/CAA J. Automatica Sinica **5**(1), 19–35 (2018)
10. Zadeh, L.A., Kacprzyk, J.: Fuzzy logic for the management of uncertainty. John Wiley & Sons Inc, New York (1992)
11. Zimmermann, H.J.: Fuzzy control. In: Fuzzy Set Theory and Its Applications, pp. 203–240. Springer (1996)

Implementation of a Device Adopting the OI (Operation Interruption) Protocol to Prevent Illegal Information Flow in the IoT

Shigenari Nakamura[1]([✉]), Tomoya Enokido[2], Lidia Ogiela[3], and Makoto Takizawa[4]

[1] Tokyo Metropolitan Industrial Technology Research Institute, Tokyo, Japan
nakamura.shigenari@gmail.com
[2] Rissho University, Tokyo, Japan
eno@ris.ac.jp
[3] Pedagogical University of Krakow, Kraków, Poland
lidia.ogiela@gmail.com
[4] Hosei University, Tokyo, Japan
makoto.takizawa@computer.org

Abstract. In the IoT (Internet of Things), the CBAC (Capability-Based Access Control) model is proposed to make sensor and actuator devices secure. Here, an owner of a device issues a capability token, i.e. a set of access rights to a subject. The subject is then allowed to manipulate resource objects in the device according to the access rights in the capability token. There is a problem a subject sb can get data from a resource object r^1 brought to another resource r^2 by getting the data from the resource r^2 even if the subject sb is not allowed to get data from the resource r^1. Here, the data in the resource r^1 illegally flow to the subject sb. In our previous studies, the OI (Operation Interruption) protocol is proposed where illegal operations are interrupted and is evaluated in the simulation. In this paper, we implement a device supporting the OI protocol and evaluate the authorization process of the OI protocol in terms of the execution time. In the evaluation, we make clear the features of the execution time of the authorization process for *get* and *put* operations in the OI protocol.

Keywords: IoT (Internet of Things) · Device security · CBAC (Capability-Based Access Control) model · Illegal information flow · Information flow control · OI (Operation Interruption) protocol · Implementation of the OI protocol

1 Introduction

Access control models [3] are used in order to realize secure information systems. Here, only an authorized subject like user is allowed to manipulate an object in an authorized operation as well as cryptography methods [13]. However, even if a subject is not allowed to get data in an object o_1, the subject can get the data by accessing another object o_2 [3]. Here, information illegally flows from the object o_1 via the object o_2 to the subject. Hence, we have to prevent illegal information flow among subjects and objects in the access control models. In the LBAC (Lattice-Based Access Control) model [14], each entity, i.e. subject or object are assigned a security class. The illegal information flow relation is defined to be an ordered relation among classes and every operation implying the illegal information flow is prohibited. In our previous studies [9,10,12], various types of protocols to prevent illegal information flow are proposed. In the paper [9], protocols to prevent illegal information flow occurring in database systems are discussed based on the RBAC (Role-Based Access Control) model [15]. In the papers [10,12], protocols to prevent illegal information flow occurring in P2PPSO (Peer-to-Peer Publish/Subscribe with Object concept) systems are proposed based on the TBAC (Topic-Based Access Control) model [12].

The IoT (Internet of Things) is composed of various types and millions of nodes including not only computers but also devices like sensors and actuators [6]. It is not easy to adopt the traditional access control models such as the RBAC [15] and ABAC (Attribute-Based Access Control) [19] models for the IoT due to the scalability of the IoT. Hence, the CBAC (Capability-Based Access Control) model named "CapBAC model" is proposed [5]. Here, an owner of a device issues a capability token to a subject like users and applications. A capability token is a set of access rights. An access right is a pair $\langle r, op \rangle$ for a resource r and an operation op. The subject is allowed to manipulate a resource r in an operation op only if a capability token including an access right $\langle r, op \rangle$ is issued to the subject.

Suppose a subject sb_1 is issued a capability token including a pair of access rights $\langle r^1, get \rangle$ and $\langle r^2, put \rangle$ of a pair of resource objects r^1 and r^2 in devices d_1 and d_2, respectively, by owners of the devices. Suppose d_1 is a sensor device and d_2 is a hybrid device equipped with both sensors and actuators. A sensor just gives sensor data to a subject. On the other hand, an actuator stores data in its resource and supports an action given by the data. A subject sb_2 is issued a capability token including an access right $\langle r^2, get \rangle$ by an owner of the device d_2. First, the subject sb_1 gets sensor data from the resource object r^1 in the sensor d_1 and then gives the data to the resource r^2 in the device d_2. Next, the subject sb_2 gets the data from the resource r^2 in the device d_2. Here, the subject sb_2 can obtain the data of the resource r^1 via the resource r^2 although the subject sb_2 is not issued a capability token including the access right $\langle r^1, get \rangle$. Here, the data of the resource r^1 flow to the subject sb_2 via the resource r^2. This is illegal information flow from the resource r^1 to the subject sb_2. In our previous studies [8,11], protocols to prevent illegal information flow in the IoT based on the CBAC model are proposed. In the OI (Operation Interruption) protocol [11], it is checked whether or not the illegal information flow occurs each time a subject issues an operation to a resource using the information flow relations defined based on the CBAC model. If the illegal information flow occurs, the operation is interrupted, i.e. not performed, at the device. Hence, every illegal information flow is prevented from occurring. The OI protocol is proposed and evaluated in terms of the number of operations interrupted in the simulation [11].

In this paper, we implement a device supporting the OI protocol to prevent illegal information flow and evaluate the authorization process of the OI protocol in terms of the execution time. An IoT device is realized in a Raspberry Pi 3 Model B+ equipped with Raspbian [2]. A device and subject are implemented as a CoAP server and CoAP client, respectively, in CoAPthon3 [17].

In the evaluation, we make clear the features of the execution time of the authorization process for *get* and *put* operations in the OI protocol. For a *get* operation, the more number of capability tokens are issued to subjects, the longer execution time of the authorization process. On the other hand, the execution time of the authorization process of a *put* operation is constant even if the number of capability tokens issued to subjects increases.

In Sect. 2, we present the system model. In Sect. 3, we discuss types of information flow relations based on the CBAC model. In Sect. 4, we overview the OI protocol. In Sect. 5, we implement the OI protocol on a Raspberry Pi 3 Model B+. In Sect. 6, we evaluate the authorization process of the OI protocol in terms of the execution time.

2   System Model

In order to make information systems secure, types of access control models [4,15,19] are widely used. Here, a system is composed of two types of entities, subjects and objects. A subject sb issues an operation op to an object o. Then, the operation op is performed on the object o. Here, only an authorized subject sb is allowed to manipulate an object o in an authorized operation op. Most of the access control models are based on ACLs (Access Control Lists) such as the RBAC (Role-Based Access Control) [15] and ABAC (Attribute-Based Access Control) [19] models. An ACL is a list of access rules specified by an authorizer. Each access rule is a tuple $\langle sb, o, op \rangle$ which is composed of a subject sb, an object o, and an operation op. This means, a subject sb is granted an access right $\langle o, op \rangle$ of an object o and an operation op. In the ACL system, if a subject sb tries to access the data of an object o in an operation op, a service provider has to check whether or not the subject sb is authorized to manipulate the object o in the operation op by using the ACL, i.e. $\langle sb, o, op \rangle$ in the ACL. In scalable systems like the IoT, the ACL gets also scalable and it is difficult to maintain and check the ACL. Hence, the RBAC and ABAC models are not suitable for the scalable distributed systems where there is no centralized coordinator and each node is an autonomous process which makes a decision by itself.

In the paper [5], the "CapBAC model" is proposed as a CBAC model to make devices securely accessed in the IoT. In the distributed CapBAC model [7], there is no intermediate entity between each pair of a subject and a device to implement the access control. Here, owners exist for each device. Let r be a computation resource object which contains data. In this paper, a term "resource" stands for a component object. A resource in a device d_k is denoted by r^k. An owner of a device first issues a capability token cap^i to a subject sb_i. CAP^i shows a set of capability tokens issued to the subject sb_i. A capability token cap^i issued to the subject sb_i is a set of access rights on the resources. An access right is a pair $\langle r, op \rangle$ of a resource object r and an operation op. Suppose a subject sb_i tries to manipulate a resource r^k in a device d_k in an operation op. Here, the subject sb_i sends the access request including the capability token cap^i to the device d_k. If the capability token cap^i is valid, the operation op is performed on the resource r^k and the subject sb_i receives the reply of the request. Otherwise, the access request is rejected and the subject sb_i receives the negative reply, i.e. the request is denied. Hence, it is easier to adopt the CBAC model to the IoT than the ACL-based models. However, it is more difficult to change access rights granted to each subject because access rights are distributed to subjects.

In the IoT, a set D of devices $d_1, \ldots, d_{dn}$ $(dn \geq 1)$ are interconnected in networks. Each device d_k holds the number nr_k (≥ 1) of resource objects. $d_k.R$ denotes a set of resources $r_1^k, \ldots, r_{nr_k}^k$ of the device d_k, i.e. $d_k.R = \{r_1^k, \ldots, r_{nr_k}^k\}$. Let wR be a set of whole resources in the system, i.e. $wR = d_1.R \cup, \ldots, \cup d_{dn}.R = \{r_1^1, \ldots, r_{nr_1}^1, \ldots, r_1^{dn}, \ldots, r_{nr_{dn}}^{dn}\}$. In this paper, we consider three types of devices, sensor, actuator, and hybrid devices. A sensor device just collects data by sensing events which occur in physical environment. An actuator device acts according to the action request from a subject. Let SB be a set of subjects $sb_1, \ldots, sb_{sbn}$ $(sbn \geq 1)$ in a system. A subject sb_i gets and puts data from a sensor s and to an actuator a, respectively. In addition, a subject sb_i issues both get and put operations to a hybrid device h like robots and cars.

3   Information Flow Relations

Through manipulating resources of devices, data are exchanged among resources and subjects. If data of a resource r flow into an entity e, i.e. subject or resource, the resource r is referred to as a *source* resource of the entity e. Let $sb.sR$ and $r.sR$ be sets of *source* resources whose data flow into a subject sb and a resource r, respectively, which are initially ϕ. For example, if a subject sb gets data from a resource r^1 in a device d_1 and puts the data to another resource r^2 in a device d_2, data of the resource r^1 flow into the subject sb and the resource r^2. Here, the sets $sb.sR$ and $r^2.sR$ include the resource r^1, i.e. $sb.sR = r^2.sR = \{r^1\}$. In another example, a sensor device d_1 gets data by sensing events and stores the data in a resource r^1. Here, the set $r^1.sR$ includes the resource r^1, i.e. $r^1.sR = \{r^1\}$.

In this section, we define types of information flow relations on resources and subjects based on the CBAC model. Let $IN(sb_i)$ be a set of resources whose data a subject sb_i is allowed to get, i.e. $IN(sb_i) = \{r \mid \langle r, get \rangle \in cap^i \wedge cap^i \in CAP^i\}(\subseteq wR)$. A subject sb_i can get data from a resource r only if the resource r is included in the set $IN(sb_i)$.

Definition 1. A resource r *flows* to a subject sb $(r \rightarrow sb)$ iff (if and only if) $r.sR \neq \phi$ and $r \in IN(sb)$.

If $r \rightarrow sb$ holds, data brought to the resource r may be brought to the subject sb. Otherwise, no data flow from the resource r into the subject sb.

Definition 2. A resource r *legally flows* to a subject sb ($r \Rightarrow sb$) iff $r \rightarrow sb$ and $r.sR \subseteq IN(sb)$.

The condition "$r.sR \subseteq IN(sb)$" means that even if the data of another resource exist in the resource r, all the data in the resource r are allowed to be brought to the subject sb.

Definition 3. A resource r *illegally flows* to a subject sb ($r \mapsto sb$) iff $r \rightarrow sb$ and $r.sR \nsubseteq IN(sb)$.

The condition "$r.sR \nsubseteq IN(sb)$" means that the data of some resource in $r.sR$ are not allowed to be brought into the subject sb.

4   An OI (Operation Interruption) Protocol

In the CBAC model of the IoT, a capability token cap^i which is a set of access rights is issued to a subject sb_i. If the capability token cap^i includes an access right $\langle r, op \rangle$, the subject sb_i is allowed to manipulate a resource r in an operation op. If data of a resource r^1 in a device d_1 brought into another resource r^2 in a device d_2 are brought into a subject sb_i which is not allowed to get the data of the resource r^1, the data of the resource r^1 illegally flow into the subject sb_i. In Sect. 3, the illegal information flow relation ($\mapsto$) is defined based on the CBAC model. In this section, we discuss the OI (Operation Interruption) protocol [11] to prevent illegal information flow based on the information flow relations among subjects and resources.

$IN(sb_i)$ is a set of resources from which a subject sb_i is allowed to get data, i.e. $IN(sb_i) = \{r \mid \langle r, get \rangle \in cap^i \land cap^i \in CAP^i\}(\subseteq wR)$. $sb_i.sR$ and $r.sR$ are sets of *source* resources whose data are brought to the subject sb_i and the resource r, respectively. For each subject sb_i and resource r, the sets $sb_i.sR$ and $r.sR$ are manipulated, which are initially ϕ. If a subject sb_i issues a *get* operation to a resource r, it is checked whether or not the subject sb_i is allowed to get all the data in resources in the set $r.sR$. If there exists at least one resource in the set $r.sR$ whose data are not allowed to be brought into the subject sb_i, i.e. $r.sR \nsubseteq IN(sb_i)$, the *get* operation is interrupted to prevent illegal information flow. The OI protocol is shown as follows:

[OI (Operation Interruption) protocol]

1. A device d_k gets data by sensing events occurring around the device d_k. Here, d_k is a sensor or hybrid device.
   a. The device d_k stores the sensor data in the resource r^k and $r^k.sR = r^k.sR \cup \{r^k\}$;
2. A subject sb issues a *get* operation on a resource r^k to a device d_k.
   a. If $r^k \Rightarrow sb$, the subject sb gets data from the resource r^k and $sb.sR = sb.sR \cup r^k.sR$;

b. Otherwise, the *get* operation is interrupted at the device d_k;

3. A subject sb issues a *put* operation on a resource r^k to a device d_k.

a. The subject sb puts the data to the resource r^k and $r^k.sR = sb.sR$;

Example 1. Figure 1 shows an example of the OI protocol where there are a pair of subjects sb_1 and sb_2, a sensor device d_1, and a hybrid device d_2. Here, a capability token cap^1 of a pair of access rights $\langle r^1, get \rangle$ and $\langle r^2, put \rangle$ is granted to the subject sb_1. On the other hand, a capability token cap^2 of an access right $\langle r^2, get \rangle$ is granted to the subject sb_2.

First, a sensor d_1 collects data by sensing events occurring around itself and stores the data to a resource $r^1(\in d_1.R)$. Here, $r^1.sR = \{r^1\}$. Next, the subject sb_1 issues a *get* operation to the sensor device d_1 to get sensor data. Here, $r^1 \Rightarrow sb_1$ holds because $r^1 \rightarrow sb_1$ and $r^1.sR(= \{r^1\}) \subseteq IN(sb_i)(= \{r^1\})$. Hence, the subject sb_1 can get the data from the resource r^1 in the sensor d_1 and $sb_1.sR = sb_1.sR(= \phi) \cup r^1.sR(= \{r^1\}) = \{r^1\}$. Then, the subject sb_1 issues a *put* operation to the hybrid device d_2 to store the data in a resource r^2. Here, $r^2.sR(= \phi) = sb_1.sR = \{r^1\}$. Finally, the subject sb_2 issues a *get* operation to the hybrid device d_2 to get the data. Here, $r^2 \mapsto sb_2$ holds because $r^2 \rightarrow sb_2$ and $r^2.sR(= \{r^1\}) \not\subseteq IN(sb_2)(= \{r^2\})$. Hence, the *get* operation of the subject sb_2 is interrupted at the hybrid device d_2 to prevent illegal information flow as shown in Fig. 1.

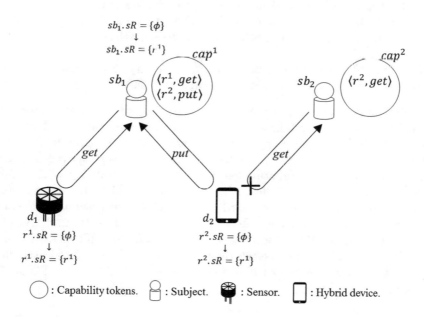

Fig. 1. OI protocol.

5   Implementation of the OI Protocol

We implement the OI protocol on a hybrid device. A hybrid device d_1 is implemented in a Raspberry Pi 3 Model B+ [1] equipped with Raspbian [2]. A subject sb_1 is implemented in Python on the hybrid device d_1. The hybrid device d_1 and the subject sb_1 are implemented as a CoAP server and CoAP client in CoAPthon3 [17], respectively.

5.1   Capability Token

In this paper, we consider a capability token which is composed of the following fields referring to a capability token example shown in the paper [7]:

- Identifier (variable size): identifier of the capability token.
- Issued-time (10 bytes): time at which the capability token is issued.
- Issuer (64 bytes): issuer of the capability token, i.e. owner of a device.
- Subject (64 bytes): subject granted the access rights in the capability.
- Device (variable size): device whose access right is in the capability token.
- Access rights: set of access rights granted to the subject. Each access right is composed of the following fields.
    - Operation (variable size): operation allowed, which is any CoAP (Constrained Application Protocol) [16] method, e.g. GET and PUT.
    - Resource (variable size): resource object of the device which is allowed to manipulate in the operation which is designated in the previous field.
- Not_before (10 bytes): time before which the capability token must not be valid.
- Not_after (10 bytes): time after which the capability token must not be valid.
- Signature (64 bytes): signature of the capability token.

Capability tokens are included in the payload field of a CoAP request. The "Issuer" field includes a public key of the issuer who issues the capability token. "Signature" field contains a signature generated with the private key of the issuer. "Subject" field includes a public key of the subject sb_1 who is granted the capability token. The signature generated with the private key of the subject sb_1 is included in the payload field of a CoAP request. The keys and signatures of issuers and subjects are generated in the ECDSA algorithm implemented as a Python package [18]. The values of "Issuer", "Signature", and "Subject" fields are encoded into Base64 form. The capability token is valid at time τ where Not_before $< \tau <$ Not_after. "Issued-time", "Not_before", and "Not_After" fields are Unix times, i.e. the number of seconds elapsed since 00:00:00 UTC on Jan. 1, 1970.

5.2   Authorization Process

Suppose a subject sb_1 issues an operation op to the hybrid device d_1 to manipulate a resource object r^1. Here, the subject sb_1 sends a CoAP request with capability tokens to the hybrid device d_1. We consider two operations, *get* and *put* on devices. If the operation *op* is *put*, the subject sb_1 includes the resource set $sb_1.sR$ in the payload field of the CoAP request. On receipt of a CoAP request with the capability tokens from a subject sb_1, the following steps of the authorization process are performed on the hybrid device d_1:

Step 1. The hybrid device d_1 checks whether or not the subject sb_1 is allowed to manipulate the resource r^1 in the operation *op*. If the subject sb_1 is issued a capability token cap^1 which satisfies the following conditions, the subject sb_1 is allowed to issue the operation *op*. Otherwise, the operation *op* is interrupted.
- The operation *op* matches the value of "Operation" field of the capability token cap^1.
- The destination and the Resource-URI option of the CoAP request match the values of "Device" and "Resource" fields of the capability token cap^1, respectively.
- The current time is larger and smaller than the values of "Not_before" and "Not_after" fields.

Step 2. If the operation *op* is *get*, the hybrid device d_1 checks whether or not the subject sb_1 is allowed to get data from every resource in $r^1.sR$. If at least one resource from which the subject sb_1 is not allowed to get data exist in $r^1.sR$, the operation *op* is interrupted.

Step 3. The public key of the subject sb_1 in the field "Subject" is obtained in each capability token. The signature generated with a private key of the subject sb_1 is included in the payload field of the CoAP request. The hybrid device d_1 checks whether or not the signature of the subject sb_1 is valid. If the signature is not valid, the operation *op* is interrupted.

Step 4. Each capability token holds the public key of its issuer and the signature generated with private key of its issuer in the "Issuer" and "Signature" fields, respectively. The hybrid device d_1 checks whether or not the signature of every capability token used in the above steps 1 and 2 is valid. If at least one capability token which is not valid exists, the operation *op* is interrupted.

After the authorization process, the hybrid device d_1 sends a CoAP response to the subject sb_1. If the authorization process is completed for a *get* operation, the hybrid device d_1 includes its resource set $r^1.sR$ in the payload field of the CoAP response. The subject sb_1 adds the resources in the set $r^1.sR$ to its *source* resource set $sb_1.sR$, i.e. $sb_1.sR = sb_1.sR \cup r^1.sR$. On the other hand, if the authorization process is completed for a *put* operation, the hybrid device d_1 updates the resource set $r^1.sR$ with $sb_1.sR$, i.e. $r^1.sR = sb_1.sR$.

6   Evaluation

In the evaluation, first, a subject sb_1 issues a *put* operation to the hybrid device d_1 to store data in the resource r^1. Here, the data are randomly generated. The data size is also randomly decided from 80 to 120 [Bytes]. Next, the subject sb_1 issues a *get* operation to the hybrid device d_1 to get data in the resource r^1.

We iterate the above measurement ten times, i.e. a pair of *put* and *get* operations are iterated ten times, and measure the execution time of every step of every authorization process. Here, we generate a scenario by which the subject sb_1 issues operations so that the authorization process is completed, i.e. no illegal information flow occurs, because the total execution time is longest in this case. If illegal information flow occurs, the authorization process aborts in the middle and the execution time is always shorter. After the measurements, we calculate the average execution time.

Let et_m be the execution time of each step m of the authorization process ($m = 1, 2, 3, 4$) presented in the preceding section. In the step 1, whether or not the subject sb_1 is allowed to issue the operation op to the resource r^1 is checked. In the step 2, whether or not the subject sb_1 is allowed to get data from every resource in $r^1.sR$ is checked. In the step 3, whether or not the signature of the subject sb_1 is valid is checked. In the step 4, whether or not the signature of every capability token used in the above steps 1 and 2 is valid is checked.

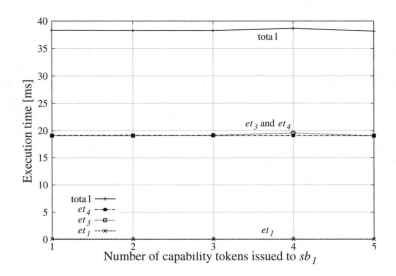

Fig. 2. Execution time for a *put* operation.

Figure 2 shows the execution time of each step in the authorization process for a *put* operation. The horizontal axis is the number of capability tokens issued to the subject sb_1, i.e. $|CAP^1|$. We generate CAP^1 and $sb_1.sR$ such that $CAP^1 = \{cap_m^1 \mid \langle r^m, put \rangle \in cap_m^1 (m = 1, \ldots, |CAP^1|)\}$ and $sb_1.sR = \{r^m \mid m = 1, \ldots, |CAP^1|\}$, respectively. For example, if $|CAP^1| = 3, CAP^1 = \{cap_1^1 (\ni \langle r^1, put \rangle), cap_2^1 (\ni \langle r^2, put \rangle), cap_3^1 (\ni \langle r^3, put \rangle)\}$ and $sb_1.sR = \{r^1, r^2, r^3\}$.

In the authorization process for a *put* operation, the step 2 is not needed. The execution time et_3 is almost the same as et_4 ($et_3 \simeq et_4$). Here, $et_1 < et_3 \simeq et_4$. If the subject sb_1 tries to put data to the resource r^1, the subject sb_1 sends the only capability token cap^1 such that $cap^1 \ni \langle r^1, put \rangle$ to the hybrid device d_1. Here, the steps 3 and 4 are performed for only one capability token cap^1. Hence, the execution times et_3 and et_4 are almost same.

The label "total" shows the total execution time of all the steps for a *put* operation, i.e. $et_1 + et_3 + et_4$. Here, the total execution time is constant as the number of capability tokens issued to the subject sb_i increases.

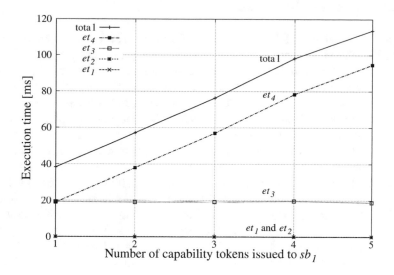

Fig. 3. Execution time for a *get* operation.

Figure 3 shows the execution time of each step in the authorization process for a *get* operation. The horizontal axis is the number of capability tokens issued to the subject sb_1, i.e. $|CAP^1|$. In order to perform every step of the authorization process, we generate CAP^1 such that $CAP^1 = \{cap_m^1 \mid \langle r^m, get \rangle \in cap_m^1 (m = 1, \ldots, |CAP^1|)\}$. Since a *put* operation is issued before, $r^1.sR = \{r^m \mid m = 1, \ldots, |CAP^1|\}$.

The execution time et_1 is almost the same as et_2 ($et_1 \simeq et_2$) for $|CAP^1|$ is 1 according to the experiment. In addition, the execution time et_3 is also almost the same as et_4 ($et_3 \simeq et_4$) for $|CAP^1|$ is 1. Here, $et_1 \simeq et_2 < et_3 \simeq et_4$. On the other hand, $et_1 \simeq et_2 < et_3 < et_4$ for $|CAP^1| \geq 2$. In the OI protocol, if the subject sb_1 tries to get data from the resource r^1, the subject sb_1 sends not only the capability token cap^1 ($\ni \langle r^1, get \rangle$) but also the other capability tokens to the hybrid device d_1 to indicate that the subject sb_1 is allowed to get data form the other resources in $r^1.sR$. In the evaluation, we assume the subject sb_1 has only one public key. This means, the values of "Subject" fields of all the capability tokens are same. Hence, the step 3 is performed for only one capability token. On the other hand, the step 4 is performed for every capability token used in the steps 1 and 2. Hence, the execution time et_4 is longer than the execution time et_3 for $|CAP^1| \geq 2$.

The label "total" shows the total execution time of all the steps for a *get* operation, i.e. $et_1 + et_2 + et_3 + et_4$. Here, the total execution time increases as the number of capability tokens issued to the subject sb_1 increases. This means, the more number of capability tokens are checked in the step 4, the longer the execution time of the authorization process of the OI protocol becomes.

7   Concluding Remarks

In order to make the IoT (Internet of Things) secure, we take the CBAC (Capability-Based Access Control) model. In this paper, three types of devices, sensor, actuator, and hybrid devices are considered. In the IoT, a subject sb can get data of a resource r^1 from another resource r^2 to which the data of the resource r^1 are brought although the subject sb is not allowed to get the data from the resource r^1. Here, illegal information flow occurs. In our previous studies, the OI (Operation Interruption) protocol where operations implying illegal information flow are interrupted is proposed to prevent illegal information flow in the IoT based on the CBAC model. In this paper, we implemented the OI protocol on a hybrid device which is realized in Raspberry Pi 3 Model B+. We evaluated the authorization process of the OI protocol in terms of the execution time. In the evaluation, we make clear the features of the execution time of the authorization process for *get* and *put* operations in the OI protocol. For *get* operations, the more number of capability tokens are issued to subjects, the longer the execution time of the authorization process becomes. On the other hand, for *put* operations, the execution time of the authorization process is constant even if the number of capability tokens issued to subjects increases.

Acknowledgements. This work was supported by Japan Society for the Promotion of Science (JSPS) KAKENHI Grant Number JP20K23336.

References

1. Raspberry Pi 3 model B+. https://www.raspberrypi.org/products/raspberry-pi-3-model-b-plus/
2. Raspbian, version 10.3, 13 Februrary 2020. https://www.raspbian.org/
3. Denning, D.E.R.: Cryptography and Data Security. Addison Wesley, Boston (1982)
4. Fernandez, E.B., Summers, R.C., Wood, C.: Database Security and Integrity. Adison Wesley, Boston (1980)
5. Gusmeroli, S., Piccione, S., Rotondi, D.: A capability-based security approach to manage access control in the Internet of Things. Math. Comput. Model. **58**(5–6), 1189–1205 (2013)
6. Hanes, D., Salgueiro, G., Grossetete, P., Barton, R., Henry, J.: IoT Fundamentals: Networking Technologies, Protocols, and Use Cases for the Internet of Things. Cisco Press, Indianapolis (2018)
7. Hernández-Ramos, J.L., Jara, A.J., Marín, L., Skarmeta, A.F.: Distributed capability-based access control for the internet of things. J. Internet Serv. Inf. Secur. **3**(3/4), 1–16 (2013)

8. Nakamura, S., Enokido, T., Takizawa, M.: Time-based legality of information flow in the capability-based access control model for the internet of things. Concurr. Comput. Pract. Exp. https://doi.org/10.1002/cpe.5944

9. Nakamura, S., Enokido, T., Takizawa, M.: A flexible read-write abortion protocol with role safety concept to prevent illegal information flow. J. Ambient Intell. Humaniz. Comput. **9**(5), 1415–1425 (2018)

10. Nakamura, S., Enokido, T., Takizawa, M.: Causally ordering delivery of event messages in p2 PPSO systems. Cogn. Syst. Res. **56**, 167–178 (2019)

11. Nakamura, S., Enokido, T., Takizawa, M.: Information flow control based on the CAPBAC (capability-based access control) model in the IoT. Int. J. Mobile Comput. Multimedia Commun. **10**(4), 13–25 (2019)

12. Nakamura, S., Enokido, T., Takizawa, M.: Information flow control in object-based peer-to-peer publish/subscribe systems. Concurr. Comput. Pract. Exp. **32**(8), e5118 (2020)

13. Ogiela, L., Ogiela, M.R.: Cognitive security paradigm for cloud computing applications. Concurr. Comput. Pract. Exp. **32**(8), e5316 (2020)

14. Sandhu, R.S.: Lattice-based access control models. Computer **26**(11), 9–19. IEEE (1993)

15. Sandhu, R.S., Coyne, E.J., Feinstein, H.L., Youman, C.E.: Role-based access control models. Computer **29**(2), 38 47. IEEE (1996)

16. Shelby, Z., Hartke, K., Bormann, C.: Constrained application protocol (COAP). IFTF Internet-draft (2013). http://tools.ietf.org/html/draft-ietf-core-coap-18

17. Tanganelli, G., Vallati, C., Mingozzi, E.: Coapthon: Easy development of COAP-based IoT applications with python. In: IEEE 2nd World Forum on Internet of Things (WF-IoT 2015), pp. 63–68 (2015)

18. Warner, B.: python-ecdsa-0.11, March 11, 2014. httpoi//github.com/ecdsa/python-ecdsa

19. Yuan, E., Tong, J.: Attributed based access control (ABAC) for web services. In: Proceedings of the IEEE International Conference on Web Services (ICWS 2005) (2005)

Simulation Results of CCM Based HC for Mesh Router Placement Optimization Considering Two Islands Model of Mesh Clients Distributions

Aoto Hirata[1], Tetsuya Oda[2(✉)], Nobuki Saito[2], Yuki Nagai[2], Masaharu Hirota[3], and Kengo Katayama[2]

[1] Engineering Project Course, Okayama University of Science (OUS),
1-1 Ridaicho, Kita-ku, Okayama 700-0005, Japan
t17p013ha@ous.jp
[2] Department of Information and Computer Engineering,
Okayama University of Science (OUS),
1-1 Ridaicho, Kita-ku, Okayama 700-0005, Japan
{oda,katayama}@ice.ous.ac.jp, {t17j033sn,t18j057ny}@ous.jp
[3] Department of Information Science, Okayama University of Science (OUS),
1-1 Ridaicho, Kita-ku, Okayama 700-0005, Japan
hirota@mis.ous.ac.jp

Abstract. Wireless mesh networks (WMNs) are one of the wireless network technologies that enables routers to communicate with each other wirelessly to create a stable network over a wide area at a low cost, and it has attracted much attention in recent years. In order to provide a lower cost and more stable network, various methods for optimizing the placement of mesh routers are being studied. In a previous work, we proposed a Coverage Construction Method (CCM) and CCM based Hill Climbing (HC) for mesh router placement problem considering normal and uniform distribution of mesh clients. In this paper, we evaluate the performance of CCM based HC for mesh router placement optimization considering two islands model of mesh clients distributions.

1 Introduction

The Wireless Mesh Networks (WMNs) [1–3] are one of the wireless network technologies that enables routers to communicate with each other wirelessly to create a stable network over a wide area at a low cost, and it has attracted much attention in recent years. The placement of the mesh router has a significant impact on cost, communication range, and operational complexity. Therefore, research is being done to optimize the placement of these mesh routers. In our previous work [4–8], we proposed and evaluated the different meta-heuristics such as Genetic Algorithms (GA) [9], Hill Climbing (HC) [10], Simulated Annealing (SA) [11], Tabu Search (TS) [12] and Particle Swarm Optimization (PSO) [13] for mesh router placement optimization. Also, we proposed a Coverage Construction

L. Barolli et al. (Eds.): EIDWT 2021, LNDECT 65, pp. 180–188, 2021.
https://doi.org/10.1007/978-3-030-70639-5_17

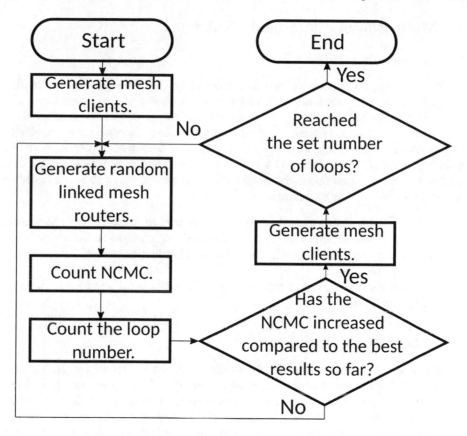

Fig. 1. Flowchart of the CCM.

Method (CCM) for mesh router placement problem [14] and CCM besed Hill Climbing (HC) method [15]. The CCM is able to rapidly create a group of mesh routers with the radio communication range of all the mesh routers linked to each other. CCM based HC was able to cover many of mesh clients generated by the normal and uniform distribution.

In this paper, we evaluate the performance of CCM based HC for mesh router placement optimization considering two islands model of mesh clients distributions. As evaluation metrics, we considered Size of Giant Component (SGC) for connection between mesh routers and Number of Covered Mesh Clients (NCMC) for mesh clients within radio communication range of SGC of mesh routers.

The structure of the paper is as follows. In Sect. 2, we defines the mesh router placement problem. In Sect. 3, we describes proposed system of the CCM and CCM based HC. In Sect. 4, we presents the simulation results and compares them with other studies. Finally, conclusions and future work are given in Sect. 5.

2   Mesh Router Placement Problem

2.1   Problem Overview

In this problem, we are given a two-dimensional continuous area where to deploy a number of mesh routers and a number of mesh clients of fixed positions. The objective of the problem is to optimize a location assignment for the mesh routers to the two-dimensional continuous area that maximizes the network connectivity and mesh clients coverage. Network connectivity is measured by the SGC of the each mesh routers link, while the NCMC is the number of mesh clients that within the radio communication range of at least one mesh router. An instance of the problem consists as follows.

- An area $Width \times Height$ where to problem area in mesh router placement. Positions of mesh routers are not pre-determined, and are to be computed.
- The mesh routers, each having its radio communication range, defining thus a vector of routers.
- The mesh clients located in arbitrary points of the considered area, defining a matrix of clients.

2.2   Mesh Clients Distribution

In this section, we describe mesh clients distributions in this paper. In our previous work [14,15], we have set the random number distribution for mesh client generation to be normal and uniform. This is because we assumed that the mesh clients would be either densely packed in the center or distributed throughout the range. In this paper, we consider two iland model as mesh clients placement. The conditional equations for the two island models are as follows

- Divide the set area into four parts and select any two of them.
- Generate a mesh client of any distributions within those two regions.

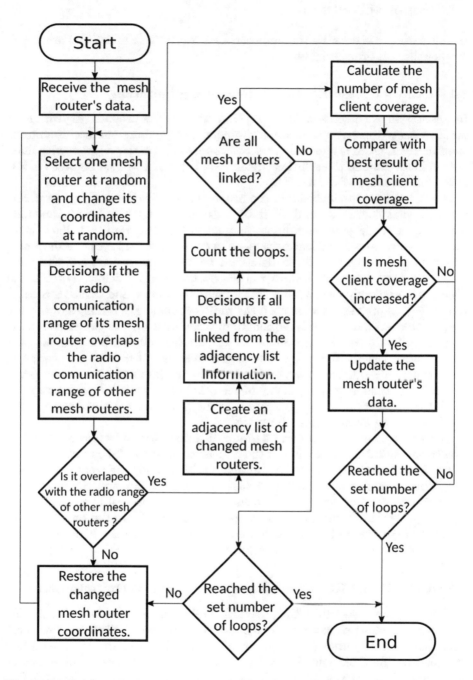

Fig. 2. Flowchart of the CCM based HC method for mesh router placement optimization.

3   Proposed Method

In this section, we describe the proposed system. In Fig. 1 and Fig. 2 are shown the flowchart of proposed system.

3.1   CCM for Mesh Router Placement Optimization

In this section, we describe a CCM [14] that our previous proposed algorithm for mesh router placement optimization problem. The proposed method is summarized in Fig. 1. CCM is a method that continues to derive the solution with maximized SGC. Among the solutions generated, the one with the highest NCMC is the final solution of CCM

First, generate mesh clients in the problem area. Next, randomly determine a single point coordinate and let it be mesh router 1. Once again, randomly determine a single point coordinate and let it be mesh router 2. Each mesh router has a radio communication range as a circle. So consider the mesh routers as circles and perform collision detection for two mesh routers. If the radio communication ranges of the two routers do not overlap, delete router 2 and once again randomly determine a single point coordinate and make it as mesh router 2. This process repeats until the radio communication ranges of the two mesh routers overlaps. If the radio communication ranges of the two mesh routers overlap, generate next mesh routers. Determine the collision between the next mesh routers generated and the one generated so far. If there is no overlap in radio communication range with any mesh routers, its mesh router is removed and generate randomly again. If any of the other mesh routers have overlapping radio communication ranges, generate next mesh routers. Continue this process until the setting number of mesh routers.

This allows for the creation of a group of routers with all radio communication ranges linked together without the derivation of connected component using Depth First Search (DFS) [16,17]. However, this method only creates a population of mesh routers at set area and does not take into account the location of the mesh clients. So, repeat this process a setting number of loop. When it is repeating, determine how many mesh clients are included in the radio communication range group of the mesh router, and the one with the highest number of coverage in the repetition is the solution.

3.2   CCM Based HC for Mesh Router Placement Optimization

In this section, we describe CCM based HC [15]. The implementation of the HC in the mesh router placement problem is shown in Fig. 2. First, we randomly select one of the routers in the group of mesh routers in the initial solution obtained by the CCM and change the coordinates of that chosen mesh router randomly. Then, decision the NCMC by the entire mesh router. If the NCMC is greater than that of the mesh router placement obtained so far, then the changed mesh router placement is the neighbor solution. If the NCMC is less than that of mesh router placement's NCMC obtained so far, the changed mesh router coordinates are restored. Repeat this process a setting number of loop. This sequence of process is the HC. However, this process alone is inadequate for the mesh router placement problem. This is because, depending on the placement of the changed mesh routers, the radio communication range of all mesh routers will not be linked. Therefore, it is necessary to create an adjacency list for a mesh router each time the mesh router placement is changed and use DFS to find out if the radio communication ranges of all the mesh routers are linked. And, NCMC is decision only when the radio communication range of all mesh routers is linked, and only when NCMC is greater than the neighbor solution, the placement of the mesh routers is the neighbor solution.

In this algorithm, in order to increase the probability that all the radio communication ranges are linked, loops the randomly change of coordinates until it overlaps with the radio communication range of one of the mesh routers. We also tightened the conditions for collision detection to cover as many clients as possible. Specifically, the collision is recognized only when the sum of the square of the difference between the x-axis of the two routers and the square of the difference between the y-axis is bigger than 12 and less than 16.

Table 1. Parameters and value for simulation

Width of problem area	32
Hight of problem area	32
Number of mesh routers	16
Radius of radio communication range of mesh routers	2
Number of mesh clients	48 total, 24 on each island
Distributions of mesh clients	Normal distribution and uniform distribution
Number of loop for CCM	3000
Number of loop for HC method	100000
Centroid in normal distribution 1	(8, 8), (24, 24)
Centroid in normal distribution 2	(10, 10), (22, 22)

Table 2. Simulation results of CCM.

	Normal distribution 1	Normal distribution 2	Uniform distribution
Best SGC	16	16	16
Average SGC	16	16	16
Best NCMC	24	27	20
Average NCMC	23	25	18

4   Simulation Results

In this section, we evaluate the proposed method. The parameters used in the simulation are shown in the Table 1. In this simulation, we consider three types of clients, two islands of uniform distribution and two types of two islands of normal distribution 15 times each. The simulation results are summarized in Table 2 and Table 3.

In Table 2 and Table 3, we show the simulation results of best SGC and avg. SGC. For each simulation results, all mesh router nodes are connected. We can see that the normal distribution 1 often target only one of the islands. However, the proposed method was able to find both islands in many cases and cover many more clients in normal distribution 2, where the coordinates of the mean are set closer to the center than in normal distribution 1.

Table 3. Simulation results of CCM based HC.

	Normal distribution 1	Normal distribution 2	Uniform distribution
Best SGC	16	16	16
Average SGC	16	16	16
Best NCMC	43	46	32
Average NCMC	26	40	25

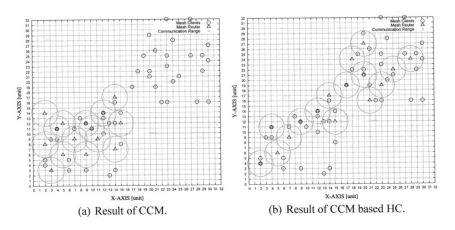

(a) Result of CCM.   (b) Result of CCM based HC.

Fig. 3. Visualization results in uniform distribution.

In Fig. 3, Fig. 4 and Fig. 5, we can see that the mesh routers group generated by the CCM are able to expand its coverage and cover more mesh clients by applying the HC.

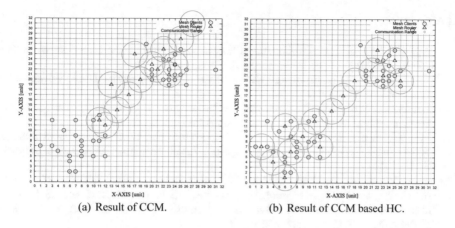

(a) Result of CCM.   (b) Result of CCM based HC.

Fig. 4. Visualization results in normal distribution 1.

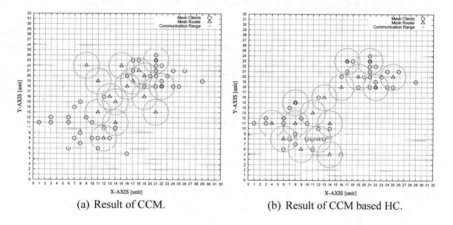

(a) Result of CCM.   (b) Result of CCM based HC.

Fig. 5. Visualization results in normal distribution 2.

5   Conclusion

In this paper, we evaluated the performance of CCM based HC for mesh router placement optimization considering two islands model of mesh clients distributions. The simulation results show that the proposed method can cover a large number of mesh clients in two island models.

In the future, we would like to apply it to the Simulated Annealing, which is an advanced version of the CCM based HC. In addition, we would like to adapt our method to the genetic algorithm which is often mentioned as a solution to the mesh router placement problem.

Acknowledgement. This work was supported by JSPS KAKENHI Grant Number 20K19793.

References

1. Akyildiz, I.F., et al.: Wireless mesh networks: a survey. Comput. Netw. **47**(4), 445–487 (2005)
2. Jun, J., et al.: The nominal capacity of wireless mesh networks. IEEE Wirel. Commun. **10**(5), 8–15 (2003)
3. Oyman, O., et al.: Multihop relaying for broadband wireless mesh networks: from theory to practice. IEEE Commun. Mag. **45**(11), 116–122 (2007)
4. Oda, T., et al.: WMN-GA: a simulation system for WMNs and its evaluation considering selection operators. J. Ambient Intell. Humaniz. Comput. **4**(3), 323–330 (2013)
5. Ikeda, M., et al.: Analysis of WMN-GA simulation results: WMN performance considering stationary and mobile scenarios. In: Proceedings of the 28th IEEE International Conference on Advanced Information Networking and Applications (IEEE AINA-2014), pp. 337–342 (2014)
6. Oda, T., et al.: Analysis of mesh router placement in wireless mesh networks using Friedman test considering different meta-heuristics. Int. J. Commun. Netw. Distrib. Syst. **15**(1), 84–106 (2015)
7. Oda, T., et al.: A genetic algorithm-based system for wireless mesh networks: analysis of system data considering different routing protocols and architectures. Soft. Comput. **20**(7), 2627–2640 (2016)
8. Sakamoto, S., et al.: Performance evaluation of intelligent hybrid systems for node placement in wireless mesh networks: a comparison study of WMN-PSOHC and WMN-PSOSA. In: Proceedings of the 11th International Conference on Innovative Mobile and Internet Services in Ubiquitous Computing (IMIS-2017), pp. 16–26 (2017)
9. Holland, J.H.: Genetic algorithms. Sci. Am. **267**(1), 66–73 (1992)
10. Skalak, D.B.: Prototype and feature selection by sampling and random mutation hill climbing algorithms. In: Proceedings of the 11th International Conference on Machine Learning (ICML-1994), pp. 293–301 (1994)
11. Kirkpatrick, S., et al.: Optimization by simulated annealing. Science **220**(4598), 671–680 (1983)
12. Glover, F.: Tabu search: a tutorial. Interfaces **20**(4), 74–94 (1990)
13. Kennedy, J., Eberhart, R.: Particle swarm optimization. In: Proceedings of the IEEE International Conference on Neural Networks (ICNN-1995), pp. 1942–1948 (1995)
14. Hirata, A., et al.: Approach of a solution construction method for mesh router placement optimization problem. In: Proceedings of the IEEE 9th Global Conference on Consumer Electronics (IEEE GCCE-2020), pp. 1–2 (2020)
15. Hirata, A., et al.: A coverage construction method based hill climbing approach for mesh router placement optimization. In: Proceedings of the 15th International Conference on Broadband and Wireless Computing, Communication and Applications (BWCCA-2020), pp. 355–364 (2020)
16. Tarjan, R.: Depth-first search and linear graph algorithms. SIAM J. Comput. **1**(2), 146–160 (1972)
17. Lu, K., et al.: The depth-first optimal strategy path generation algorithm for passengers in a metro network. Sustainability **12**(13), 1–16 (2020)

Proposal and Evaluation of a Tabu List Based DQN for AAV Mobility

Nobuki Saito[1], Tetsuya Oda[1(✉)], Aoto Hirata[2], Yuki Nagai[1],
Masaharu Hirota[3], and Kengo Katayama[1]

[1] Department of Information and Computer Engineering,
Okayama University of Science (OUS),
1-1 Ridaicho, Kita-ku, Okayama 700-0005, Japan
{t17j033sn,t18j057ny}@ous.jp, {oda,katayama}@ice.ous.ac.jp
[2] Engineering Project Course, Okayama University of Science (OUS),
1-1 Ridaicho, Kita-ku, Okayama 700-0005, Japan
t17p013ha@ous.jp
[3] Department of Information Science, Okayama University of Science (OUS),
1-1 Ridaicho, Kita-ku, Okayama 700-0005, Japan
hirota@mis.ous.ac.jp

Abstract. The Deep Q Network (DQN) is one of the methods of the deep reinforcement learning algorithm, which is a deep neural network structure used to estimate Q-values in Q-learning methods. The authors have previously designed and implemented a DQN-based mobility control methods for Autonomous Aerial Vehicle (AAV) In this paper, we propose and evaluate a DQN based on tabu list strategy for AAV mobility control. For evaluation, we simulate were conducted for the mobility control of AAV in a staircase environment using normal DQN and tabu list based DQN. The simulation results showed that a tabu list based DQN was a better solution than the normal DQN.

1 Introduction

In recent years, Unmanned Aerial Vehicleds (UAV) are expected to be used not only for aerial photography and transportation, but also in various other fields, such as human search and rescue, inspection, surveying and observation, crime prevention, and agriculture. The Autonomous Aerial Vehicle (AAV) [1], which have the ability to operate autonomously, are also expected to be used in a variety of fields similar to UAVs. However, existing autonomous flight systems are designed for outdoor use and rely on Global Navigation Satellite System (GNSS) and other location information. Therefore, autonomous movement control is essential to achieve operations that are independent of the external environment, including indoors and in non-GNSS environments such as tunnels and underground.

© The Author(s), under exclusive license to Springer Nature Switzerland AG 2021
L. Barolli et al. (Eds.): EIDWT 2021, LNDECT 65, pp. 189–200, 2021.
https://doi.org/10.1007/978-3-030-70639-5_18

The authors are studying Wireless Sensor and Actuator Networks (WSANs) that can act autonomously in consideration of disaster monitoring. A WSAN consists of wireless network nodes, all of which have the ability to sense events (sensors) and perform actions (actuators) based on the sensing data collected by the sensors. The application areas of WSAN include AAVs, Autonomous Underwater Vehicles (AUVs), Heating, Ventilation, and Air Conditioning (HVAC), Internet of Things (IoT), ambient intelligence, ubiquitous robotics, and etc. WSAN nodes in these applications are nodes with integrated sensors and actuators that have high processing power, high communication capability, high battery capacity, and may include other functions such as mobility.

In our previous works, we designed and implemented a Deep Q-Network (DQN) based AAV testbed [2] and simulation system for behavioral control methods of actuator nodes in WSANs based on DQN [3–7]. DQN is a type of deep reinforcement learning, in which the Q-values in the Q-learning algorithm are estimated by a deep learning algorithm. Deep reinforcement learning is a function approximation method that uses deep neural networks for the value function and policy function in reinforcement learning. DQN which uses a convolutional neural network (CNN) as a function approximation for Q-leaning, is a deep reinforcement learning method proposed by Mnih et al. [8,9]. DQN combines the methods of neural fitting Q-iteration [10,11] and experience replay [12], shares the hidden layer of the action value function for each action pattern, and can stabilize learning even with nonlinear functions such as CNN [13,14]. However, there were some points where learning was difficult to progress for problems with complex operations and rewards, or problems where it takes a long time to obtain a reward.

In this paper, we proposed a DQN based on tabu list strategy [15], and evaluated it in a staircase environment for autonomous movement control of an AAV. The structure of the paper is as follows. In Sect. 2, we show the DQN based AAV testbed implementation. In Sect. 3, we describes proposed method of the tabu list based DQN. In Sect. 4, we presents the simulation results. Finally, conclusions and future work are given in Sect. 5.

2   Implementation of an AAV Testbed

This section describes the designed AAV testbed for autonomous mobility control and the DQN for controlling AAV.

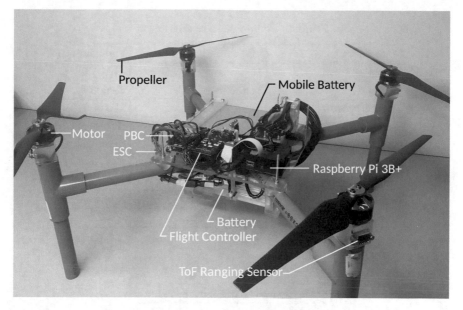

Fig. 1. The implemented AAV.

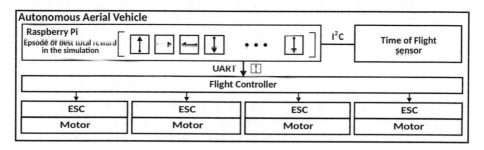

Fig. 2. AAV control system.

2.1 Design of a Quadrotor

The design of AAV makes use of a quadcopter, which is a type of multicopter. Multicopters are high maneuverable and can go into places that are difficult for humans to enter, such as disaster areas and danger zones. It also has the advantage of not requiring space for takeoffs and landings and being able to stop at mid-air during the flight, therefore enabling activities at fixed points. The quadrotor is a type of rotary-wing aircraft that uses four rotors for takeoff and propulsion. They are also less expensive to manufacture and operate with less power than hexacopters and octocopters.

Figure 1 shows a photograph of the implemented AAV testbed. The quadrotor frame is mainly composed of a polyvinyl chloride (PVC) pipe and acrylic plate. The parts for connecting the battery, motor, sensor, etc. to the frame

Table 1. Components of quadrotor.

Component	Model	Manufacture
Propeller	15 × 5.8	Hobby king
Morter	MN3508 700 KV	T-motor
Electric speed controller	F45A 32bitV2	T-motor
Flight controller	Pixhawk 2.4.8	Hobby ant
Power distribution board	MES-PDB-KIT	Lynxmotion
Li-Po battery	22.2v 12000 mAh XT90	YoWoo
Mobile battery	Pilot Pro 2 23000 mAh	Poweradd
ToF ranging sensor	VL53L0X	Kookye
Raspberry Pi 3	Model B Plus	ABOX

Table 2. Size of quadrotor.

Size (including propeller)	Values
Length [cm]	87
Width [cm]	87
Height [cm]	30
Diagonal [cm]	107
Weight [g]	4259

were created using an optical 3D printer. Figure 2 shows a AAV control system. The Flight Controller (FC) is a component that calculates the optimum motor rotation speed for flight based on the information sent from the built-in acceleration sensor and gyro sensor. The Electronic Speed Controller (ESC) is a part that controls the rotation speed of the motor in response to commands from FC. The Raspberry Pi uses telemetry communication to send commands for AAV mobility. In addition, multiple Time of Flight (ToF) range sensors using Inter-Integrated Circuit (I^2C) communication and General-Purpose Input/Output (GPIO) are used to acquire and save flight data. Output and save values for the process of movement when simulating a single indoor path environment using DQN. The movement process during the DQN simulation is stored values and reproduces the simulation results. Table 1 shows the components used in the quadrotor. Table 2 shows the size and weight of the quadrotor including the propeller.

2.2   DQN for AAV Mobility Control

We present the design and implementation of proposed simulation system based on DQN for AAV mobility control. The DQN for moving control of AAV structure is shown in Fig. 3. The proposed simulating system is implemented by Rust programming language [16, 17].

In this work, we use the Deep Belief Network (DBN), where computational complexity is smaller than CNN for DNN part in DQN. The environment is set as v_i. At each step, the agent selects an action a_t from the action sets of the AAV and observes a communication and sensing coverage v_t from the current state. If a_t is an action to shields including walls and floors, or to coordinates that are added to the tabu list, the agent selects action again.

The change of the AAV reward r_t was regarded as the reward for the action. For a reinforcement learning, we can complete all of these AAV sequences m_t as Markov decision process directly, where sequences of observations and actions $m_t = v_1, a_1, v_2, \ldots, a_{t-1}, v_t$. Likewise, it uses a method known as experience replay in which it store experiences of the agent at each timestep, $e_t = (m_t, a_t, r_t, m_{t+1})$ in a dataset $D = e_1, \ldots, e_N$, cached over many episodes into a Experience Memory. Defining the discounted reward for the future by a factor γ, the sum of the future reward until the end would be $R_t = \sum_{t'=t}^{T} \gamma^{t'-t} r_{t'}$. T means the termination time-step of the AAV. After running experience replay, the agent selects and executes an action according to an ϵ-greedy strategy. Since using histories of arbitrary length as inputs to a neural network can be difficult, Q function instead works on fixed length format of histories produced by a function ϕ. The target was to maximize the action value function $Q^*(m, a) = \max_\pi E[R_t | m_t = m, a_t = a, \pi]$, where π is the strategy for selecting of best action. From the Bellman equation, it is equal to maximize the expected value of $r + \gamma Q^*(m', a')$, if the optimal value $Q^*(m', a')$ of the sequence at the next time step is known.

$$Q^*(m', a') = E_{m' \sim \xi}[r + \gamma \max_{a'} Q^*(m', a') | m, a] \tag{1}$$

Not using iterative updating method to optimize the equation, it is common to estimate the equation by using a function approximator. Q-network in DQN was such a neural network function approximator with weights θ and $Q(s, a; \theta) \approx Q^*(m, a)$. The loss function to train the Q-network is:

$$L_i(\theta_i) = E_{s,a \sim \rho(.)}[(y_i - Q(s, a; \theta_i))^2]. \tag{2}$$

The y_i is the target, which is calculated by the previous iteration result θ_{i-1}. $\rho(m, a)$ is the probability distribution of sequences m and a. The gradient of the loss function is shown in Eq. (3):

$$\nabla_{\theta_i} L_i(\theta_i) = E_{m,a \sim \rho(.);s' \sim \xi}[(y_i - Q(m, a; \theta_i)) \nabla_{\theta_i} Q(m, a; \theta_i)]. \tag{3}$$

The initial weights values are assigned as Normal Initialization [18]. The input layer is using AAV and the position of events, total reward values in Experience Memory and AAV patterns. The hidden layer is connected with 256 rectifier units in Rectified Linear Units (ReLU) [19]. The output Q values are AAV mobile patterns.

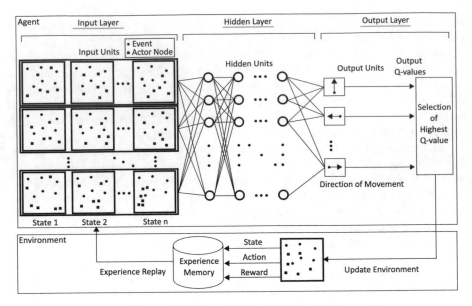

Fig. 3. DQN for AAV mobility control.

We consider tasks in which an agent interacts with an environment. In this case, the AAV moves step by step in a sequence of observations, actions and rewards. We took in consideration the mobility of AAV. For an AAV are considered 7 mobile patterns (up, down, forward, back, left, right, stop).

3   Proposed Method

We describes proposed method of the tabu list based DQN. Reward value is decided using Eq. 4 in this paper.

$$r = \begin{cases} 10 & (if\ (x_{current} = x_{global\ destinations}) \wedge \\ & (y_{current} = y_{global\ destinations}) \wedge \\ & (z_{current} = z_{global\ destinations})). \\ 3 & (else\ if\ ((x_{before} < x_{current}) \wedge (x_{current} \leq x_{local\ destinations})) \vee \\ & ((x_{before} > x_{current}) \wedge (x_{current} \geq x_{local\ destinations})) \vee \\ & ((y_{before} < y_{current}) \wedge (y_{current} \leq y_{local\ destinations})) \vee \\ & ((y_{before} > y_{current}) \wedge (y_{current} \geq y_{local\ destinations})) \vee \\ & ((z_{before} < z_{current}) \wedge (z_{current} \leq z_{local\ destinations})) \vee \\ & ((z_{before} > z_{current}) \wedge (z_{current} \geq z_{local\ destinations}))). \\ -1 & (else). \end{cases} \quad (4)$$

If the reward value is positive, the prohibited area will be added to the tabu list according to the rules shown in Algorithm 1. Tabu rule addition method is shown in Fig. 4. The tabu list will hold the added prohibited area until the end of the episode, and will be initialized for each episode.

Algorithm 1. Tabu List for DQN.

Require: The coordinate with the highest evaluated value in the section is (x, y, z).

1: **if** $(x_{before} \leq x_{current}) \wedge (x_{current} \leq x)$ **then**
2:   $tabulist \Leftarrow ((x_{min} \leq x_{before}) \wedge (y_{min} \leq y_{max}) \wedge (z_{min} \leq z_{max}))$
3: **else if** $(x_{before} \geq x_{current}) \wedge (x_{current} \geq x)$ **then**
4:   $tabulist \Leftarrow ((x_{before} \leq x_{max}) \wedge (y_{min} \leq y_{max}) \wedge (z_{min} \leq z_{max}))$
5: **else if** $(y_{before} \leq y_{current}) \wedge (y_{current} \leq y)$ **then**
6:   $tabulist \Leftarrow ((x_{min} \leq x_{max}) \wedge (y_{min} \leq y_{before}) \wedge (z_{min} \leq z_{max}))$
7: **else if** $(y_{before} \geq y_{current}) \wedge (y_{current} \geq y)$ **then**
8:   $tabulist \Leftarrow ((x_{min} \leq x_{max}) \wedge (y_{before} \leq y_{max}) \wedge (z_{min} \leq z_{max}))$
9: **else if** $(z_{before} \leq z_{current}) \wedge (z_{current} \leq z)$ **then**
10:   $tabulist \Leftarrow ((x_{min} \leq x_{max}) \wedge (y_{min} \leq y_{max}) \wedge (z_{min} \leq z_{before}))$
11: **else if** $(z_{before} \geq z_{current}) \wedge (z_{current} \geq z)$ **then**
12:   $tabulist \Leftarrow ((x_{min} \leq x_{max}) \wedge (y_{min} \leq y_{max}) \wedge (z_{before} \leq z_{max}))$
13: **end if**

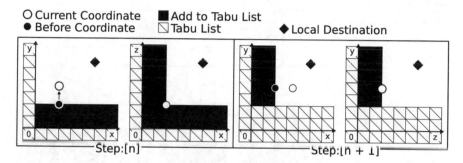

Fig. 4. Tabu rule addition method.

4 Performance Evaluation

The Table 3 shows the parameters used in the simulation. Figure 5 show a snap-shots of the problem area in simulation scenario, and Fig. 6 show a problem area for simulation scenario based Fig. 5. Table 4 and Fig. 7 show the partitioning of the search space, and the setting of the local destinations in this simulation. The section number 6 in Fig. 7 is staircase space, and section number 7 is the global destinations.

Figure 8 shows the change in reward value for the action in each iteration in the Worst, Median, and Best episodes with the total reward value of normal DQN (which does not take the tabu list into account) and tabu list based DQN. The episodes with the best total reward values both show an increasing trend in total reward values. Normal DQN is not possible to reach the global destination because it is possible to explore the already explored space again and obtain the reward, and because it is difficult to explore widely in the space due to the random determination of the direction of movement. In contrast, the reward value in the Best episode of DQN considering the tabu list is rapidly increasing from around 1700 iterations. This indicates that the AAV has reached its global

Table 3. Simulation parameters of DQN.

Parameters	Values
Number of episode	10000
Number of iteration	2000
Number of hidden layers	3
Number of hidden units	15
Initial weight value	Normal initialization
Activation function	ReLU
Action selection probability (ϵ)	$0.999 - (t/\text{Number of episode})$ ($t = 0, 1, 2, \ldots$, Number of episode)
Learning rate (α)	0.04
Discount rate (γ)	0.9
Experience memory size	300×100
Batch size	32
Number of AAV	1

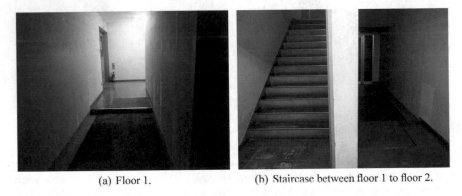

(a) Floor 1.   (b) Staircase between floor 1 to floor 2.

Fig. 5. Snapshots of the problem area in simulation scenario.

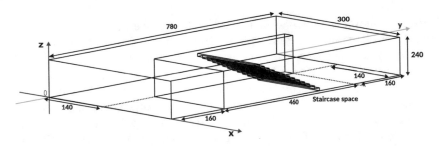

Fig. 6. Problem area for simulation.

Table 4. Destination for section.

Section no	x	y	z
1	150	80	100
2	75	535	100
3	75	710	100
4	150	535	100
5	225	535	100
6	225	80	340
7	150	80	340

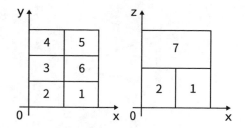

Fig. 7. Segmented section of the target space.

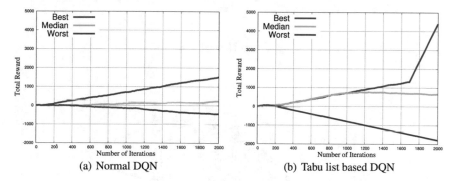

(a) Normal DQN   (b) Tabu list based DQN

Fig. 8. Results of total rewards.

destination. In the `Median` episode, the total reward value earned by DQN that took the tabu list into account tended to increase, while the total reward value earned by DQN that did not take the tabu list into account was close to flat, suggesting that they were not able to make the transition to a higher reward value.

Figure 9 shows the trajectories of movement in three-dimentional space for normal DQN and tabu list based DQN in the episode with the best total reward. The simulation results show that the performance of the tabu list based DQN better than normal DQN. Because, the normal DQN explores the initial position near, while the tabu list based DQN explores more extensively and is able to reach the destination. This indicates that the use of a tabulated list prohibits repeated search in the same space and restricts the direction of randomly selected movement, thereby eliminating or reducing the point that the search space is limited due to random movement and the point that repeated rewards are obtained in the same search space when the tabu list is not used.

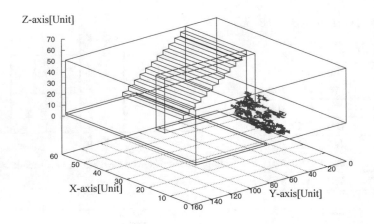

(a) Normal DQN

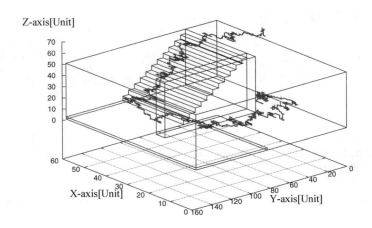

(b) Tabu list based DQN

Fig. 9. Visualization results.

5 Conclusion

In this paper, we proposed a tabu list based DQN for AAV mobility control.
For evaluation for tabu list DQN, we conducted using DQN with and without
considering the tabu list for a staircase environment. The simulation results
show a tabu list used DQN can reach a better solution than normal DQN by
prohibiting repeated search in the same space and restricting randomly selected
movement directions.

In the future, we would like to improve the tabu list based DQN for AAV control by conducting simulations with different scenarios.

Acknowledgement. This work was supported by Grant for Promotion of Okayama University of Science (OUS) Research Project (OUS-RP-20-3).

References

1. Stöcker, C., et al.: Review of the current state of UAV regulations. Remote Sens. **9**(5), 1–26 (2017)
2. Saito, N., Oda, T., Hirata, A., Hirota, Y., Hirota, M., Katayama, K.: Design and implementation of a DQN based AAV. In: Proceedings of the 15th International Conference on Broadband and Wireless Computing, Communication and Applications (BWCCA-2020), pp. 321–329 (2020)
3. Oda, T., Obukata, R., Ikeda, M., Barolli, L., Takizawa, M.: Design and implementation of a simulation system based on deep Q-network for mobile actor node control in wireless sensor and actor networks. In: Proceedings of the 31th IEEE International Conference on Advanced Information Networking and Applications Workshops (IEEE WAINA-2017) (2017)
4. Oda, T., Kulla, E., Cuka, M., Elmazi, D., Ikeda, M., Barolli, L.: Performance evaluation of a deep Q-network based simulation system for actor node mobility control in wireless sensor and actor networks considering different distributions of events. In: Proceedings of the 11th International Conference on Innovative Mobile and Internet Services in Ubiquitous Computing (IMIS-2017), pp. 36–49 (2017)
5. Oda, T., Elmazi, D., Cuka, M., Kulla, E., Ikeda, M., Barolli, L.: Performance evaluation of a deep Q-network based simulation system for actor node mobility control in wireless sensor and actor networks considering three-dimensional environment. In: Proceedings of the 9th International Conference on Intelligent Networking and Collaborative Systems (INCoS-2017), pp. 41–52 (2017)
6. Oda, T., Kulla, E., Katayama, K., Ikeda, M., Barolli, L.: A deep Q-network based simulation system for actor node mobility control in WSANs considering three-dimensional environment: a comparison study for normal and uniform distributions. In: Proceedings of the 12th International Conference on Complex, Intelligent, and Software Intensive Systems (CISIS-2018), pp. 842–852 (2018)
7. Toyoshima, K., Oda, T., Hirota, M., Katayama, K., Barolli, L.: A DQN based mobile actor node control in WSAN: simulation results of different distributions of events considering three-dimensional environment. In: Proceedings of the 8th International Conference on Emerging Internet, Data & Web Technologies (EIDWT-2020), pp. 197–209 (2020)
8. Mnih, V., Kavukcuoglu, K., Silver, D., Rusu, A.A., Veness, J., Bellemare, M.G., Graves, A., Riedmiller, M., Fidjeland, A.K., Ostrovski, G., Petersen, S., Beattie, C., Sadik, A., Antonoglou, I., King, H., Kumaran, D., Wierstra, D., Legg, S., Hassabis, D.: Human-level control through deep reinforcement learning. Nature **518**, 529–533 (2015)
9. Mnih, V., Kavukcuoglu, K., Silver, D., Graves, A., Antonoglou, I., Wierstra, D., Riedmiller, M.: Playing atari with deep reinforcement learning, pp. 1–9 (2013). arXiv:1312.5602v1
10. Lei, T., Ming, L.: A robot exploration strategy based on Q-learning network. In: IEEE International Conference on Real-Time Computing and Robotics (RCAR-2016), pp. 57–62 (2016)

11. Riedmiller, M.: Neural fitted Q iteration - first experiences with a data efficient neu-
    ral reinforcement learning method. In: The 16th European Conference on Machine
    Learning (ECML-2005). Lecture Notes in Computer Science, vol. 3720, pp. 317–328
    (2005)
12. Lin, L.J.: Reinforcement learning for robots using neural networks. Technical
    report, DTIC document (1993)
13. Lange, S., Riedmiller, M.: Deep auto-encoder neural networks in reinforcement
    learning. In: Proceedings of the 2010 International Joint Conference on Neural
    Networks (IJCNN-2010), pp. 1–8 (2010)
14. Kaelbling, L.P., Littman, M.L., Cassandra, A.R.: Planning and acting in partially
    observable stochastic domains. Artif. Intell. **101**(1–2), 99–134 (1998)
15. Glover, F.: Tabu search - part I. ORSA J. Comput. **1**(3), 135–206 (1989)
16. The Rust Programming Language. https://www.rust-lang.org/. Accessed 14 Oct
    2019
17. Takano, K., Oda, T., Kohata, M.: Design of a DSL for converting rust program-
    ming language into RTL. In: Proceedings of the 8th International Conference on
    Emerging Internet, Data & Web Technologies (EIDWT-2020), pp. 342–350 (2020)
18. Glorot, X., Bengio, Y.: Understanding the difficulty of training deep feedforward
    neural networks. In: Proceedings of the 13th International Conference on Artificial
    Intelligence and Statistics (AISTATS-2010), pp. 249–256 (2010)
19. Glorot, X., Bordes, A., Bengio, Y.: Deep sparse rectifier neural networks. In: Pro-
    ceedings of the 14th International Conference on Artificial Intelligence and Statis-
    tics (AISTATS-2011), pp. 315–323 (2011)

Performance Comparison of Replication Protocols for Low Demand Files in MANET

Takeru Kurokawa and Naohiro Hayashibara[✉]

Kyoto Sangyo University, Kyoto, Japan
{i1986061,naohaya}@cc.kyoto-su.ac.jp

Abstract. Mobile ad-hoc network consists of a collection of mobile devices that are interconnected to each other. It has attracted attention in various areas, such as information sharing and intelligent transportation. In particular, information sharing on MANET in an emergency is crucial for victims. In this situation, information on relief supplies and safety confirmation should be replicated to keep it on the network as long as possible. On the other hand, the cost of replicated information should be taken into account since the storage of each mobile device is limited. In this paper, we compare the performance of replication protocols, including the cuckoo search based replication protocol for low demand files that has been proposed. We also discuss the impact of file access models on file availability, the success ratio of file requests, and the costs for storage and communication.

1 Introduction

Mobile ad-hoc network (MANET) is widely used by many applications such as the information sharing system in an emergency [11], indoor location tracking [1] and intelligent transportation in a smart city.

Now, we suppose the information sharing in the case of emergency (e.g., disaster). In this case, there is no network infrastructure available. So, MANET that consists of mobile phones, is an alternative to a network infrastructure (e.g., wifi access points) for sharing information among victims. Victims in a shelter upload information on disaster relief supplies, medical support, livelihood support, and confirmation of individual safety.

The number of replication of data (files) is often in proportion to the number of requests in typical replication schemes, for example, Path replication [3] and Owner replication [9]. Thus, files that are frequently required by users create many replicas. For instance, information on disaster relief supplies and medical, livelihood support is required by many people and replicated actively. Contrary to this, files with low demand create a small number of replicas. This type of file could be disappeared quickly in MANETs by node's leaving or churn. On the other hand, those files could be important for specific individuals, for example, the safety information on family and relatives.

L. Barolli et al. (Eds.): EIDWT 2021, LNDECT 65, pp. 201–211, 2021.
https://doi.org/10.1007/978-3-030-70639-5_19

In general, the number of replicas is correlated with the availability of the files. The availability of information increases as the number of its replicas increases. For improving the availability of low demand files, the total number of replicas must be increased. As a result, the cost of the storage of each node and the network becomes high. This is a serious problem in MANETs.

This work aims to improve the availability of low demand files without imposing a heavy load on the devices and the network. Now, we assume a mobile ad-hoc network that consists of mobile devices, and information is shared on it in the case of emergency (e.g., disaster). In this case, we need to take into account the storage cost in each device because it is strictly limited. We also make low demand (but still important) information available on the network as long as possible at the same time.

For this purpose, we have proposed a novel approach for data replication on MANETs based on Cuckoo search (CSPR) [7] in the previous work. Cuckoo search [13] is a meta-heuristic algorithm inspired by the egg-laying habits of cuckoos. It is known as an efficient solution for nonlinear global optimization problems, and it is proven that it guarantees the global convergence by using Lévy flight [4].

Roughly speaking, we use Cuckoo search for allocating the replicated files to nodes in MANET. Each node evaluates other nodes that are close to it based on the residual of their storage and battery, then sends copies of files, instead of eggs, to the best nodes to improve file availability, especially for low demand files. On the other hand, we need to consider the tradeoff between file availability and the costs for replication since the storage and the bandwidth are limited in MANET.

In this paper, we measure the availability of files, the success ratio of file requests, the storage cost, and the communication cost of CSRP in different scenarios to evaluate that CSRP is suitable for low demand files in MANET. Moreover, we also compare several existing replication protocols with CSRP.

2   Related Work

Owner replication [9] is a simple replication protocol that replicates data to the node that issues a search request for it. It means that one replicated data corresponds to the search query. Thus, the cost of replicating data is low. However, this protocol may require many messages for searching for data.

Path replication [3] creates replicas of the data and allocates them to all nodes along the path from the requesting node to the providing node in a search query for it. This protocol reduces the search traffic (i.e., the number of messages) by a factor of three compared to Owner replication. On the other hand, it imposes a higher storage cost than Owner replication because of the more significant number of replicated data.

Kageyama and Shibusawa proposed a replication protocol (KS protocol) for improving the availability of low demand files in super-node based P2P systems [6]. Note that the storage of each super-node has large enough to store files. It replicates the high demand files using Owner replication. In addition, it computes a demand forecast for low demand files and decides the number of replicas to be created.

3   System Model

We assume that a MANET consists of a set of nodes $N = \{n_1, n_2, ..., n_{|N|}\}$, which are mobile devices (e.g., smartphones, laptop computers, wearable devices). Each node has a unique identifier. Each pair of nodes can communicate with each other. It means that a node in the network has a multi-hop path to any other node. We can obtain the remaining amount of battery RB_i^t and storage RS_i^t of a node N_i at time t. We also define the capacity of battery CB_i and storage CS_i of a node N_i. Moreover, A_i^t is defined as a duration (minutes) that N_i joins the network at time t. We conducted simulations by a cycle-based simulator. Thus, A_i^t is represented as the number of cycles.

A MANET we assume is a dynamic distributed system. Nodes leave from the network by going out of the communication range of any other node, turning off the devices, and so on. In addition, nodes can join the network as new nodes. Once a node leaves the network, it joins it again with another identifier without any data and logs in the previous execution. It means that we assume fail-stop failure. We also assume a membership service such as [2,8,10] to manage the status of each node.

Both CSRP protocol and KS protocol determine target files that should be replicated based on the access frequency of each file. Basically, it is monitored while in execution. In our simulation, we assume that it obeys Pareto distribution.

The topology of a MANET is assumed to be a random geometric graph.

4   Cuckoo Search Based Replication Protocol

This work aims to improve the availability of low demand files in MANETs. Kurokawa and Hayashibara proposed the Cuckoo Search based Replication Protocol (CSRP) [7]. It is a modified version of KS protocol [6] with a replica allocation mechanism using Cuckoo Search [13]. In the beginning of the protocol, each node affixes a label of *high demand* or *low demand* to each file in its storage. A file d_i is labeled as low demand if the Eq. 1 holds, otherwise the file is labeled as high demand.

$$|D_i^t| \leq \log_{10} |N^t| \cdot \alpha \tag{1}$$

$|N^t|$ indicates the number of nodes at time t and $|D_i^t|$ is the total number of files of d_i (i.e., including the original file and replicated ones) where a set of files D_i^t including the original file d_i and its replicas and i is an identifier of the file. α is a given parameter and indicates a weight factor.

Then it executes the replica allocation protocol, which consists of two procedures, the replication procedure for high demand files and low demand files. The former one uses Owner replication for high demand files. On the other hand, the latter one uses a replica allocation protocol based on Cuckoo search [13] for low demand files. Cuckoo search is for a maximization problem of the value of the object function. We use this mechanism to select the best nodes that hold replicated low demand files for improving the availability of those files.

4.1   Replica Allocation Mechanism Based on Cuckoo Search

The aim of the mechanism is to improve the availability of low demand files in a MANET without having a high cost of storage and communication among nodes. We need to select eligible nodes to hold replicated files as long as possible. Cuckoo search is for selecting eligible nodes to hold replicated files based on the evaluation of nodes' survivability.

We suppose that each node N_i has two parameters, the capacity of battery CB_i and storage CS_i. Moreover, it has three variables, the remaining amount of battery RB_i^t and storage RS_i^t, and the duration A_i^t at time t. The proposed protocol evaluates the nodes in the network based on the following equation and computes the evaluation value $E(i)$ of each node N_i.

$$E(i) = ln\left(\left(\alpha \times \frac{RB_i^t}{CB_i} + \beta \times \frac{RS_i^t}{CS_i}\right) \times A_i^t\right) \qquad (2)$$

α and β are weights for the evaluation where α, $\beta = [0, 1]$. To simplify the capacity of battery and storage, CB_i and CS_i are defined as 100.0. Thus, RB_i^t and RS_i^t are variables from 0.0 to 100.0. $E(i)$ corresponds to the objective function $f(x)$ in Cuckoo search.

File availability A^t at time t is defined as follows.

$$A^t = \frac{|D^t|}{|D_{original}| + |D_{replicated}|} \qquad (3)$$

$|D_{original}|$ and $|D_{replicated}|$ are the total number of the original files and that of the replicated ones, respectively. The total number of files available at time t is denoted as D^t where $|D^t| = |D_{original}^t| + |D_{replicated}^t|$.

We assume that files are disappeared from the network by nodes' leaving. A node that has a large enough amount of battery is not likely to leave the network because of the battery shortage. The proposed mechanism tends to allocate replicated data into such a node. On the other hand, it takes into account the residual space of storage. It computes $E(i)$ for each node n_i and searches nodes with the values of $E(i)$ as high as possible by Cuckoo search. It contributes to improve the availability of low demand files and to reduce the average storage cost of mobile devices.

Cuckoo search algorithm uses Lévy walk that is known as an efficient algorithm for a wide-area search. Since the topology of a MANET is usually represented as a graph, we use the Lévy walk algorithm for unit disk graphs [12].

CSRP stops replicating d_j if d_j does not have any request for P cycles from the latest request. This is because the storage requirement of a MANET that consists of mobile devices is very severe. We call P the *protection period* that is a system-wide parameter. It could affect file availability and the costs for replication.

5 Performance Evaluation

The evaluation aims to clarify the impact of the file access distribution and the protection period P in CSRP on file availability, the success ratio of file requests, and the file access efficiency. We also measure the cost of storage and communication to improve file availability.

We also compare the performance of CSRP with that of Owner replication [9], Path replication [3], and KS replication protocol [6].

We implemented these protocols and the proposed one on PeerSim simulator [5] and used it for the evaluation. PeerSim is a cycle-based simulator for Peer-to-Peer systems implemented in Java.

5.1 Scenarios, Environment and Configuration

We assume two models for file access; A) Normal distribution, B) Pareto distribution. The latter indicates that most of the files are labeled as low demand. We also assume three configurations of CSRP regarding the parameter of the protection period $P = \{100, 200, 300\}$ for each file access model. Thus, we conducted simulation runs in six scenarios A-1, 2, 3 and B-1, 2, 3. Each simulation run is executed for 500 cycles (time unit).

We suppose that the number of nodes $|N| = 2000$, and they are located uniformly at random in the 1000×1000 field. Each node is interconnected with the neighbor nodes that are within a certain communication range of it. Hence the topology of a MANET is a random geometric graph.

In the beginning, the storage residual of each node is set to 100, and the battery residual of it is decided randomly from 50 to 100.

We assume that 50 individual files are randomly located in nodes in the network at the beginning, and no more file is uploaded after that. The main purpose of the simulation is to observe the availability of these files and the impact of replication of them by different replication protocols on the storage and the network.

5.2 Performance Criteria

We conducted simulation runs to measure the following criteria.

File Availability: It indicates how many files are accessible on the network.
Search Success Ratio: It indicates how much search requests succeed to find out their target files.

File Access Efficiency: We measure average hops from the source node to the
destination that hold a target file. The access efficiency is low if the number
of hops is smaller. Thus, it indicates the efficiency of file access.

Storage Cost: It indicates how much replicated data consumes the storage of
devices in the network.

Communication Cost: It indicates how many packets are required for replication. In MANETs, each node communicates other nodes in a multi-hop
fashion. We measure the number of hops that are required for replica
allocation.

We investigate the trade-off between file availability and the costs for storage
and communication, and compare the performance of CSRP with other replication schemes.

5.3   File Availability

The primary purpose of data replication is to improve file availability. It indicates
how many files are available compared to the total amount of files stored. File
availability A^t is represented in Eq. 3. In MANET, files can be disappeared by
nodes' leave. Thus, file availability is an essential criterion to evaluate replication
protocols.

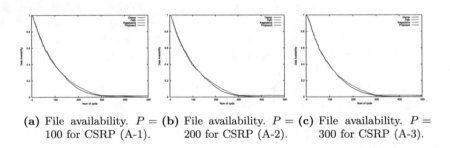

(a) File availability. $P =$ (b) File availability. $P =$ (c) File availability. $P =$
100 for CSRP (A-1).   200 for CSRP (A-2).   300 for CSRP (A-3).

Fig. 1. File availability of different replication protocols. File access obeys Normal
distribution.

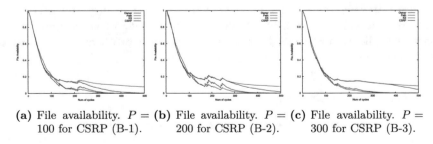

(a) File availability. $P =$ (b) File availability. $P =$ (c) File availability. $P =$
100 for CSRP (B-1).   200 for CSRP (B-2).   300 for CSRP (B-3).

Fig. 2. File availability of different replication protocols. File access obeys Pareto
distribution.

Figure 1 and 2 show the availability of files of different protocols in different access models. Each of them includes three results correspond different configurations of $P = 100$, 200 and 300 in CSRP, respectively.

According to Fig. 1 and 2, CSRP and KS are efficient for improving the availability of low demand files. KS protocol is an ideal protocol for replicating low demand files because it replicates files without considering the storage consumption of each node. CSRP drops the availability from 100 cycles after P. Although some file archiving system is necessary, CSRP is able to get closer to KS protocol regarding the availability by configuring P.

5.4   Success Ratio of File Requests and Access Efficiency

Table 1 to 6 show the success ratio of file requests and the average hops from the source node to the destination that holds the target file.

According to the results, the success ratio of file requests is high on both CSRP and KS protocol regardless of the file access distribution. The efficiency of CSRP and KS protocol regarding the success ratio and the access efficiency is emphasized in the situation that most of the files are low demand (Scenario B).

CSRP is able to get closer to KS protocol regarding the average hops by configuring P. CSRP with $P = 300$ is almost the same as KS protocol (see Table 5 and 6).

Table 1. Scenario A-1 $P = 100$ for CSRP.

Protocol	Success ratio	Avg. hops
Owner	81%	7.07
Path	86%	6.36
KS	100%	5.78
CSRP	94%	6.62

Table 2. Scenario B-1. $P = 100$ for CSRP.

Protocol	Success ratio	Avg. hops
Owner	52%	13.05
Path	58%	11.92
KS	100%	8.52
CSRP	76%	10.33

Table 3. Scenario A-2. $P = 200$ for CSRP.

Protocol	Success ratio	Avg. hops
Owner	83%	7.41
Path	85%	6.56
KS	100%	5.72
CSRP	100%	5.74

Table 4. Scenario B-2. $P = 200$ for CSRP.

Protocol	Success ratio	Avg. hops
Owner	58%	12.13
Path	64%	10.39
KS	100%	8.32
CSRP	94%	9.52

Table 5. Scenario A-3. $P = 300$ for CSRP.

Protocol	Success ratio	Avg. hops
Owner	83%	6.99
Path	87%	6.34
KS	100%	5.78
CSRP	100%	5.67

Table 6. Scenario B-3. $P = 300$ for CSRP.

Protocol	Success ratio	Avg. hops
Owner	50%	12.17
Path	55%	11.41
KS	100%	8.48
CSRP	99%	8.55

5.5   Storage Cost

The storage cost is crucial in MANET. Improvement of file availability usually imposes storage costs because file availability is proportional to the number of replicated files. In other words, there exists a tradeoff between them. We measure the average residual of storage of each node and the cumulative storage usage.

Figure 3 to 4 show the cumulative storage occupancy required by replication protocols with different access models.

Path replication consumes the storage compared to Owner replication. CSRP and KS protocol are in between them in Scenario A (see Fig. 3a, 3b, 3c). On the other hand, CSRP and KS protocol tend to increase the cumulative storage occupancy in Scenario B because the ratio of low demand files is larger than that in Scenario A. Obviously, CSRP stops increasing the storage occupancy because it quits replicating a file in P cycles from the last request.

There is a tradeoff between storage cost and file availability. According to the simulation result, we observed that CSRP requires 1.5 times the storage cost to improve file availability by 46% in Scenario B.

We show the guideline to configure the parameter P in CSRP by the measurement of the storage cost.

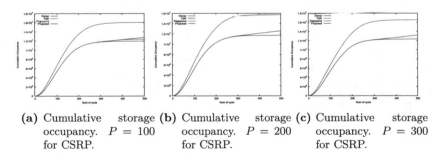

(a) Cumulative   storage occupancy. $P = 100$ for CSRP.

(b) Cumulative   storage occupancy. $P = 200$ for CSRP.

(c) Cumulative   storage occupancy. $P = 300$ for CSRP.

Fig. 3. File availability of different replication protocols. File access obeys Normal distribution.

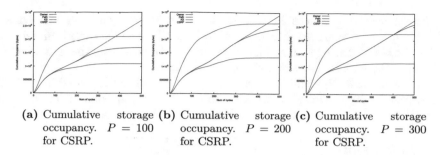

(a) Cumulative storage occupancy. $P = 100$ for CSRP.

(b) Cumulative storage occupancy. $P = 200$ for CSRP.

(c) Cumulative storage occupancy. $P = 300$ for CSRP.

Fig. 4. File availability of different replication protocols. File access obeys Pareto distribution.

5.6   Communication Cost

We compare the cost for communication on replication in four protocols including CSRP.

Mobile devices mostly consume their battery power through communication. Thus, this cost is also crucial in MANETs in terms of energy consumption.

Figure 5 to 6 show the cumulative storage occupancy required by replication protocols with different access models.

The communication cost of CSRP and KS protocol in Scenario B starts increasing earlier than that in Scenario A. It means that the ratio of low demand files influences the communication cost.

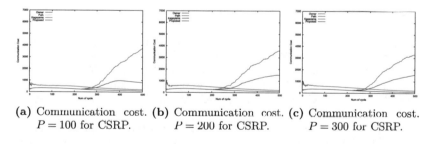

(a) Communication cost. $P = 100$ for CSRP.

(b) Communication cost. $P = 200$ for CSRP.

(c) Communication cost. $P = 300$ for CSRP.

Fig. 5. Communication cost of different replication protocols.

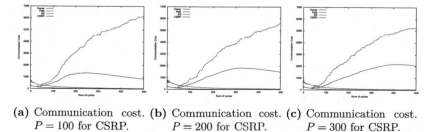

(a) Communication cost. $P = 100$ for CSRP.

(b) Communication cost. $P = 200$ for CSRP.

(c) Communication cost. $P = 300$ for CSRP.

Fig. 6. Communication cost of different replication protocols.

Although both protocols have a substantial cost for replicating low demand files, KS protocol aggressively replicates such files in a whole execution, whereas CSRP stops replicating if there is no request for P cycles from the latest request for saving the cost for storage and communication.

Although CSRP is slightly worse than KS protocol in file availability, CSRP is more than 60% efficient than KS protocol regarding communication cost, especially in the situation that most of files are in low demand.

6 Conclusion

We measured file availability, the success ratio of file requests, the average number of hops, storage cost, and communication cost of CSRP with the preservation period P. We also compared CSRP with other replication protocols regarding those criteria and discuss the tradeoff between file availability, file access efficiency, and the costs for storage and communication.

According to the simulation results, the gap between CSRP and KS protocol regarding file availability, the success ratio of file requests, and the average number of hops reduce with preserving the superiority of CSRP regarding communication efficiency if P is adequately configured. Although the storage efficiency of CSRP reduces to improve file availability, the cumulative storage occupancy of CSRP is bounded.

Since CSRP could improve file availability and file access efficiency without imposing high storage and network cost, it is a suitable replication protocol for low demand files in MANET.

References

1. Ali, A., Latiff, L.A., Fisal, N.: GPS-free indoor location tracking in mobile ad hoc network (MANET) using RSSI. In: 2004 RF and Microwave Conference (IEEE Cat. No.04EX924), pp. 251–255 (2004). https://doi.org/10.1109/RFM.2004.1411119
2. Briesemeister, L., Hommel, G.: Localized group membership service for ad hoc networks. In: Proceedings. International Conference on Parallel Processing Workshop, pp. 94–100 (2002)
3. Cohen, E., Shenker, S.: Replication strategies in unstructured peer-to-peer networks. SIGCOMM Comput. Commun. Rev. **32**(4), 177–190 (2002). https://doi.org/10.1145/964725.633043, http://doi.acm.org/10.1145/964725.633043
4. He, X., Wang, F., Wang, Y., XS., Y.: Nature-inspired algorithms and applied optimization. In: Studies in Computational Intelligence, vol. 744. Springer (2018)
5. Jelasity, M., Montresor, A., Jesi, G.P., Voulgaris, S., Arteconi, S., Hales, D., Marcozzi, A., Picconi, F.: Peersim (2016). http://peersim.sourceforge.net/
6. Kageyama, J., Shibusawa, S.: Replication that prevents low-demand files from disappearing in P2P network. In: Proceedings of The First Forum on Data Engineering and Information Management (2009) (in Japanese)
7. Kurokawa, T., Hayashibara, N.: Data replication based on cuckoo search in mobile ad-hoc networks. In: Barolli, E.T., Hellinckx, L.P. (eds.) Proceedings of International Conference on Advances on Broad-Band Wireless Computing, Communication and Applications (BWCCA 2019), Lecture Notes in Networks and Systems, vol. 97, pp. 199–209 (2019). https://doi.org/10.1007/978-3-030-33506-9_18

8. Liu, J., Sacchetti, D., Sailhan, F., Issarny, V.: Group management for mobile ad hoc networks: Design, implementation and experiment. In: Proceedings of the 6th International Conference on Mobile Data Management. Association for Computing Machinery, New York, NY, USA (2005). https://doi.org/10.1145/1071246.1071276

9. Lv, Q., Cao, P., Cohen, E., Li, K., Shenker, S.: Search and replication in unstructured peer-to-peer networks. In: Proceedings of the 16th International Conference on Supercomputing, ICS 2002, New York, NY, USA, pp. 84–95. ACM (2002). https://doi.org/10.1145/514191.514206, http://doi.acm.org/10.1145/514191.514206

10. Osman, H., Taylor, H.: Managing group membership in ad hoc m-commerce trading systems. In: 2010 10th Annual International Conference on New Technologies of Distributed Systems (NOTERE), pp. 173–180 (2010)

11. Sakano, T., Kotabe, S., Komukai, T., Kumagai, T., Shimizu, Y., Takahara, A., Ngo, T., Fadlullah, Z.M., Nishiyama, H., Kato, N.: Bringing movable and deployable networks to disaster areas: development and field test of MDRU. IEEE Network **30**(1), 86–91 (2016). https://doi.org/10.1109/MNET.2016.7389836

12. Shinki, K., Nishida, M., Hayashibara, N.: Message dissemination using lévy flight on unit disk graphs. In: Proceedings of the 31st IEEE International Conference on Advanced Information Networking and Applications (AINA 2017). Taipei, Taiwan ROC (2017)

13. Yang, X., Deb, S.: Cuckoo search via lévy flights. In: 2009 World Congress on Nature Biologically Inspired Computing (NaBIC), pp. 210–214 (2009). https://doi.org/10.1109/NABIC.2009.5393690

A Study on Comparative Evaluation of Credit Card Fraud Detection Using Tree-Based Machine Learning Models

Thitiwat Ruangsakorn[1]([✉]) and Song Yu[2]

[1] Graduate School of Engineering, Fukuoka Institute of Technology, 3-30-1 Wajiro-Higashi, Higashi-Ku, Fukuoka 811-0295, Japan
mjm19202@bene.fit.ac.jp
[2] Department of System Management, Fukuoka Institute of Technology, 3-30-1 Wajiro-Higashi, Higashi-Ku, Fukuoka 811-0295, Japan
song@fit.ac.jp

Abstract. Credit card fraud is a severe problem that distresses financial companies and cardholders around the world and is becoming more and more serious along with the development of technology. The loss every year due to these fraudulent acts is billions of dollars.

Fraud detection has been an interesting topic in machine learning. In this study, we focus on the comparative evaluation of results by using the tree-based machine learning models (decision tree, random forest, and XGBoost) to detect fraudulent card behavior. In addition, we apply the SMOTE technique to handle imbalance data. Numerical tests show that the accuracy for decision tree, random forest and XGBoost are 96.82%, 97.06%, and 98.35%, respectively. Hence, we conclude that XGBoost performs superior to the other algorithms.

1 Introduction

In the last decade, the rapid growth of internet makes it more convenient for users to enjoy the advantages of e-commerce and online transaction. On the other hand, credit card fraud is growing and becoming a serious problem. Fraudulent behavior can be found in various areas of financial transactions such as credit cards, insurance fraud, tax evasion, etc. Some financial organizations lose 5% of revenue to fraud each year.

Detecting fraud is usually a challenging task because of the following features of the fraud: uncommonness, concealment, change over time and organization. The criminals committing fraud are the minority and they try to blend in and conceal their activities. They constantly try to find new methods to avoid getting caught and change their behavior over time. Moreover, the fraudsters are working together and organize the crime in networks. Hence, it is making fraud harder to detect. For that reason, distinguishing fraudulent transactions using traditional techniques is tedious and uneconomical [1].

Under such circumstance, machine learning technology is considered as one of the most hopeful and powerful techniques in handling fraud identification. Some previous studies formulate fraud detection problems as classification models of machine learning.

L. Barolli et al. (Eds.): EIDWT 2021, LNDECT 65, pp. 212–219, 2021.
https://doi.org/10.1007/978-3-030-70639-5_20

In [2], the authors conducted comparative analyses of fraudulent activity identification on credit cards using support vector machine, k-nearest neighbor, naïve bayes, and logistic regression.

Recently, there are new machine learning models that improve classifying performance in classification problems. One of them is XGBoost [3], a tree-based model.

An issue with imbalanced classification is that there are a couple of instances of the minority class for a model to adequately learn with the decision boundary. Synthetic Minority Over-sampling Technique (SMOTE) was introduced to handle this imbalanced data [4].

This study applies those new techniques to predict whether an online transaction is fraudulent or non-fraudulent. We compare the performances of 3 tree-based machine learning models (decision tree, random forest, and XGBoost) and check whether the performance can be improved in detecting fraud. The simulation results show that XGBoost outperforms to the other algorithms.

The rest of this paper is organized as follows: Sect. 2 illustrates the 3 tree-based machine learning models for detecting fraud. Then, we conduct simulation to detect fraudulent transaction using the 3 models and logistic regression, and compare model performances in Sect. 3. Finally, the conclusion and future work of this study are discussed in Sect. 4.

2 Models

There are various kind of models that can deal with the classification problems. Since the tree-based models have the advantage of data interpretation, in this study, we consider using tree-based classification models: decision tree, random forest, and XGBoost to solve the fraud problem. We will briefly introduce the 3 models in this section.

2.1 Decision Tree

Decision tree is the basis of tree-based machine learning models, and is widely used in machine learning classification problems. The model arranges information in a tree-like structure as per the name.

As shown in Fig. 1, a decision tree consists of nodes, edges, and leaves. Nodes are for assessing attributes, edges are for branching by the value of the chosen attribute, and leaves label classes where a specific class is attached to each leaf. To determine which attribute is the best for classifying the data, we considered to use Gini impurity to measure the impurity score of classes for each node. A lower impurity score indicates the better classification ability of the selected attribute. The equation of Gini impurity is shown as follows:

$$Gini = 1 - \sum_{k=1}^{K} p_k^2 \tag{1}$$

where K is the class of the target variable, and p_k is the proportion of the same class inputs which present in a particular group. Decision trees and visual cryptography was applied on credit card fraud detection problems by implementation of this approach in fraud detection systems [5].

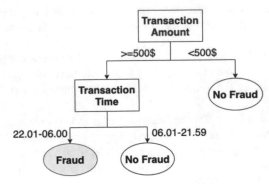

Fig. 1. Example of a decision tree

Figure 1 illustrates a decision tree with the concept of detecting the likely fraudulent transactions, where the attribute of the transaction amount was chosen to be the initiate node by the impurity score. A transaction amount of more than $500 tends to be a fraudulent transaction and vice versa. Then, the next nodes were added by repeating the step of the impurity score calculation. Therefore, with this process, the transaction with an amount of more than $500 and an action time between 22.01–06.00 will classify as a likely fraudulent transaction.

2.2 Random Forest

Random forest comprises thousands of individual decision trees that work as an ensemble method to create a more accurate and stable prediction on a new sample. As shown in Fig. 2, the learning process of random forest is initiated by randomly sampling features from a dataset and forming a tree for each sample. After training each individual tree to predict the target feature, all results are combined to make the prediction by majority voting.

Random forest has the ability to detect fraudulent transactions with more accuracy than expected. A study developed to detecting credit card fraud using random forest gave the accuracy as more than 90% [6].

2.3 XGBoost

XGBoost stands for eXtream Gradient Boosting, which is also a tree-based algorithm. Boosting is an ensemble method with the essential goal of reducing bias and variance. The process of the learning method is shown in Fig. 3. The objective is to make weak trees successively by having each new tree center around the weakness (misclassified data) of the past one. After a weak learner is added, the feature weights are readjusted. The entire shaping of a strong model after convergence is due to the auto-correction after every new learner is added.

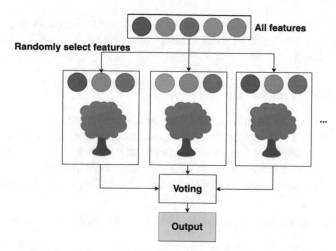

Fig. 2. Process of the learning method for random forest

For detecting credit card fraud, some studies used the XGBoost model to detect customer transactions and outperformed the logistic regression, support vector machine, and random forest [7].

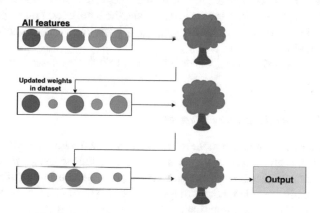

Fig. 3. Process of the learning method for XGBoost

3 Simulation Results

In this section, we conduct simulations using the models introduced in the previous section. The first part describes the dataset for training the models. The procedure of preparing data for the experiment is also explained. Then, the second part of this section presents the results of the simulations, and compare the performances of the models.

The simulations were performed by the Python programming language with a Jupyter notebook environment. The device has an Intel Core i5-8259U CPU @ 2.3 GHz processor and 8.0 GB of RAM.

3.1   Dataset and Prepocessing

The dataset for the tests were acquired from the Kaggle website [8]. It consists of 590, 540 transactions and 433 features for each transaction. An example of data is shown in Fig. 4. Other features in the dataset contain data such as datetime, transaction amount, product code, address, and device type. The most important target feature in the dataset is "IsFraud" which represents whether the transaction is fraud or non-fraud.

TransactionID	isFraud	TransactionDT	TransactionAmt	ProductCD	card1	card2	card3	card4	card5	card6
2987000	0	86400	68.5	W	13926	NaN	150.0	discover	142.0	credit
2987001	0	86401	29.0	W	2755	404.0	150.0	mastercard	102.0	credit
2987002	0	86469	59.0	W	4663	490.0	150.0	visa	166.0	debit

Fig. 4. Example of the dataset

Most of the feature columns in the dataset are categorical data. Since machine learning models need to process the input as integer data, the label encoding is used to convert the categorical data into integers. In this technique, each label was assigned to a unique integer based on alphabetical ordering.

After several numerical tests, we decide to split 80% of dataset into a training set and 20% into a test set.

3.2   Simulation Results and Comparison

To verify the performance of the models, the following two metrics are considered. The first is accuracy, the proportion of the number of predictions to the number of samples. And then, a confusion matrix, which is a set of performance indicators for machine learning classification problem where output can be two or more classes. Table 1 shows a table with four different combinations of predicted and actual values.

Based on the values of the four metrics: accuracy, sensitivity, specificity, and precision, the model performance could be evaluated. With the values from Table 1, these four metrics can be calculated by the equations below:

Table 1. Confusion matrix table

Actual sample	Predicted no	Predicted yes
No (Not Fraud)	True Negative	False Positive
Yes (Fraud)	False Negative	True Positive

$$Accuray = \frac{TP + TN}{TP + TN + FP + FN} \qquad (2)$$

$$Sensitivity(recall) = \frac{TP}{TP + FN} \qquad (3)$$

$$Specificity = \frac{TN}{TN + FP} \qquad (4)$$

$$Precision = \frac{TP}{TP + FP} \qquad (5)$$

In the four metrics above, accuracy and sensitivity are more important metrics than the other two to evaluate the model performance. Accuracy indicates the total ratio of correct predictions. On the other hand, sensitivity is the proportion of true positive to sum of true positive and false negative, which means sensitivity measures how many fraudulent transactions were detect as fraud.

Table 2. Comparison of results for the tree-based models

Metrics	Classifiers (%)		
	Decision tree	Random forest	XGBoost
Accuracy	96.82	97.06	98.35
Sensitivity	59.64	21.31	57.35
Specificity	98.20	99.87	99.89
Precision	55.29	87.51	95.01
Computation time (minutes)	2.03	95.16	3.14

In Table 2, the results show that (1) all the three models obtain satisfactory accuracy and specificity. (2) However, sensitivity is quite low in every model. (3) XGBoost has the best performance across the evaluation metrics except for the value of sensitivity. (4) Computation time for Random forest is much longer than the other two models.

As mention in Sect. 1, only a very small portion of the transaction is fraudulent. Such an imbalance of data is considered to be the reason of the poor performance in sensitivity. To improve the value to sensitivity, we introduce the SMOTE (Synthetic Minority Over-sampling Technique) to handle the imbalanced data. SMOTE synthesizes new samples of the minority class by choosing data that is close in the component space, attracting a line between the data in the element space, and drawing another example at a point along that line.

As shown in Table 3, sensitivity is improved greatly in all the models. And the computation times for Decision tree and XGBoost are still moderate ones.

Table 3. Comparison of results for tree-based models with the SMOTE technique applied

Metrics	Classifiers (%)		
	Decision tree	Random forest	XGBoost
Accuracy	97.79	97.82	98.03
Sensitivity	98.01	96.81	96.68
Specificity	98.00	96.88	99.38
Precision	97.58	98.80	99.36
Computation time (minutes)	4.27	193.42	6.49

In the related previous work, the logistic regression model was performed with their dataset and the results show that logistic regression had the best performance. Therefore, we kept attention on this model. Nevertheless, after being performed on our dataset with the SMOTE technique applied, the performance of logistic regression did not go as well as expected. The results are shown in Table 4 and Table 5.

Table 4. Comparison of results for tree-based models and logistic regression

Metrics	Classifiers (%)			
	Decision tree	Random forest	XGBoost	Logistic regression
Accuracy	96.82	97.06	98.35	96.40
Sensitivity	59.64	21.31	57.35	0.11
Specificity	98.20	99.87	99.89	99.99
Precision	55.29	87.51	95.01	45.46
Computation time (minutes)	2.03	95.16	3.14	1.39

Table 5. Comparison of results for tree-based models and logistic regression with the SMOTE technique applied

Metrics	Classifiers (%)			
	Decision tree	Random forest	XGBoost	Logistic regression
Accuracy	97.79	97.82	98.03	67.43
Sensitivity	98.01	96.81	96.68	53.77
Specificity	98.00	96.88	99.38	81.09
Precision	97.58	98.80	99.36	74.00
Computation time (minutes)	4.27	193.42	6.49	3.20

4 Conclusion and Future Work

In this paper, we applied decision tree, random forest, XGBoost, and logistic regression to detect fraudulent credit card transactions. From the comparison of simulation results, we conclude as follows.

- The XGBoost model outperforms the decision tree and other classification learning methods for detecting credit card fraud.
- The SMOTE method provides a solution for imbalance data in prediction.

For future work, we would like to utilize the other learning methods to improve performance in detecting fraudulent credit card transactions.

References

1. Karim, A.Z., Said, J., Bakri, H.M.H.: An exploratory study on the possiblity of assets misappropriation among royal Malaysian police officials. In: International Accounting and Business Conference (IABC), pp. 625–631 (2015)
2. Adepoju, O., Wosowei, J., Lawte, S., Jaiman, H.: Comparative evaluation of credit card fraud detection using machine learning techniques. In: Global Conference for Advancement in Technology (GCAT) (2019)
3. Chen, T., Cuestrin, C.: XGBoost: a scalable tree boosting system. In: 22nd ACM SIGKDD International Conference, pp. 785–794 (2016)
4. Chawla, N.V., Bowyer, K.W., Hall, L.O., Kegelmeyer, W.P.: SMOTE: systhetic minority over-sampling technique. J. Artif. Intell. Res. 16, 321–357 (2002)
5. Patil, S., Somavanshi, H., Gaikwad, J., Deshmane, A., Badgujar, R.: Credit card fraud detection using descision tree induction algorithm. Int. J. Comput. Sci. Mob. Comput. (IJCSMC) 4, 92–95 (2015)
6. Kumar, S.M., Soundarya, V., Kavitha, S., Keerthika, E.S., Aswini, E.: Credit card fraud detection using random forest algorithm. In: 3rd International Conference on Computing and Communications Technologies (ICCCT), pp. 149–153 (2019)
7. Zhang, Y., Tong, J., Wang, Z., Gao, F.: Customer transaction fraud detection using Xgboost model. In: International Conference on Computer Engineering and Application (ICCEA), pp. 554–558 (2020)
8. IEEE Computational Intelligence Society: IEEE-CIS Fraud Detection (2020). On: https://www.kaggle.com/c/ieee-fraud-detection

Considering Cross-Referencing Method for Scalable Public Blockchain

Takaaki Yanagihara and Akihiro Fujihara[✉]

Chiba Institute of Technology, 2-17-1 Tsudanuma, Narashino, Chiba 275-0016, Japan
s1972042YU@s.chibakoudai.jp, akihiro.fujihara@p.chibakoudai.jp

Abstract. We propose a cross-referencing method for enabling multiple P2P network domains to manage their own public blockchains and periodically exchange among one another the state of the latest fixed block in the blockchain with hysteresis signatures. We evaluated the effectiveness of the proposed method from a theoretical viewpoint by introducing the tamper-resistance-improvement ratio. We confirmed that the tamper-resistance-improvement ratio was high, meaning that it is possible to achieve a more scalable public blockchain balanced with decentralization and tamper resistance with the proposed method.

1 Introduction

Since Bitcoin [1] appeared in 2009, blockchain technology has been gaining considerable public attention. The word "blockchain" was born from a dialogue between the Bitcoin inventor, Satoshi Nakamoto, and a cryptographer and Bitcoin developer Hal Finney on the cryptography mailing list [2]. Blockchain is a write-once-read-many database with a block structure and works as a timestamp server that uses a cryptographic hash function to link blocks in a long chain in a time-series order. This database is not a new idea as noted in the Bitcoin white paper [3], but blockchain has become iconic image for related technology. Bitcoin discloses all transactions written in a blockchain. Thus, it becomes possible for a large indefinite number of nodes to build consensus on the transactions. Because this consensus algorithm was essentially a new idea, compared with previous algorithms in distributed systems, it is called Nakamoto Consensus.

In Bitcoin and other public blockchains, core nodes that manages a blockchain frequently transfer messages to share transaction and block data with each other on the peer-to-peer (P2P) networks worldwide. There is a delay of several to several tens of seconds when blocks are transferred distant nodes. This delay requires the blockchain system to lengthen the average block generation time to avoid branching the blockchain. In Bitcoin, the average block generation time is controlled by the difficulty of the proof of work (PoW) [4] around 10 min, which is sufficiently longer than the latency for a block to be transferred to most nodes in the network [5]. Due to this long waiting time for block generation, the transaction volume that can be processed in a unit time decreases drastically, which is known as the blockchain scalability problem.

L. Barolli et al. (Eds.): EIDWT 2021, LNDECT 65, pp. 220–231, 2021.
https://doi.org/10.1007/978-3-030-70639-5_21

Several technologies are used to solve this problem. Off-chain scaling technologies, such as Lightning Network [6], are well known. However, off-chain transactions do not leave records on the blockchain, which goes against the original idea of Bitcoin to maximize their auditability. Bitcoin has been used as a payment means of conducting illegal transactions in darknet markets. There have been many reports on the managers and users of these markets being arrested [7–9]. These arrests are attributable to Bitcoin's disclosure of all transactions with tamper resistance, which are available as legal evidence. If off-chain scaling technologies become widespread, the number of transactions that cannot be audited by governments can be easily created and off-chain services might become hotbeds of illegal transactions in darknet markets and money laundering by criminal organizations. Thus, when we consider the use of blockchain technology on the basis of law and ethics, it is ultimately necessary to solve the scaling problem on-chain.

One method of solving this problem involves autonomously forming domains between core nodes that are geographically close to each other and managing a unique blockchain for each domain [10–12]. This method, however, has a problem in that the decentralization and tamper resistance of blockchains degraded as the number of core nodes in each domain decreases.

We propose a cross-referencing method as a solution to the scalability problem with which multiple domains periodically share their blockchain states via central core nodes (CCNs) with each other and save them as a hysteresis signature [13] into the cross-reference part of the latest block in their blockchains. We theoretically evaluated how much the tamper-resistance-improvement ratio R increases. We conducted Monte Carlo simulations to compare the hash powers when our cross-referencing method is and is not used. The R was high, meaning that it is possible to archive a more scalable public blockchain balanced with decentralization and tamper resistance with the proposed method.

2   Related Work

There are two methods that are used to solve the scalability problem on-chain: to increase the block size or shorten the time interval of block generation. The former is being tested on the Bitcoin SV (BSV) Scaling Test Network (STN) [14, 15]. In Bitcoin Core (BTC), the maximum block size is 1 MB and the transaction processing speed is said to be 7 transactions per second (TPS) on average. In BSV STN, however, the average size of generated blocks reaches 98.01 MB and the transaction processing speed is 632 TPS on the latest 144 blocks (when we accessed on August 8, 2020) by removing the upper limit of the block size [15].

The latter method is also being used with bloXroute [16]. The key idea of bloXroute is to create a new network (Layer 0) called a Blockchain Distribution Network on top of the peer-to-peer (P2P) network (Layer 1) to shorten the block-propagation time. In Bitcoin, the time interval of block generation is about ten minutes, which needs to be controlled to be long enough compared with the block-propagation time on the P2P network. Therefore, the faster the block propagation becomes, the shorter the time interval of block generation will be.

An hysteresis signature [13] enhances the tamper resistance of ordinary electronic signatures by adding a nested structure with the other electronic signatures. A typical structure of a hysteresis signature is shown in Fig. 2. A nested structure is naturally created by signing the content including the previous signature. By repeating the nested structure, the electronic signatures are chained to create a time-series context. For example, let the previous hysteresis signature be S_{n-1} and let the content to be signed be the block data of m domains in total $(B_{D_1}, \cdots, B_{D_m})$. In this case, the hysteresis signature is created by signing the concatenation of the summary of the previous hysteresis signature $H(S_{n-1})$ and the hashed content $H(B_{D_1}), \cdots, H(B_{D_m})$. The created signature S_n is also added as a part of the hysteresis signature, as shown in Fig. 2. Ordinary electronic signatures are tampered with due to leakage of the private key, and it is also impossible to detect tampering. In hysteresis signatures, on the other hand, since the signature is signed in the nested structure, it is necessary to tamper with all nested signatures after the tampered content, which makes tampering significantly more difficult. This is similar to the block structure in the public blockchain. If a contradiction in a hysteresis signature is found, it is possible to detect that the content has been partially tampered.

BBc-1 [17] is a type of distributed ledger technology, in which a distributed system manages a consistent ledger, but it does not handle any public blockchain. The system stores private transaction data in a tamper-resistant manner by using a hysteresis signature. There is a reference implementation of the node, and the effectiveness of the system is confirmed. In BBc-1, all the transaction data have a hysteresis signature, which is exchanged across domains to increase tamper resistance. In addition, the energy cost is low compared with a typical public blockchain, such as Bitcoin, because of the absence of a PoW. However, it is difficult to theoretically estimate how much tamper resistance is improved by introducing a hysteresis signature.

Atomic Swap is a technique that allows the exchange of cryptocurrencies recorded on different blockchains without needing to trust any third party [18]. This technique is also useful for coexisting multiple blockchains and their native cryptocurrencies.

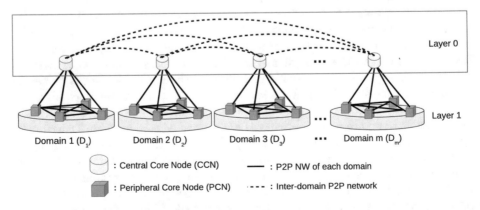

Fig. 1. Two-layer P2P network used in this study

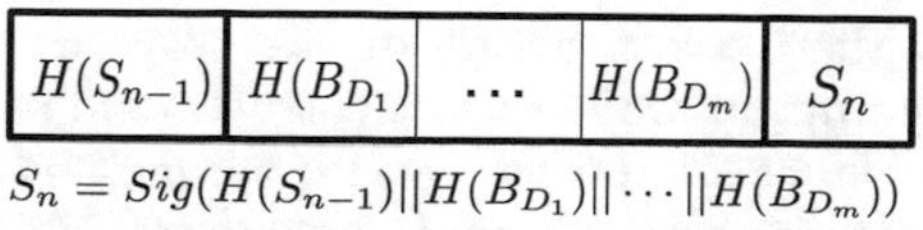

$$\boxed{H(S_{n-1})}\ \boxed{H(B_{D_1})}\ \cdots\ \boxed{H(B_{D_m})}\ \boxed{S_n}$$

$$S_n = Sig(H(S_{n-1})||H(B_{D_1})||\cdots||H(B_{D_m}))$$

Fig. 2. Hysteresis signature

3 Proposed Method

The structure of the P2P network used in this study is shown in Fig. 1. The whole networks consists of two layers of P2P networks, *i.e.*, layers 0 and 1. In Layer 1, multiple P2P networks of typical public blockchains that share transaction and block data are assumed. Each P2P network in Layer 1 is called a *domain*, and it is assumed that there is a set of core nodes in each domain. There are two types of core nodes: CCNs and peripheral core node (PCNs). It is assumed that at least one CCN is selected as a leader in each domain beforehand. In this study, we assumed that the number of domains is m and the CCNs of multiple domains $(D_1, D_2, \cdots, D_m)$ have a prior agreement to share the block data and domain hysteresis signatures to use our cross-referencing method. In Layer 0, the CCNs are connected to each other to form another P2P network, which is disconnected from that of Layer 1. For simplicity, we consider the case in which the number of CCNs for each domain is one. it is also possible to generalize the case in which the number of CCNs is more than one. Note that the addition of Layer 0 over Layer 1 is common with bloXroute [16], but the difference is that Layer 0 is also another P2P network with our proposed method, while Layer 0 with bloXroute is a faster network transport layer of both the transaction and block data.

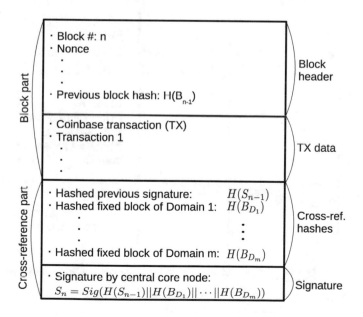

Fig. 3. Block structure of proposed method

The structure of the block structure of the proposed method is shown in Fig. 3. The difference from an ordinary block structure in public blockchains and that with our method is that the structure with our method has a cross-reference part. In this part, an hysteresis signature is created by signing the concatenation of the summary of the previous hysteresis signature $H(S_{n-1})$ and the hashed content $H(B_{D_1}), \cdots , H(B_{D_m})$. The created signature S_n is the same as that shown in Fig. 2. The block data with the cross-reference part are shared between CCNs via the P2P network in Layer 0.

The timeline of the proposed cross-referencing method in a normal case meaning that no stop failure occurs in CCNs is shown in Fig. 4. The timeline is divided into three phases. The details of the execution flow of each phase are explained below.

1. In Phase 1, a CCN in a domain notifies the other CCNs to start cross referencing by sending a message. Then, each CCN transfers a message including the l-confirmed block to other CCNs, where l is a positive integer and the l-confirmed block means the block approved l blocks before the latest block. Phase 1 finishes if all the CCNs collect all the l-confirmed block data by sharing them with each other.
2. In Phase 2, each CCN first generates a hysteresis signature, as shown in Fig. 2. The CCN then sends a request message with the hysteresis signature to PCNs in the same domain to mine the block, having the cross-reference part. After independently mining the latest block, a PCN sends a message with the mined block back to the CCN, and the CCN checks whether the block is properly

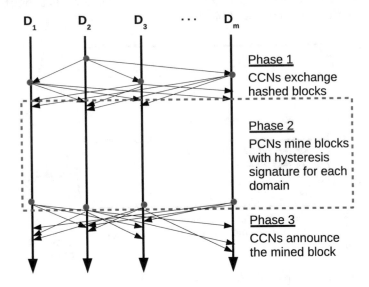

Fig. 4. Timeline of cross-referencing method in normal case where no stop failure occurs in CCNs

mined. If the block is not properly mined, the CCN waits until the properly mined block is received. If the block is properly mined, Phase 2 finishes.
3. In Phase 3, each CCN broadcasts the mined block to announce that the cross-referencing method has successfully finished.

We designed a distributed algorithm [19] for our cross-referencing method. The details are shown in Fig. 5. We also considered a distributed algorithm that is tolerant to t-stop failures (CCNs in m domains do not respond because they experience stop failure or reject the execution of cross referencing) as shown in Fig. 6.

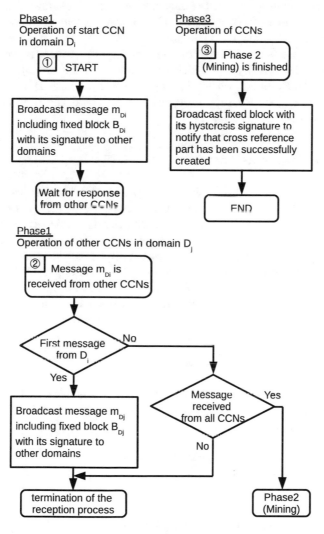

Fig. 5. Algorithm 1 (Phase 2 is omitted because it consists of usual mining process on Layer 1)

There are three assumptions with these algorithms to work properly.

1. The P2P network in Layer 0 is synchronous and its structure should be a complete graph.
2. All the CCNs are reliable meaning that they execute the cross-referencing method following the algorithm to share the requested block data with each other.
3. In Algorithm 1, none of the CCNs experience any stop failure in Algorithm 2, they allow t-stop failures, which means that our cross-referencing method works even if at most t CCNs do not cooperate to share the block data.

The efficiency of these distributed algorithms can be evaluated by measuring communication and time complexity. The communication complexity in Algorithm 1 is $O(m^2 \cdot b)$, where m is the number of CCNs (equivalent to the number of domains), and b is the bit size of the message to send between CCNs. The communication complexity in Algorithm 2 is $O((t+1)(m^2 \cdot b))$. Assume that the time complexity in Algorithm 1 is $T_1 + T_2 + T_3$, where $T_i (i = 1, 2, 3)$ is the waiting time taken for Phase i in Algorithm 1. Then, the time complexity in Algorithm 2 is $(t+1)T_1 + T_2 + T_3$.

4   Theoretical Evaluation

Our cross-referencing method should improve the tamper resistance of a blockchain. We defined and theoretically evaluated R. Let the total number of core nodes in the distributed system be N and the hash rate of core node i be h_i, where the hash rate is the number of times that a cryptographic hash function can be computed per unit time. In this case, tamper resistance can be estimated by the maximum hash rate because the node with the highest hash rate has the highest probability of generating blocks, so it continues mining. On the other hand, most nodes with relatively smaller hash rates have a smaller probability of generating blocks, so they tend to quit mining. Therefore, the tamper resistance is proportional to the following value, $i.e.$,

$$\max\{h_1, h_2, \cdots, h_N\}. \tag{1}$$

Suppose that the total number of core nodes in the m-th domain is D_m. The tamper resistance of each domain is given by the maximum hash power of the nodes for each domain, $i.e.$,

$$A_1 = \max\{h_{11}, \cdots, h_{1D_1}\}, \tag{2}$$

$$A_2 = \max\{h_{21}, \cdots, h_{2D_2}\}, \tag{3}$$

$$\vdots$$

$$A_m = \max\{h_{m1}, \cdots, h_{mD_m}\}. \tag{4}$$

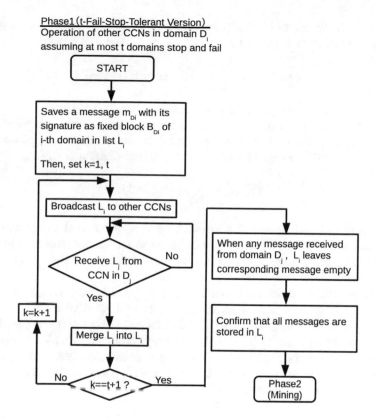

Fig. 6. Algorithm 2 (Phase 1 only; other phases are common to those in Algorithm 1)

In the normal case in which there is only one domain having $N = D_1 + D_2 + \cdots + D_m$ core nodes, the tamper resistance of the domain without our cross-referencing method can be estimated by

$$A = \max\{A_1, A_2, \cdots, A_m\} \tag{5}$$

On the other hand, by applying our cross-referencing method between m domains, tamper resistance can be estimated by the sum of all the highest hash rates in the domains, *i.e.*,

$$B = \sum_{i=1}^{m} A_i, \tag{6}$$

because the cross referencing is recorded in all the blockchains of m domains. To tamper with the block data, it is necessary to remove all hysteresis signatures in the cross-reference parts in the corresponding blockchains of m domains for tampering.

Therefore, the R in domain R_i combined with cross referencing is given by

$$R_i = \frac{B}{A_i} > 1 \qquad (i = 1, \cdots, m). \tag{7}$$

In addition, R, where our cross-referencing method provides the domain with the highest hash rate, is given by

$$(R_i \geq)R = \frac{B}{A} > 1 \qquad (i = 1, \cdots, m).$$ (8)

As described above, it is possible to estimate R for all domains.

To estimate a typical R, we conducted Monte Carlo simulations in which the hash rate h_{ij} (i is the domain number, and j is the serial number of nodes in a domain) is randomly assigned following the Pareto distribution, i.e.,

$$P(h_{ij}) = \frac{\alpha}{h_{ij}^{1+\alpha}}(h_{ij} > 1),$$ (9)

where α is a scale parameter. This distribution is often used to explain wealth distribution in economics. The rate represents the total amount of computational resources, which is proportional to the amount of capital of the miner, so using the Pareto distribution is considered appropriate.

We assume that the total number of core nodes is 10,000 and that the number of core nodes in each domain is uniform, i.e., 10, 100, and 1000 nodes. Example simulation results are shown in Figs. 7 and 8. In general, as m increases, the peak of the histogram of R shifts toward the higher position. It can also be confirmed that the dispersion of the distribution increases as m increases. A comparison of Figs. 7 and 8 indicates that the peak of the distribution of R shifts toward the smaller position as α becomes smaller.

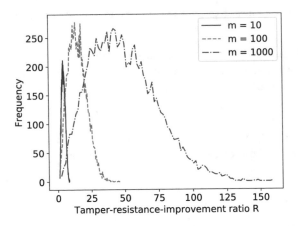

Fig. 7. Simulation results of probability density function of tamper-resistance-improvement ratio R when scale parameter $\alpha = 2$

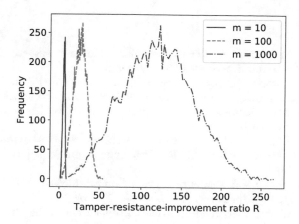

Fig. 8. Simulation results of probability density function of R when $\alpha = 3$

5   Conclusion

We proposed a cross-referencing method for enabling multiple domains of P2P networks to manage their own blockchains and periodically exchange among each other the state of the latest confirmed block in the blockchain with a hysteresis signature.

The effectiveness of the proposed method was examined from the theoretical perspective by introducing the tamper-resistance-improvement ratio R. In general, as the number of domains increases, the peak of the distribution of R shifts toward the higher position. we confirmed that the dispersion of this distribution increases as the number of domains increases.

We assumed that the hash rate obeys the Pareto distribution and the comparison of the scale parameter of the Pareto distribution $\alpha = 2, 3$ showed that as α decreases, the peak of the distribution of R shifts to the smaller position.

We are currently developing a program of CCNs for doing our cross-referencing method as a reference implementation of the communication protocol between CCNs. The program is open to the public at our Github website [20]. Due to the page limitation of this paper, we have not presented experimental results, but we are will present them elsewhere.

For future work, we will determine if the proposed cross-referencing method can be successfully applied to various situations, for example, when the starting CCN for cross referencing is not fixed and frequently changes, and when multiple cross referencing requests are sent from multiple domains at the same time.

Acknowledgements. This work was partially supported by the Japan Society for the Promotion of Science (JSPS) through KAKENHI (Grants-in-Aid for Scientific Research) Grant Numbers 17K00141, 17H01742, and 20K11797.

References

1. Nakamoto, S.: Bitcoin: a peer-to-peer electronic cash system (2008). https://bitcoin.org/bitcoin.pdf. Accessed 15 Nov 2020
2. A record of when the words "block chain" were first used. https://www.metzdowd.com/pipermail/cryptography/2008-November/014827.html. https://satoshi.nakamotoinstitute.org/emails/cryptography/6/#selection-37.36-37.47. Accessed 15 Nov 2020
3. Massias, H., Avila, X.S., Quisquater, J.-J.: Design of a secure timestamping service with minimal trust requirement. In: 20th Symposium on Information Theory in Benelux (1999)
4. Back, A.: Hashcash - A Denial of Service Counter-Measure (2002). http://www.hashcash.org/papers/hashcash.pdf. Accessed 15 Nov 2020
5. Decker, C., Wattenhofer, R.: Information propagation in the bitcoin network. In: IEEE P2P 2013 Proceedings, Trento, pp. 1–10 (2013)
6. Poon, J., Dryja, T.: The Bitcoin Lightning Network: Scalable Off-Chain Instant Payments (2016). https://lightning.network/lightning-network-paper.pdf. Accessed 15 Nov 2020
7. FBI: Manhattan U.S. Attorney Announces Seizure of Additional $28 Million Worth of Bitcoins Belonging to Ross William Ulbricht, Alleged Owner and Operator of "Silk Road" Website (2013). https://archives.fbi.gov/archives/newyork/press-releases/2013/manhattan-u.s.-attorney-announces-seizure-of-additional-28-million-worth-of-bitcoins-belonging-to-ross-william-ulbricht-alleged-owner-and-operator-of-silk-road-website. Accessed 15 Nov 2020
8. The US Department of Justice: AlphaBay, the Largest Online 'Dark Market,' Shut Down (2017). https://www.justice.gov/opa/pr/alphabay-largest-online-dark-market-shut-down. Accessed 15 Nov 2020
9. The US Department of Justice: South Korean National and Hundreds of Others Charged Worldwide in the Takedown of the Largest Darknet Child Pornography Website, Which was Funded by Bitcoin (2019). https://www.justice.gov/opa/pr/south-korean-national-and-hundreds-others-charged-worldwide-takedown-largest-darknet-child. Accessed 15 Nov 2020
10. Fujihara, A.: Proposing a system for collaborative traffic information gathering and sharing incentivized by blockchain technology. In: Advances in Intelligent Networking and Collaborative Systems, pp. 170–182. Springer (2019)
11. Fujihara, A.: PoWaP: proof of work at proximity for a crowdsensing system for collaborative traffic information gathering. Internet Things 10, 100046 (2019)
12. Fujihara, A.: Proposing a blockchain-based open data platform and its decentralized oracle. In: Advances in Intelligent Systems and Computing, vol. 1035, pp. 190–201. Springer (2019)
13. Susaki, S., Matsumoto, T.: Alibi establishment for electronic signatures. Trans. Inf. Process. Soc. Jpn. 43(8), 2381–2393 (2002). (in Japanese)
14. Bitcoin SV (Satoshi Vision). https://github.com/bitcoin-sv/bitcoin-sv. Accessed 15 Nov 2020
15. Bitcoin Scaling Test Network. https://bitcoinscaling.io/. Accessed 15 Nov 2020
16. Klarman, U., Bsu, S., Kuzmanovic, A., Sirer, E.G.: bloXroute: a scalable trustless blockchain distribution network. BloXkroute Labs, Whitepaper, version 1.0 (2018). https://bloxroute.com/wp-content/uploads/2018/03/bloXroute-whitepaper.pdf. Accessed 15 Nov 2020

17. Saito, K., Kubo, T.: BBc-1: Beyond Blockchain One (2018). https://beyond-blockchain.org/public/bbc1-design-paper.pdf. Accessed 15 Nov 2020
18. Atomic swap, Bitcoin Wiki. https://en.bitcoin.it/wiki/Atomic_swap. Accessed 15 Nov 2020
19. Kshemkalyani, A.D., Singhal, M.: Distributed Computing: Principles, Algorithms, and Systems. Cambridge University Press, Cambridge (2011)
20. Yanagihara, T., Fujihara, A.: Experimental implementation of the cross-referencing method for scalable public blockchain. https://github.com/cit-fujihalab/crossref_BC. Accessed 15 Nov 2020

On Reducing Measurement Load on Control-Plane in Locating High Packet-Delay Variance Links for OpenFlow Networks

Nguyen Minh Tri[1]([✉]), Nguyen Viet Ha[1], Masahiro Shibata[1], Masato Tsuru[1], and Akira Kawaguchi[2]

[1] Kyushu Institute of Technology, Kitakyushu, Japan
{tri.nguyen-minh414,nguyen.viet-ha503}@mail.kyutech.jp,
{shibata,tsuru}@cse.kyutech.ac.jp
[2] The City College of New York of The City University of New York, New York, USA
akawaguchi@ccny.cuny.edu

Abstract. We previously proposed a method to locate high packet-delay variance links for OpenFlow networks by probing multicast measurement packets along a designed route and by collecting flow-stats of the probe packets from selected OpenFlow switches (OFSs). It is worth noting that the packet-delay variance of a link is estimated based on arrival time intervals of probe packets without measuring delay times over the link. However, the previously used route scheme based on the shortest path tree may generate a probing route with many branches in a large network, resulting in many accesses to OFSs to locate all high delay variance links. In this paper, therefore, we apply an Eulerian cycle-based scheme which we previously developed, to control the number of branches in a multicast probing route. Our proposal can reduce the load on the control-plane (i.e., the number of accesses to OFSs) while maintaining an acceptable measurement accuracy with a light load on the data-plane. Additionally, the impacts of packet losses and correlated delays over links on those different types of loads are investigated. By comparing our proposal with the shortest path tree-based and the unicursal route schemes through numerical simulations, we evaluate the advantage of our proposal.

1 Introduction

The OpenFlow technology was proposed more than a decade ago and is becoming widespread as a replacement solution for traditional network not only in data centers but also in enterprise networks and wide area networks. The ongoing prevalence of cloud computing and contents delivery networking requires flexible traffic engineering on a network connecting globally-distributed data-centers, which is often centrally managed by OpenFlow [1,2]. By decoupling the

L. Barolli et al. (Eds.): EIDWT 2021, LNDECT 65, pp. 232–245, 2021.
https://doi.org/10.1007/978-3-030-70639-5_22

control-plane and data-plane, OpenFlow lets network operators configure, manage, monitor, secure, and optimize network resources very quickly via dynamic, automated programs. On the data-plane, switches forward packets based on per-flow rules installed and records the statistical information (flow-stats) of each flow. On the control-plane, a controller manages switches in the network by installing appropriate rules into each switch and collecting flow-stats from each switch.

Passive measurement by collecting per-link (from a physical input/output port of switch) traffic information via SNMP is commonly used to monitor and detect performance degraded links in traditional networks. However, in the edge-cloud computing for emerging IoT technologies, since a "link" between two nodes is not always physical but sometimes virtual, a per-link passive measurement cannot detect the performance degeneration of such virtual links. In OpenFlow network, by collecting the flow-stats from switches through the OpenFlow monitoring messages or by monitoring the OpenFlow-standard operating messages themselves, passive measurement approaches can operate in a per-flow manner without extra loads on the data-plane. However, there is a trade-off between the measurement accuracy and the load incurred on the control-plane, and thus some research efforts have been made. For example, [3] can calculate the network utilization by only using FlowRemoved and PacketIn messages of Open-Flow standard, but it cannot trace quickly changed links. In [4], the authors proposed a dynamic algorithm to balance the request frequency and accuracy.

Active measurement by probing packets is essential for flexibly and promptly monitoring any desired part of the entire network. The status of all links on a specific measurement route could be actively but aggregately monitored. However, probing at a high sending rate for a long duration can impose a greater load on the data-plane, and thus some research efforts have been made on how to reduce the load while still retaining reliability and precision. In [5], an infrastructure that focuses on reducing the flow entries and the number of probe packets in the round-trip time (RTT) monitoring is proposed. In [6], a measurement scheme that can cover all links in both directions while minimizing flow entries on switches is presented. For datacenter networks, an effective probe matrix is designed to locate real-time failures in [7]. However, those methods rely on unicast probing in an end-to-end (among servers or beacons) manner and generally suffer from a concentration of many overlapped probing paths traversing a small number of bottleneck links near the sender of the probing packets.

Based on those existing works, we previously presented a monitoring framework that combines an active measurement by probing multicast packets from a measurement host and a passive measurement by collecting the flow-stats from selective switches. Then, on that, we proposed a method to estimate the packet delay variance from the arrival time interval of packets at each switch and locate high packet-delay variance links. The variance of the packet delays on a link or on an end-to-end path can be clearly defined as a statistical value and can represent a degree of the packet delay variations or fluctuations. However, the term "packet delay variation" is sometimes related with jitter [8] and sometimes defined by slightly different ways. We focus on the packet delay variance (the variance of the packet delays). Since links with a high packet delay variance are likely congested or physically unstable, it is of importance to monitor and locate such links in network performance management. Note that instead of directly measuring link delays and calculating delay variance, our method estimates the packet delay variance on a directional link or a directional segment between two ports (e.g., upper and lower ports of a link) based on the variances of packet arrival intervals monitored at each of those two ports. Differently from our previous work [9], the contribution of this paper is that we adopt an essential extension on a better route scheme [10] and provide an in-depth evaluation on how to reduce the load on the control-plane (i.e., the number of accesses to OFSs) while maintaining an acceptable measurement accuracy with a light load on the data-plane.

2   Monitoring Framework

2.1   Overview

The monitoring framework is based on that we previously proposed to monitor and locate high-loss links using multicast probing on OpenFlow networks [11]. It is for OpenFlow-based full-duplex networks consisting of the OpenFlow controller (OFC) and OpenFlow switches (OFS), with the measurement host (MH) that sends a series of multicast probe packets traversing all links in the network. An MH is directly connected to an OFS (called "measurement node"). The input port of the measurement node connecting the MH is called "root port". Probing packets are launched at the MH toward the root port, traversed input ports of some OFSs, and finally discarded at some input ports (called "leaf port"). A measurement path from the root port to a leaf port is called "terminal path". Note that the present method for estimation on delay variance requires an extension of flow entry and Flow-Stats Reply message to monitor the statistics of packet arrival time intervals on a specific flow [9].

First, as illustrated in Fig. 1, a measurement request from an MH is sent to the OFC. Then, the OFC obtains network topology, calculates probe packet routes, and installs them to OFSs. Probe packets are routed along a multicast tree route so that each probe packet passes through each link once and only once in each direction of each link separately to monitor bidirectional full-duplex links in both directions. The number of directed links on a terminal path is the terminal path length. A sequence of adjacent directed links along a path is called "segment" as a part of a terminal path. After that, the MH starts actively sending the probe packets. The packet arrival time interval at an individual input port on each OFS are passively recorded as flow-stats at each OFS and then, if needed, is collected by the OFC. Finally, the OFC calculates the packet-delay variance on a link (or a segment) between two ports and compared with a threshold to detect a high packet-delay variance link (or segment).

To reduce the loads on both the data-plane and the control-plane incurred by the measurement, two technical components, a flexible design of multicast measurement routes to cover all links in the active probing and a dynamic decision on

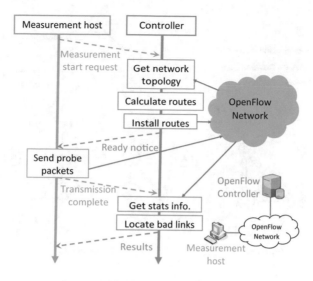

Fig. 1. Measurement process to locate performance degenerate links [11]

the sequential access order to switches for collecting the flow-stats, are required. In this paper, we focus on the former, i.e., the route scheme. The shorted-path tree based route scheme proposed in [9] suffers from generating many terminal paths so that it needs a large number of accesses to locate high packet-delay variance links. Therefore, as explained in the following subsection, we adopt a better route scheme based on [10] with fewer terminal paths to reduce the load on the control-plane while keeping an acceptable path length to minimize the load on the data-plane.

2.2   The Backbone-and-Branch Tree Route Scheme (BBT)

The proposed route scheme is called the backbone-and-branch tree route scheme (BBT). In this subsection, the BBT scheme is briefly explained as the example in Fig. 2. The Eulerian cycle algorithm is used to build backbone paths in the original undirected graph (network). Since an Eulerian cycle exists if and only if the graph consists of only even-degree vertices, first we need to remove all links between couples of odd-degree vertices (nodes) temporarily, called "omitted links", see dashed lines in Fig. 2a. Then, we generate a backbone cycle by using the Eulerian cycle algorithm to cover all remaining undirected links. From the generated backbone cycle, we can build one or two backbone paths. To avoid too long terminal paths, the BBT T2 with two halves of the Eulerian cycle as backbone paths is used in this paper, see the bold lines in Fig. 2.

After building backbone paths, we divide each backbone path into multiple backbone segments with almost the same length. At the end node of each segment, called branch node, the reverse direction segment of route on the backbone path is added as extension of the route toward the measurement node, called

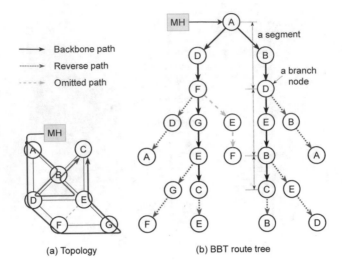

Fig. 2. Example of BBT route design

the reverse path, see the doted lines in Fig. 2b. By this way, both directions of each full-duplex link should be traversed by a measurement path. The reverse path has the same length with its backbone segment but the opposite direction. Finally, we integrate additional paths of temporally omitted links into the route tree, see the dashed line in Fig. 2b. Those operations eventually construct a route tree consisting of multiple terminal paths for multicast measurements.

3   Estimate Packet Delay Variance from Arrival Intervals

To estimate the packet delay variance, a simple and direct method is measuring packet delay times of samples (i.e., probe packets in our case) and computing their unbiased variance. However, the packet delay time measurement requires matching and subtracting the arrival times of a same packet monitored at two different OFSs. Thus, the list of arrival times of all probing packets should be moved from a place to another, which induces a considerable load on the control and/or data planes. In our method, instead of direct measurement of per-packet delay times, each OFS monitors the arrival time intervals of two adjacent packets in a series of probe packets and computes their statistics to record locally and incrementally, which can be performed within each OFS independently and does neither require to store a long list of per-packet information nor to exchange it between OFSs or the OFC. After the above measurement of probe packets is finished, the OFC collects the arrival time interval statistics at each input port of OFSs and estimates the packet delay variance between two ports using the collected statistics by using an estimation method explained below.

The sequence diagram of probe packets is shown in Fig. 3. Let Q_k^j and D_{k+1}^j be the queuing delay at the kth OFS of the jth probe packet and the transmission

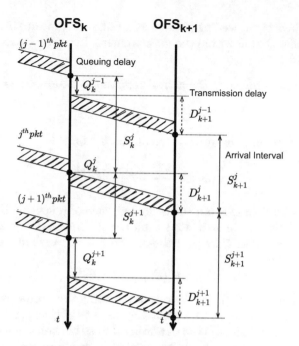

Fig. 3. Sequence diagram of packets probing

delay (including the propagation delay) of the jth probe packet from the kth OFS to the $(k+1)^{th}$ OFS, respectively. The arrival time interval S_{k+1}^j of the $(j-1)^{th}$ and j^{th} probe packets at the $(k+1)^{th}$ OFS is

$$S_{k+1}^j = S_k^j + (Q_k^j + D_{k+1}^j) - (Q_k^{j-1} + D_{k+1}^{j-1}) \tag{1}$$

where S_k^j is the arrival time interval of the $(j-1)^{th}$ and j^{th} probe packets at the kth OFS. Note that, the arrival time interval S_1^j of jth packet at the first OFS is equal to the initial sending time interval at the MH because we can assume no queuing delay between the MH and the first OFS.

Since we assume the bandwidth of each link and the probe packet size do not change in time, we have $D_{k+1}^j = D_{k+1}^{j-1}$, and thus the arrival time intervals of packets only depend on the queuing delays of OFSs (mainly at egress/output ports of OFSs) and the initial sending interval S_1^j.

$$S_{k+1}^j = S_k^j + Q_k^j - Q_k^{j-1} \tag{2}$$

The following ideal preconditions are defined to estimate the delay variance from the arrival time intervals. We will discuss the impact of a deviation from those conditions (i.e., correlated delays) later.

- Queuing delays of a packet at different links (different output ports) along the measurement path are independent within the measurement duration. Hence S_k^j and $Q_k^j - Q_k^{j-1}$ are independent.

- Queuing delays of succeeding packets at a link over time are independent and identically distributed within the measurement duration. Hence Q_k^j and Q_k^{j-1} are independent.

Therefore, the variance of the arrival intervals is expressed as follows

$$V[S_{k+1}] \cong V[S_k] + 2V[Q_k] \tag{3}$$

From Eq. 3, the queuing delay variance at the kth OFS is

$$V[Q_k] \cong \frac{V[S_{k+1}] - V[S_k]}{2} \tag{4}$$

This means that, in general, the delay variance of a specific link or segment between two OFSs can be estimated from the difference of the arrival interval variances of those OFSs. Note that the arrival interval variance at each OFS is simply computed by

$$V[S_k] = E[(S_k)^2] - (E[S_k])^2 \tag{5}$$

where $E[(S_k)^2]$ and $(E[S_k])^2$ can be computed using the sample mean incrementally, that is, we do not need to store a long list of $\{S_k^j | j = 1, 2, \ldots\}$.

One problem to consider is the packet loss. Possible holes in a series of probe packets due to packet losses should be considered and removed in the process of monitoring the arrival time intervals. If the jth probe packet is lost somewhere and an OFS receives the $(j-1)^{th}$ and $(j+1)^{th}$ packets but not receive the jth packet, then OFS discards the arrival time interval between $(j-1)^{th}$ and $(j+1)^{th}$ packets and does not count it in the statistics. To detect such holes by lost packets at OFS, the MH embeds a sequence number into ID field of IP header of each probe packet.

Another problem is called the "narrow interval"; meaning that the jth packet arrives at the kth OFS before the $(j-1)^{th}$ packet departs from the OFS, i.e., two succeeding packets stay in the same queue. If two adjacent packets arrive at an OFS closely and meet similar congestion levels, Q_k^{j-1} and Q_k^j are similar and thus have a positive correlation. This problem will decrease our estimation's accuracy on delay variance $V[Q_k]$. A simple solution is to enlarge the initial sending time interval at the MH, although it will prolong the measurement duration. In our simulation, we adopt this approach.

4   Locate High Packet Delay Variance Links

In the first step of the location process, the OFC queries OFSs that have leaf ports to collect the information on arrival time intervals at those ports and estimates the delay variance of each terminal path using the information at the leaf ports and the root port by (4). If the delay variances of all terminal paths are less than the threshold h, that means the network do not include any high delay variance link. Here, h is a design parameter that represents the target delay variance quality of links to maintain, which depends on the target applications.

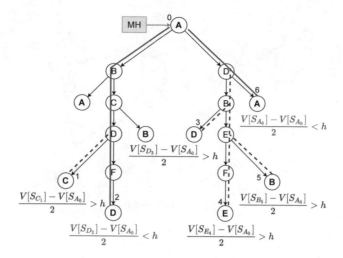

Fig. 4. Example of the order of accesses in locating high delay variance links

If the delay variance of a terminal path exceeds h, this terminal path is likely to include one or more high delay variance links.

Then, by considering the correlation among terminal paths in terms of delay variance, OFC can narrow the search scope, i.e., the expected locations of high delay variance links. For example, if a terminal path is high delay variance and there are no other high delay variance terminal paths, the high delay variance links are located within a segment between the leaf port and the nearest branch port on the considered terminal path. The dashed line on the left part in Fig. 4 shows an example of this case in which S_{X_i} represents the arrival time intervals of probe packets received at the port i of the OFS X. Here, to locate high delay variance links, the ports along this segment should be queried by OFC in a binary-search manner. Eventually, the delay variance of each high delay variance link is measured based on the difference between the delay variance at the link's upper and lower ports.

If there are multiple terminal paths whose delay variance values exceed threshold h, the port that is most commonly shared by those paths and nearest to the root among them is queried first to collect the arrival time interval of probe packets at that port. By considering the sub-trees separated by that port, the same procedure can be performed on each sub-tree recursively. An example of this case is shown in the right part of Fig. 4. Here, the next queried port is the ingress port of the OFS D.

5   Simulation Evaluation

5.1   Simulation Settings

We evaluate the search performance of our proposal by numerical simulation on a real-world network topology in a topology database [12], illustrated in Fig. 5.

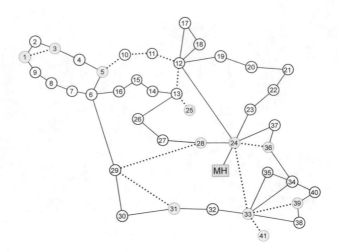

Fig. 5. Renater network topology

In the simulation, we compare the newly proposed route BBT T2 with the previous shortest path tree-based route (Model 2) and the unicursal route (the route has only one terminal path with a maximum length). Their information of terminal paths is in Table 1.

Table 1. Number of terminal paths and path lengths of route schemes

	Paths	Average	Min	Max
Unicursal	1	108	108	108
Model 2	26	5.5	3	12
BBT T2	8	19.6	8	28

Paths: Number of terminal paths
Average: Average length of terminal paths. Min: The minimum length. Max: The maximum length.

In the simulation, the parameters relating with packet delay times are set on each link as follows. A baseline static delay time of a link is set to a randomly selected fixed value from a range of [10.0, 20.0] (ms). An additional dynamic delay (queuing delay) of each output port of OFS is a random variable with an exponential distribution that is independent of each other. The mean value (the expectation) of this random variable of dynamic delay is randomly selected from a range of

- [5.0, 10.0] (ms) for each of a specific number of high delay variance links,
- [2.0, 4.0] (ms) for each of 10% moderate delay variance links,
- [0.2, 1.0] (ms) for each of other little delay variance links.

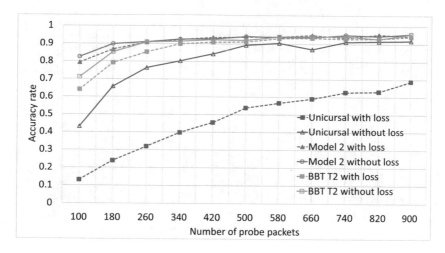

Fig. 6. Accuracy rate for locating 2 high-delay variance links with packet loss

We assume a random light loss rate in range of [0, 0.01] on every link. The threshold h of high delay variance is 25. The initial sending interval of probe packets is 150 (ms) to avoid correlated delays among the adjacent packets and the narrow interval problem. The number of probe packets varies from 100 to 900. All resulting values are averaged over 1,000 measurement instances.

5.2   Simulation Results

Figure 6 shows the measurement accuracy depending on the number of probe packets. The measurement accuracy is defined as the ratio of the number of measurements in which all 2 high-delay variance links are correctly located to the total number of measurements (1,000 in our setting). We compare the results in two scenarios: with and without the packet loss. The estimation accuracy of delay variance relies on the number of probe packets. Therefore, the packet loss has a strong impact on the measurement accuracy. We see that a route scheme with longer terminal paths needs many probe packets to operate accurately. This is because each packet loss on an upstream link of measurement paths will make a hole of recorded arrival time interval on all remaining links, making the estimated value smaller or larger than the true value; these are "underestimation" or "overestimation", respectively. The underestimation leads OFC to skip over high delay variance links. On the other hand, the overestimation leads OFC to unnecessarily and mistakenly seek high delay variance links. Additionally, accumulated errors over multiple links in a long segment will also create the underestimation or overestimation and result in a decrease of accuracy. The unicursal route suffers from a significantly low accuracy due to its long terminal path.

Figure 7 shows the number of the required accesses from OFC to OFSs until the high delay variance link location process ends in case of 2 high delay variance

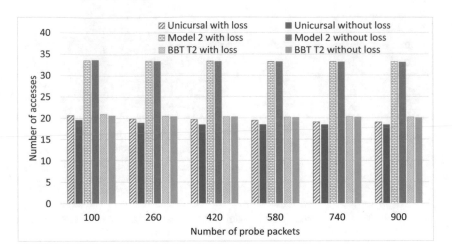

Fig. 7. Number of accesses for locating 2 high-delay variance links with packet loss

links, depending on the number of probe packets. Since the location process does not know the number of high delay variance links, the process lasts until it judges there is no other high delay variance links. Note that the results of the location process are not always correct because of errors in estimation. The shortest path tree-based route (Model 2) suffers from a larger number accesses due to a large number of terminal paths.

From results in Figs. 6 and 7, there is a trade-off between the load on the data-plane (the number of required probe packets for a certain estimation accuracy) and the load on the control-plane (required accesses). However, our proposed BBT T2 route can clearly balance the trade-off.

Although we assume the queuing delays on different ports along a terminal path are independent, a positive correlation may happen especially between high delay variance links. We examine this situation in case that 2 high delay variance links are positively correlated on the packet delays with the correlation coefficient $\rho = 0.5$ with the existence of packet losses. Figures 8 and 9 compare the performance with and without the correlation of delays between 2 high delay variance links.

From Fig. 8, the accuracy of the unicursal route (with a very long terminal path) is improved in the correlated case. This is because the positive delay correlation between ports within a segment makes the estimated value of the delay variance of the segment larger than the true value. This overestimation may introduce a fail-safe checking and increase the accuracy rate while increasing unnecessary accesses as shown in Fig. 9. Whereas, in Model 2 and BBT T2 with shorter terminal paths, the probability that these 2 corrected high delay variance links are positioned on the same terminal path is small. Therefore, the overestimation is small, and the results are similar both in the correlated case and the uncorrelated (independent) case as shown in Figs. 8 and 9. This suggests

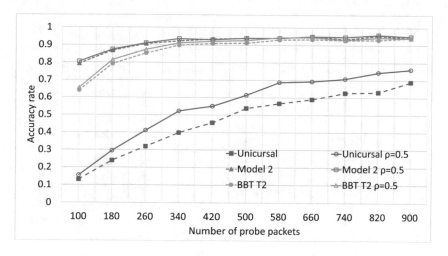

Fig. 8. Accuracy rate for locating 2 correlated high-delay variance links

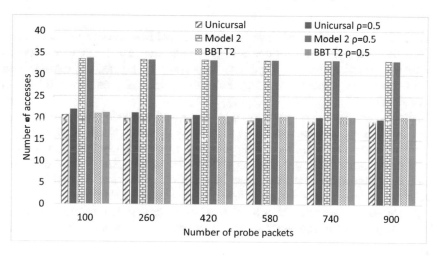

Fig. 9. Number of accesses for locating 2 correlated high-delay variance links

the resiliency of BBT T2 route scheme in correlated delays situations to some extent. As the result, in this simulation scenario with packet losses and correlated delays, BBT T2 scheme can locate 2 high delay variance links at an accuracy rate 0.9 with a small number (580) of the probe packets on the data-plane and a small number (20) of the switch accesses on the control-plane.

To have a detail view about the impact of packet losses and correlated delays, we investigate the locating process of the unicursal route with a long single terminal path. As above discussed, errors in estimation lead to the underestimation and overestimation. In the underestimation, the OFC may skip out high delay variance links because estimated values are smaller than true delay variances.

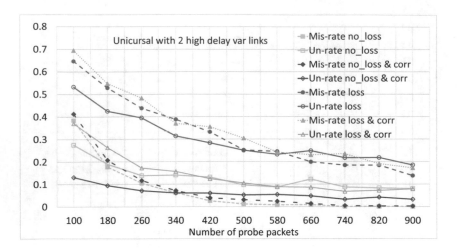

Fig. 10. Impact of the packet loss and delay correlation

On the other words, high delay variance links could be undetected. In the simulation, we count these situations and calculate the rate (called un-rate) over 1000 measurement instances. Similarly, "mis-rate" is the ratio of cases in which the OFC mistakenly detects a normal link into a high delay variance link to total measurements.

Figure 10 compares the error rates (un-rate and mis-rate) in four cases (with and without the packet loss and the positive delay correlation). With the packet loss, both mis-rate and un-rate are significantly increased, resulting in a low accuracy of measurement. About the impact of correlation, since correlated delays happen in 2 high delay variance links, the mis-rates are not changed with and without the correlation and enough small in case of no packet loss. On the other hand, because of overestimation on the delay variances, the un-rate is accidentally but significantly decreased in the correlation case regardless of the packet loss.

6   Concluding Remarks

Based on our previously proposed framework for monitoring and locating links with a high packet-delay variance in an OpenFlow network, in this paper, we apply an Eulerian cycle-based route scheme that can control the number of terminal paths and the lengths of them. Compared with the shortest path tree-based route in [9], the new proposal can reduce the number of accesses to switches by reducing the number of terminal paths while keeping a high accuracy rate by limiting the lengths of terminal paths. The proposed route scheme has been shown to be resilient with impacts of the packet losses and correlated delays through simulation results.

As a topic for future work, we will strive to adaptively optimize schemes for the multicast probe packet route by using the information on past measurement results to further reduce the number of accesses to OFSs.

Acknowledgements. These research results have been achieved by the "Resilient Edge Cloud Designed Network (19304)," NICT, and by JSPS KAKENHI JP20K11770, Japan.

References

1. Jain, S., Kumar, A., Mandal, S., et al.: B4: Experience with a globally-deployed software defined WAN. In: Proceedings ACM SIGCOMM 2013, pp. 3–14 (2013)
2. Hong, C.-Y., Kandula, S., Mahajan, R., et al.: Achieving high utilization with software-driven WAN. In: Proceedings ACM SIGCOMM 2013, pp. 15–26 (2013)
3. Yu, C., Lumezanu, C., Zhang, Y., et al.: FlowSense: monitoring network utilization with zero measurement cost. Lect. Notes Comput. Sci. **7799**, 31–41 (2013)
4. Chowdhury, S.R., Bari, M.F., Ahmed, R., Boutaba, R.: PayLess: a low cost network monitoring framework for software defined networks. In: Proceedings of 2014 IEEE NOMS, pp. 1–9 (2014)
5. Atary, A., Bremler-Barr, A.: Efficient round-trip time monitoring in OpenFlow networks. In: Proceedings of IEEE INFOCOM, pp. 1–9 (2016)
6. Shibuya, M., Tachibana, A., Hasegawa, T.: Efficient active measurement for monitoring link-by-link performance in OpenFlow networks. IEICE Trans. Commun. **E99B**(5), 1032–1040 (2016)
7. Peng, Y., Yang, J., Wu, C., et al.: deTector: a topology aware monitoring system for data center networks. In: Proceedings of the 2017 USENIX Annual Technical Conference, pp. 55–68 (2017)
8. Demichelis, C., Chimento, P.: IP packet delay variation metric for IP performance metrics (IPPM). The Internet Engineering Task Force, IETF-RFC (2002)
9. Tri, N.M., Nagata, S., Tsuru, M.: Locating delay fluctuation-prone links by packet arrival intervals in openflow networks. In: Proceedings of the 20th Asia-Pacific Network Operations and Management Symposium, pp. 1–6 (2019)
10. Tri, N.M., Shibata, M., Tsuru, M.: Effective route scheme of multicast probing to locate high-loss links in OpenFlow networks. J. Inf. Process., 9 (2021)
11. Tri, N.M., Tsuru, M.: Locating deteriorated links by network-assisted multicast proving on OpenFlow networks. In: Proceedings of the 24th IEEE Symposium on Computers and Communications, pp. 1–6 (2019)
12. The Internet Topology Zoo, 14 May 2020. http://www.topology-zoo.org/

Data Privacy Preservation Algorithm on Large-Scale Identical Generalization Hierarchy Data

Waranya Mahanan[1][(✉)] and Juggapong Natwichai[2]

[1] College of Arts, Media and Technology, Chiang Mai University,
Chiang Mai, Thailand
waranya.m@cmu.ac.th
[2] Data Engineering Laboratory, Computer Engineering Department,
Faculty of Engineering and Data Analytics and Knowledge Synthesis for Healthcare,
Chiang Mai University, Chiang Mai, Thailand
juggapong@eng.cmu.ac.th

Abstract. In the data bursting era, an enormous amount of data can be collected. Though, such data can benefit both the data owners and the public, the privacy of collected data is an important concern. To guarantee the privacy of data, the k-anonymous method is applied before publishing the dataset. The dataset needs to have an identical value of at least k records in order to protect the data privacy which could cause data losses. The optimal k-anonymity is concerned with minimizing the data losses and also preserve data privacy. The generalization lattice is created to map all the generalization schemes and use them to find the optimal answer. The larger number of attributes means the larger number of nodes in the generalization lattice. Thus, in the larger number of attributes, the k-anonymous algorithm takes more computation resources and time to determine the answer. Although, due to the limited computation resources and time, the existing optimal k-anonymity algorithms only find the optimal answer on the small dataset. Therefore, in this paper, we design the optimal k-anonymity algorithm with the incremental attribute concept. At m attribute, our algorithm process only the nodes which previously satisfy the k-anonymity from the $m - 1$ attributes. Thus, our algorithm can find the optimal answer at a large number of attributes without reaching the memory limit problem compared with the existing k-anonymity algorithms due to it determines only some necessary nodes in the lattice.

1 Introduction and Motivation

The collected data is now considered the most valuable asset, thus data privacy preservation is also an increasing interest. One of the well-known privacy preservation methods is k-anonymity [7], the privacy preserved dataset or the k-anonymous dataset must have the value of the record same as at least k-1 other records. For making the dataset satisfy the k-anonymity condition, the

L. Barolli et al. (Eds.): EIDWT 2021, LNDECT 65, pp. 246–255, 2021.
https://doi.org/10.1007/978-3-030-70639-5_23

generalization process is used. Some values in the dataset are generalized to the more general value using the generalization hierarchy [9]. During the generalization process of the k-anonymity, some data is lost in order to form the k-anonymous dataset. Thus, the optimal k-anonymity is considered to obtain the data k-anonymous dataset and also minimize the data losses [8]. The optimal k-anonymity can be addressed by using the generalization lattice, the lattice which indicates all the generalization schemes [2]. The optimal k-anonymity answer is the node of the generalization lattice that satisfies the k-anonymity condition and loses the smallest data.

Generally, the often-used datasets in the k-anonymity method have a diverse data type in each attribute. Therefore, there is some dataset that contains only a single data type. This type of data use only a generalization hierarchy for all attributes in the k-anonymity process, so-called an identical generalization hierarchy (IGH) data [5]. An example of the IGH dataset, shown in Table 1(a), is the interest score from 0–5 that the users give to each movie The generalization lattice of such dataset, using the generalization hierarchy in Fig. 1, shown in Fig. 2 . Each node in the lattice represents a generalization scheme. For example, the node $\{2, 0, 1\}$ means that the value of movie M1, M2, and M3 is generalized to level 2, 0, and 1 of the generalization hierarchy. The root node $\{0, 0, 0\}$ in the lowest level represents the generalized data in which M1, M2, and M3 are generalized to level 0 or they are not generalized. While at the highest level, the leaf node, $\{2, 2, 2\}$, represents the generalized dataset which all attribute generalized to the highest level. Some data losses have occurred during the generalization process. One often-used data loss metric of the k-anonymity process is the precision [10]. For IGH data, the precision of the nodes in the same generalization lattice level is always the same. And the node that generalized to the higher level always has higher precision than the node in the lower level. The optimal k-anonymity of IGH is the k-anonymous node which has the lowest precision. Thus, the optimal answer will always in the lowest level found the k-anonymous nodes due to each attribute using an identical generalization hierarchy [5].

The IGH data frequently have a large number of attributes which could take longer time and more resources to process the optimal answer. For example, from the dataset with 3 attributes in Table 1(a) using the generalization hierarchy in Fig. 1, the generalization lattice contains 27 nodes, while at the number of attributes at 10, the number of nodes is exponentially increased to 59,049 nodes. The existing k-anonymous algorithms can process only a small number of attributes because they need to create and process all nodes in the lattice in order to get the optimal answer. Thus, in this paper, we improve the optimal k-anonymity algorithm for IGH data by using the attribute incremental technique. The algorithm starts processing the k-anonymity at 3 attributes. Then, an attribute is increased by one at each iteration until reaches the objective number of attributes. At each iteration, the algorithm increases and processes only the nodes which are a k-anonymous node in the previous iteration. Thus, the proposed algorithm processes only a small number of nodes compared with the full

Table 1. An example of an *IGH* dataset.

ID	Attributes		
	M1	M2	M3
1	4	0	1
2	4	0	1
3	1	1	1
4	1	4	3
5	1	5	5
6	0	4	4
7	0	4	3
8	0	4	5

(a) Original dataset with 3 attributes [5]

ID	Attributes				
	M1	M2	M3	M4	M5
1	4	0	1	0	1
2	4	0	1	0	4
3	1	1	1	2	3
4	1	4	3	1	2
5	1	5	5	1	4
6	0	4	4	2	0
7	0	4	3	3	0
8	0	4	5	4	2

(b) Original dataset with 5 attributes

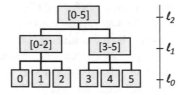

Fig. 1. The generalization hierarchy of an *IGH* dataset [5]

search of the existing algorithms. The proposed algorithm can find the optimal answer of the larger-scale dataset compared with the existing algorithms.

2   IGH Data on Increasing Number of Attributes

Since the optimal k-anonymity answer can be determined using the generalization lattice. The number of nodes in the generalization lattice, N, can be calculated by Eq. 1, where H is the number of levels of the generalization hierarchy and m is the number of attributes.

$$N = \binom{H}{1}^{m} \tag{1}$$

For example, from the IGH dataset in Table 1(a) and the generalization hierarchy in Fig. 1, the number of nodes of the generalization lattice is $\binom{3}{1}^{3} = 27$, while, at the number of attributes at 9, the number of nodes in the lattice is exponentially increased to $\binom{3}{1}^{9} = 19{,}683$. With a higher number of nodes, more resources and time are needed in order to get the optimal answer.

Therefore, some characteristics of generalization lattice on an incremental number of attributes are as follows.

1. A node of the generalization lattice at the number attributes at m can be referred to H nodes at the number of attributes at $m + 1$. As example in

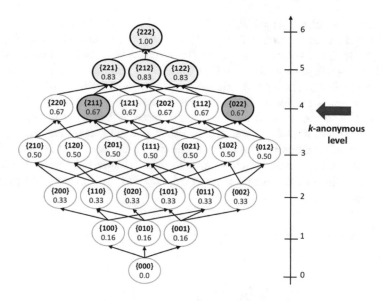

Fig. 2. The generalization lattice of the number of attribute at 3 [5]

Fig. 2, the node {0, 2, 1} can be referred to the node {0, 2, 1, 2}, {0, 2, 1, 1}, and {0, 2, 1, 0} after adding an attribute to the dataset, so-called the referred nodes.

2. At the number of attributes at m, if a node is not a k-anonymous node, then the referred nodes at the higher number of attributes are also not a k-anonymous node. For example, the node {0, 2, 1} is not a k-anonymous node, so its referred nodes {0, 2, 1, 2}, {0, 2, 1, 1}, and {0, 2, 1, 0} are also not a k-anonymous nodes.

3. At the number of attributes at $m-1$, some referred nodes of the k-anonymous nodes can be a k-anonymous node at the number of attributes at m. To be illustrated, the nodes {0, 2, 2, 2}, the referred nodes of k-anonymous node {0, 2, 2}, is a k-anonymous nodes while the {0, 2, 2, 1} and {0, 2, 2, 0} is not a k-anonymous nodes.

Furthermore, the general characteristic of the optimal k-anonymity of IGH data, the optimal answer is always the k-anonymous node in the lowest level found k-anonymous node, is still intact [5].

3   Incremental Optimal-IGH Algorithm

From the characteristics of the optimal k-anonymity on the increasing number of attributes, in this paper, an incremental optimal k-anonymity algorithm for IGH data is proposed. The algorithm is invented based on the characteristic of the increments of the attributes.

The key idea of the algorithm is to determine only the nodes which satisfy the k-anonymity in a lower number of attributes. As illustrated in Algorithm 1, the algorithm first start to evaluate the k-anonymity of the dataset with the first 3 attributes by performing the Optimal-IGH [5] algorithm. Then, iteratively determine the k-anonymity only the referred nodes of such k-anonymous nodes with an additional attribute. If the node satisfies the k-anonymity, the node will append to the list and use it as the k-anonymous nodes for the next iteration. The algorithm will iteratively determine the k-anonymous nodes until reach the total number of attributes. The optimal answer is the node in the k-anonymous level, the lowest level found k-anonymous nodes. If there are more than one k-anonymous nodes in the k-anonymous level, the DM [1, 11] is used to provide an optimal answer.

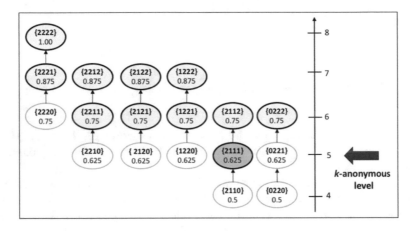

Fig. 3. The referred nodes at the number of attributes at 4

To be illustrated, from the dataset which contains 5 attributes in Table 1(b), the algorithm will determine the optimal answer of the first 3 attributes by executing the Optimal-IGH algorithm. As shown in Fig. 2, the k-anonymous nodes are presented in shaded. The k-anonymous level, the lowest level found k-anonymous nodes, is now set at 4, due to the optimal nodes will always among nodes in the k-anonymous level. The algorithm will next determine the optimal nodes of the dataset with the number of attributes at 4 by processing only the referred nodes of the k-anonymous nodes from the previous step, shown in Fig. 3. According to the characteristic of k-anonymity of the increasing number of attributes mentioned earlier, the k-anonymous nodes from a lower number of attributes with an additional highest generalized attribute are also the k-anonymous, Thus, node {2, 2, 2, 2}, {2, 2, 1, 2}, {2, 1, 2, 2}, {1, 2, 2, 2}, {2, 1, 1, 2} and {0, 2, 2, 2} are automatically considered the k-anonymous nodes. The k-anonymous level is now at the 6. Then, in each referred node's path, the algorithm will iteratively determine the k-anonymity only the path

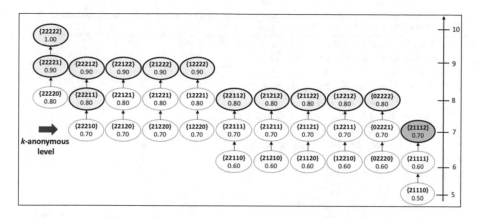

Fig. 4. The referred nodes at the number of attributes at 5

which contains a node at the k-anonymous level. After processing all nodes, the optimal answer of the number of attributes at 4 is the node {2, 1, 1, 1}, and the k-anonymous level is now set at 5. Next, an attribute is added and the k-anonymous level is set at 7, due to the k-anonymous nodes is now the node {2, 1, 1, 1, 2}. At the number of attributes at 5, all referred paths that have a node in the k-anonymous level will be processed. The optimal answer is the node {2, 1, 1, 1, 2} with the lowest data loss. In addition, if there is more than one node in the k-anonymous level, the DM will be calculated among such nodes in order to get only an optimal answer.

4   Results and Discussion

We compare the performance of our algorithm with the other fastest known k-anonymous algorithms, Extended-OIGH [5], Flash [4], and Depth-First-Search (DFS). Three real-world datasets are used to evaluate the performance of the algorithms, MovieLens [6], Jester [3], and Taxi [12]. The experiments were performed with 2.9 GHz Intel Core i5 with 8 GB memory running Mac OS X implemented based on Java SE 8.

The performance of the algorithms on the MovieLens, Jester and Taxi dataset are illustrated in Fig. 5, 6 and 7, respectively. The charts show the execution time on the various number of attributes from 3 to 15. The result shows that the execution time of the Incremental-OIGH algorithm is slightly increased when the number of attributes is increased, while the other algorithms are exponentially increased. At the lower number of attributes, the execution time of Incremental-OIGH is higher than the other algorithms because it must be iteratively increased an attribute to the lattice and processed the k-anonymity of each. While the other algorithms process only the lattice with such a number of attributes. Therefore, due to the Incremental-OIGH algorithm only evaluating the necessary nodes of the increased attribute, the execution time of it is lower

---

Algorithm 1: *Incremental-OIGH*

Input: Dataset D, Generalization Hierarchy h, k
Output: Optimal k-anonymity OP

1 **begin**
2 | $m \leftarrow$ Number of attributes of D
3 | $H \leftarrow$ Highest level of h
4 | $D_3 \leftarrow$ First 3 attributes of dataset D
5 | $KN \leftarrow Optimal\text{-}IGH(D_3)$
6 | $L \leftarrow$ Lowest level of nodes in KN
7 | $nL \leftarrow L + H$
8 | **for** $a = 4$ *to* m **do**
9 | | $tempKN \leftarrow \{\}$
10 | | **foreach** N *in* KN **do**
11 | | | $newN \leftarrow$ Node N with an additional attribute transformed to level H
12 | | | $tempKN.append(newN)$
13 | | | **for** $h = 0$ *to* $H - 1$ **do**
14 | | | | $newN \leftarrow$ Node N with an additional attribute transformed to level h
15 | | | | **if** $newN.level \leq nL$ **then**
16 | | | | | **if** $CheckAnonymity(newN)$ **then**
17 | | | | | | $tempKN.append(newN)$
18 | | | | | **end**
19 | | | | **end**
20 | | | **end**
21 | | **end**
22 | | $KN \leftarrow tempKN$
23 | | $L \leftarrow$ Lowest level of nodes in KN
24 | | $nL \leftarrow L + H$
25 | **end**
26 | $OP \leftarrow$ minimum DM value among k-anonymous nodes KN
27 | **return** OP
28 **end**

---

than the other algorithms at the number of attributes at 13, 10, and 12 on the MovieLens, Jester, and Taxi dataset, respectively. Thus, at the higher number of attributes, the Incremental-OIGH algorithm can find the optimal answer faster than the other well-known k-anonymity algorithms.

For comparing the performance of Incremental-OIGH algorithm on each dataset, the result shown in Fig. 8. The execution time is increased when the number of attributes is increased, while the execution time of the Incremental-OIGH on the MovieLens is the lowest, followed by Jester and Taxi. The reason is that the height of the generalization hierarchy is different on each dataset, for the MovieLens the height of the generalization hierarchy is 2, while they are 4 and 5 on the Jester and Taxi dataset. The higher height of the generalization

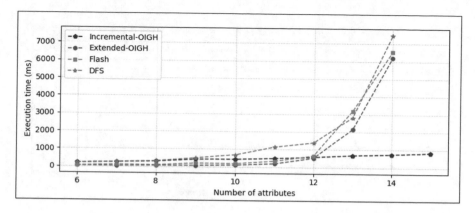

Fig. 5. Performance of the algorithms on MovieLens dataset.

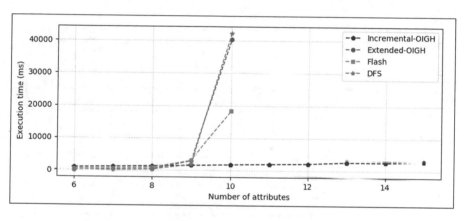

Fig. 6. Performance of the algorithms on Jester dataset.

hierarchy means the higher number of nodes in the generalization lattice which could cause a higher execution time.

The Incremental-OIGH can find the optimal k-anonymity on a large number of the attribute, due to its process only the referred nodes on the increasing number of attributes. As shown in Fig. 8, the Incremental-OIGH algorithm can process the dataset with 1,000 attributes without reaching the resource limit. While the other algorithms reach the resource limit at only the number of attributes at 20.

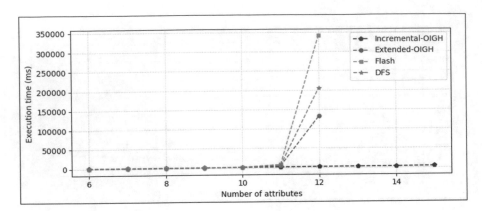

Fig. 7. Performance of the algorithms on Taxi dataset.

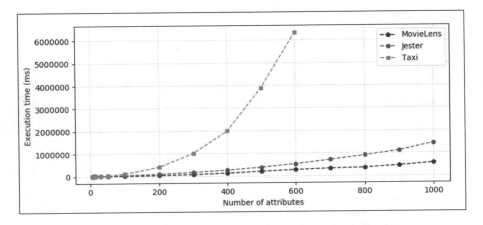

Fig. 8. Performance of the Incremental-OIGH algorithms.

5   Conclusion

This paper proposed a k-anonymity algorithm for preserving the data privacy for the Identical Generalization Hierarchy (IGH) data on the large-scale dataset. The algorithm evaluates the optimal k-anonymity answer by increasing an attribute to the generalization lattice one by one. The algorithm processes only the referred nodes of the k-anonymous nodes from a lower number of attributes. Thus, it does not have to create and evaluate all nodes in the generalization lattice for getting the optimal answer. The result shows that the Incremental-OIGH can evaluate the optimal k-anonymity on a large number of attributes without reaching the memory limit and can perform at approximately 90% greater number of attributes than the other well-known k-anonymity algorithms.

References

1. Bayardo, R.J., Agrawal, R.: Data privacy through optimal k-anonymization. In: Proceedings of the 21st International Conference on Data Engineering, ICDE 2005, pp. 217–228. IEEE Computer Society, Washington, DC, USA (2005)
2. El Emam, K., Dankar, F., Issa, R., Jonker, E., Amyot, D., Cogo, E., Corriveau, J.P., Walker, M., Chowdhury, S., Vaillancourt, R., Roffey, T., Bottomley, J.: A globally optimal k-anonymity method for the de-identification of health data. J. Am. Med. Informatics Assoc.: JAMIA **16**, 670–82 (2009)
3. Goldberg, K., Roeder, T., Gupta, D., Perkins, C.: Eigentaste: a constant time collaborative filtering algorithm. Inf. Retrieval **4**(2), 133–151 (2001)
4. Kohlmayer, F., Prasser, F., Eckert, C., Kemper, A., Kuhn, K.A.: Flash: efficient, stable and optimal k-anonymity. In: 2012 International Conference on Privacy, Security, Risk and Trust and 2012 International Conference on Social Computing, pp. 708–717 (2012). https://doi.org/10.1109/SocialCom-PASSAT.2012.52
5. Mahanan, W., Art Chaovalitwongse, W., Natwichai, J.: Data anonymization: a novel optimal k-anonymity algorithm for identical generalization hierarchy data in IoT. Serv. Oriented Comput. Appl. (2020). https://doi.org/10.1007/s11761-020-00287-w
6. Purvank: uber ride reviews dataset (2017). https://www.kaggle.com/purvank/uber-rider-reviews-dataset. Accessed 10 Nov 2020
7. Samarati, P.: Protecting respondents identities in microdata release. IEEE Trans. Knowl. Data Eng. **13**(6), 1010–1027 (2001)
8. Sweeney, L.: Achieving k-anonymity privacy protection using generalization and suppression. Int. J. Uncertain. Fuzziness Knowl.-Based Syst. **10**(5), 571–588 (2002)
9. Sweeney, L.. k -anonymity: a model for protecting privacy **10**(5), 1–14 (2002)
10. Sweeney, L.A.: Computational disclosure control: A primer on data privacy protection. Ph.D. thesis, Massachusetts Institute of Technology, Cambridge, MA, USA (2001). AAI0803469
11. Wong, R.C.W., Li, J., Fu, A.W.C., Wang, K.: (α, k)-anonymity: an enhanced k-anonymity model for privacy preserving data publishing. In: Proceedings of the 12th ACM SIGKDD International Conference on Knowledge Discovery and Data Mining, KDD Ó6, pp. 754–759. ACM, New York, NY, USA (2006)
12. Zheng, Y.: T-drive trajectory data sample. T-Drive sample dataset (2011). https://www.microsoft.com/en-us/research/publication/t-drive-trajectory-data-sample/

Routing Control and Fault Recovery Strategy of Electric Energy Router Under the Framework of Energy Internet

Yunfei Du, Xianggen Yin[✉], Jinmu Lai, Zia Ullah, Zhen Wang, and Jiaxuan Hu

State Key Laboratory of Advanced Electromagnetic Engineering, and Technology,
Huazhong University of Science and Technology, Wuhan 430074, China
xgyin@hust.edu.cn

Abstract. The inclusion of advanced technologies such as emerging energy internet and system perception via data fusion in the electric power systems enhances the overall performance and reliability of the power system. However, accurate path selection, congestion circumvention, and fault recovery in electric energy router (EER) within the emerging energy internet are the key challenges. This paper proposes the routing control strategy based on the minimum loss path (MLP) and the fault recovery strategy based on the multi-source cooperative power supply (MCPS) for accurate path selection, congestion circumvention, and fault recovery in electric energy router (EER). The proposed MLP strategy offers accurate path recognition having smaller power loss in congestion circumvention. Moreover, the MCPS strategy can guarantee the power supply of various loads and maximize the normal load to restore power during the fault. Meanwhile, two priority ranking methods, incentive reasonable quotation and incentive transaction volume realized by loss transfer, are proposed. Finally, the feasibility and superiority of the overall strategy are verified by simulation and comparison.

1 Introduction

Worldwide, the electric power demand is increasing rapidly due to load expansion that causes the rise of the energy crisis and influences developed countries' economies. The utilization of renewable energy sources recently gained remarkable attention due to lower prices and clean environmental characteristics. The penetration of renewable energy sources such as photovoltaic, wind turbine, bioenergy, and other renewable energy resources (RERs) to the power grid is reported to meet the energy demands [1]. The uncertain output of RERs and dynamic load variation are the key challenges to the control RES integrated power grid. At present, the most effective way to use RERs is to collect, store and consume the produced power locally [2, 3], which increases the electric power system performance in terms of technical benefits such as voltage profile improvement and power loss reduction [4–7] and enhance the economic benefits via energy cost minimization [8]. The high penetration of RERs into to power system influences the power flow and power quality [9], fault detection, and isolation, which make it difficult for the maximum utilization of RERs. However, using and restricting the power flow with information flow jointly can realize the maximum utilization of RESs via real-time information sharing of RERs power generation and transmission of generated power. In this regard, emerging energy internet technology provides a feasible scheme for the effective utilization of RERs. In fact, energy internet is a neo-type of information and energy integration internet [10, 11]. It takes the main network controlled by the power dispatching center as the wide-area network (e-WAN), the micro network controlled by the electric energy router (EER) as the local area network (e-LAN), and can realize the two-way on-demand transmission and active power control, so it can adapt to the access of RERs to the greatest extent. EER is the key element in the future energy internet which contains the integration of power electronic and communication technology. EER provides plug and play AC and DC interfaces for RERs and can independently realize the islanding operation or grid connection of e-LAN according to the load and fault demand, so as to ensure the real-time energy balance of the energy internet. At the same time, under the centralized control of the dispatching center, the mutual restriction of information and energy flow can realize efficient power transmission and accurate routing [12–14].

The power flow control in the traditional power grid are like spider webs—pulling on one strand causes all the strands to move. The access of EER makes the bidirectional active control of power flow possible. Therefore, the research on the routing control strategy between EERs will completely subvert the traditional power flow control method [12]. In reference [15] the relevant concepts of virtual circuit transmission using the information internet [16], aiming at minimizing power loss, and designs a routing strategy based on the open shortest path first (OSPF) to realize the minimum cost energy interaction among EERs; however, the important factor energy congestion was considered in the reference [15]. Moreover, it should be argued that it was assumed that the power is determined and will be unchanged in the whole transmission path; nevertheless, the power change along the path and affect the optimal path selection.

Investigating the problem of energy congestion, some studies improve the congestion and path selection, such as a time-division multiplexing method [17] congestion avoidance method is proposed [18]. However, while transmitting the same power, the line close to the full current causes to produce greater loss [19], which means that full current should be avoided as far as possible, and the loss of congestion line and shunt lines should be considered at the same time in order to achieve the absolute minimum loss. In addition, since EER is derived from the router concept, the existing literature lacks the consideration of the particularity of energy internet compared with the internet. For example, the priority of routers in the internet is the same, but EER in the energy internet should prioritize because it will affect the income of trading users [20]. Also, the energy internet has very high requirements for power supply reliability [21]; therefore, a fault recovery strategy must need to be required.

In view of the above highlighted issues, this paper designs the EER routing control and fault recovery strategy, the main innovations are as follows:

1) We introduced new strategies of routing control and fault recovery, and two types of EER's transaction priority ranking methods called minimum loss path (MLP) and multi-source cooperative power supply (MCPS) are proposed, which are incentive reasonable quotation and incentive transaction volume.

2) A routing control strategy based on the minimum loss path (MLP) can effectively ensure the minimum power loss both in path selection and congestion avoidance and significantly improve the user's income.

3) A fault recovery strategy based on the multi-source cooperative power supply (MCPS) can meet the key load power consumption and restore the power supply of normal loads to the maximum extent.

In Sect. 2, a neo-type energy internet framework based on EER is proposed. In Sect. 3, the routing control and fault recovery strategy under the framework are proposed. In Sect. 4, the feasibility and superiority of the proposed strategy are verified by simulation. In Sect. 5, the whole paper is summarized.

2   Neo-Type Energy Internet Framework Based on EER

This paper designs a new energy internet framework based on the EER using the existing basic architecture of Information Internet as shown in Fig. 1, which can be divided into three layers from top to bottom as follows:

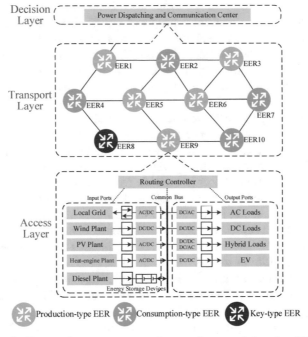

Fig. 1. Proposed a neotype energy internet framework based on EER.

1) *Decision layer.* The power dispatching and communication center works as a key element in the decision layer responsible for providing trading platform. Once the

transaction is completed, the optimal power transmission path to be selected through the routing control or fault recovery strategy considering multiple factors such as the user's needs, trading cost, equipment capacity, scheduling feasibility, and fault status.

2) *Transport layer.* It's important element is EER. According to the dispatching instructions from the decision layer, it can dispatch the e-WAN composed of multiple EERs, and complete the relay protection and control to ensure the safe, reliable, economic, and controllable transmission of power. Considering the characteristics of load and output, EER can be divided into three types: Production-type EER, which only connects to the power generation device (PGD), or simultaneously connects the PGD and load, but the output of the PGD is greater than the load demand, so the power is sold during the trading period. Consumption-type EER, which only connects to the load or simultaneously connects load and PGD, but the load demand is greater than the output. Key-type EER, which connects to the key load, and has the same trading priority with other EER when the energy internet operates stably. However, when a fault occurs, especially production-type EER is disconnected from the e-WAN, the total load demand will be greater than the total output. It is necessary to raise the priority of the key-type EER to the highest level to ensure the power consumption continuity so as to reduce loss or protect the user's safety as much as possible.

3) *Access layer.* It is the interface facing power terminal equipment downward, which makes detailed agreement on the access standards and specifications of the terminal, and is used to realize the access of PGD, energy storage, and load equipment. It provides power production and consumption services for the transport layer upward and sends a notification to the decision layer when the power terminal faults.

3 Routing Control and Fault Recovery Optimization Strategy

According to the energy internet framework based on EER proposed in Fig. 1, this paper designs the EER routing control and fault recovery optimization strategy, as shown in the overall architecture in Fig. 2. Furthermore, we introduce the priority ranking method, routing control strategy based on MLP, and fault recovery strategy based on MCPS, respectively.

3.1 Transaction Priority Ranking Method

The accurate priority ranking of the trading pairs can significantly reduce the power loss; therefore, two priority ranking methods are proposed in this paper:

1) *Incentive reasonable quotation.* As shown in Fig. 3(a), the transaction pairs' priority is sorted according to the bid difference in descending order, and the trading pair with a large bid difference is executed first. In order to incent the production-type EER to reduce and the consumption-type EER to improve quotation within the acceptable range. Since the final price is determined by the clearing price, to a certain extent, a more reasonable quotation will not decrease the user's income. On the contrary, it can decrease the power loss during stable operation or meet the load demand during fault recovery by preferential trading to improve the income.

2) *Incentive transaction volume.* As shown in Fig. 3(b), the trading pair's priority is sorted according to the transaction volume in descending order, and the trading pair with a large trading volume is executed first. In order to incent EER to increase their transaction volume as far as possible within the acceptable range. Similar to the above analysis, to a certain extent, the appropriate increase of transaction volume will reduce the loss and improve the income through preferential trading.

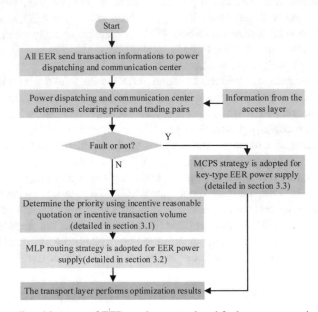

Fig. 2. The overall architecture of EER routing control and fault recovery optimization strategy.

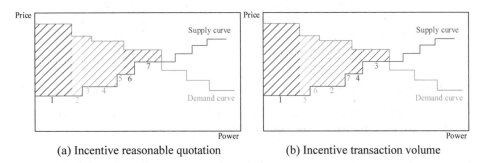

Fig. 3. Two priority ranking methods of trading pairs.

3.2 MLP Based Routing Control Strategy

The routing control strategy based on the MLP can be divided into two aspects: path selection and congestion avoidance. Firstly, the path with the minimum power loss is selected from multiple paths to transmit power; secondly, in order to avoid the power congestion caused by the capacity limitation of lines, the transmission path should be changed to share the power transmission pressure of the line until there is no congestion. The following two aspects are introduced in detail.

3.2.1 MLP Selection

In the real operation of the energy internet, the previous routing control strategy considers that once the transaction power P_{trans} is determined, it will remain unchanged in the whole transmission path, but in fact, P_{trans} will gradually decrease along the path due to the power conversion loss $P_{i.\text{loss}}$ at the EER i input and output ports and the power transmission loss $P_{[i \to j].\text{loss}}$ at the line between two adjacent EER i and j. As shown in Fig. 4, the trading volume of production-type EER s and consumption-type EER d is P_{trans}, and a certain power transmission path between them is $p_1 = s \to l \to \ldots \to n \to d$. The MLP selection method is given as follows:

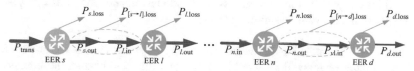

Fig. 4. Schematic diagram of power conversion loss $P_{i.\text{loss}}$ and power transmission loss $P_{[i \to j].\text{loss}}$.

1) *Solve $P_{i.\text{loss}}$ and $P_{[i \to j].\text{loss}}$.* For EER i, $P_{i.\text{loss}}$ and output power $P_{i.\text{out}}$ are

$$P_{i.\text{loss}} = P_{i.\text{in}} \cdot (1 - \eta_{\text{in}} \cdot \eta_{\text{out}}) \tag{1}$$

$$P_{i.\text{out}} = P_{i.\text{in}} - P_{i.\text{loss}} \tag{2}$$

where η_{in} and η_{out} are the power conversion efficiency of input and output port, respectively. The power transmission loss $P_{[i \to j].\text{loss}}$ at the line between EER i and j is

$$P_{[i \to j].\text{loss}} = \frac{R_{[i \to j]}}{U_{[i \to j]}^2}\left[\left(P_{i.\text{out}} + P_{[i \to j].\text{exist}}\right)^2 - P_{[i \to j].\text{exist}}^2\right] \tag{3}$$

where $R_{[i \to j]}$ is the resistance of the line $[i \to j]$, $U_{[i \to j]}$ is the line voltage class, and $P_{[i \to j].\text{exist}}$ is the existing power on the line $[i \to j]$.

2) $V_{p1} = \{s, l, \ldots, n, d\}$ represents the set of all EER on the power transmission path p_1, and L_{p1} represents the set of all lines in p_1. V_{p1} and L_{p1} is stored in the routing tables in advance. Inquiring routing table to determine the source as s and the next hop as l, and then the method in step *1)* was used to solve $P_{s.\text{loss}}$ and $P_{[s \to l].\text{loss}}$. The output power $P_{d.\text{out}}(P_{s.\text{in}})$ of d is finally obtained by constantly inquiring the routing table.

3) Once P_{trans} is determined, $P_{s.\text{in}} = P_{\text{trans}}$, then $P_{i.\text{loss}}$, $P_{[i \to j].\text{loss}}$ and EER d output power $P_{d.\text{out}}(P_{s.\text{in}})$ can be calculated, the total loss $P_{p1.\text{loss}}$ on the transmission path p_1 is

$$P_{p1.\text{loss}} = \sum_{i \in V_{p1}} P_{i.\text{loss}} + \sum_{[i \to j] \in L_{p1}} P_{[i \to j].\text{loss}} = P_{s.\text{in}} - P_{d.\text{out}} \tag{4}$$

4) The same method is used to solve the corresponding loss of other transmission paths, then the power loss of each path is sorted, and the MLP is selected as the optimal power transmission path.

3.2.2 Minimum Loss Congestion Avoidance

In order to avoid power congestion in the lines, the power transmission path should be changed to share the transmission pressure of these lines. The transaction pair cannot be executed until there is no line over-load. Assuming that the transmission line between EER i and j is congested, a minimum loss congestion avoidance method is proposed to avoid congestion and ensure a minimum loss.

1) Assume that the single line between EER i and j cannot guarantee the normal transaction, so two or more lines must be used. The first step checks the connectivity of other lines to EER i and j. In order to reduce the additional power loss caused by EER port startup or shutdown and prolong the EER service life, the number of enabled bypass lines n_{circ} should be reduced as much as possible, as initially can be set as 1, and select the bypass line with the minimum loss to start.
2) The optimization objective is to minimize the loss:

$$\min \sum_{m=1}^{n_{\text{circ}}} P_{\text{circm}.\text{loss}}(P_{\text{circm}}) + P_{[i \to j].\text{loss}}\left(P_{[i \to j]}\right) \tag{5}$$

where P_{circm} is the power distributed by bypass line m, $P_{\text{circm}.\text{loss}}$ is power loss on line m, $P_{[s \to l]}$ is the power distributed by line $[s \to l]$, and $P_{[s \to l].\text{loss}}$ is power loss on line $[s \to l]$.

The power balance constraints is

$$\sum_{m=1}^{n_{\text{circ}}} P_{\text{circm}} + P_{[i \to j]} - P_{i.\text{out}} = 0 \tag{6}$$

The Lagrange function is constructed as follows:

$$L = \sum_{m=1}^{n_{\text{circ}}} P_{\text{circm}.\text{loss}}(P_{\text{circm}}) + P_{[\to j].\text{loss}}\left(P_{[i \to j]}\right) - \lambda \left(\sum_{m=1}^{n_{\text{circ}}} P_{\text{circm}} + P_{[i \to j]} - P_{i.\text{out}}\right) \tag{7}$$

The Lagrangian minimum condition is as follows:

$$\frac{dP_{\text{circm}.\text{loss}}(P_{\text{circm}})}{dP_{\text{circm}}} = \frac{dP_{[i \to j].\text{loss}}(P_{[i \to j]})}{dP_{[i \to j]}} = \lambda, \ m = 1, 2, \ldots, n_{\text{circ}} \tag{8}$$

The existence of multiple bypass lines escalates the Eq. (8) complexity; thus, it is necessary to solve the complex nonlinear Eq. (8) using the iteration method.

3) According to Eq. (8) determine the power distributed by the original and the bypass lines and check the power congestion again. If the congestion still exists, then continue to increase n_{circ}, and perform recalculation until all lines between EER i and j are enabled.

4) In a few extreme cases, even if all the lines between EER i and j are enabled and the congestion still exists, it is necessary to find the last hop EER h of EER i by querying the routing table and update the i in Eqs. (5)–(8) to h. After updating, the calculation is repeated from step 2).

5) If there is no congestion, the calculation is finished. According to the results, the optimized bypass lines and corresponding distributed power is determined to realize the minimum loss congestion avoidance.

3.3   CPS Based Fault Recovery Strategy

MCPS based fault recovery strategy investigates production-type EER is off grid in the faulty conditions, where the power dispatching and communication center needs to raise the trading priority of key-type EER to the highest level to ensure its power consumption. If it is still powered by a single production-type EER, it may be difficult to meet the key load demand fully. In order to fully guarantee the key-type EER i demand, the fault recovery strategy based on MCPS is adopted after the fault.

1) All production-type EERs supply power to EER i. The path selected by each EER is the MLP obtained using the method in Sect. 3.2.1.

2) The optimization objective is to minimize the loss:

$$\min \sum_{m=1}^{n_{\text{multi}}} P_{pm.\text{loss}}(P_{pm}) \tag{9}$$

where P_{pm} is the power allocated by the corresponding production-type EER for path m, $P_{pm.\text{loss}}$ is the power loss at path m. The power balance constraints is

$$\sum_{m=1}^{n_{\text{multi}}} (P_{pm} - P_{pm.\text{loss}}) - P_{i.\text{out}} = 0 \tag{10}$$

The Lagrange function is constructed as follows:

$$L = \sum_{m=1}^{n_{\text{multi}}} P_{pm.\text{loss}}(P_{pm}) - \eta \left(\sum_{m=1}^{n_{\text{multi}}} (P_{pm} - P_{pm.\text{loss}}(P_{pm})) - P_{i.\text{out}} \right) \tag{11}$$

The Lagrangian minimum conditions is as follows:

$$\frac{dP_{pm.\text{loss}}(P_{pm})}{dP_{pm}} = \frac{\eta}{1+\eta}, m = 1, 2, \ldots, n_{\text{multi}} \tag{12}$$

where $P_{i.\text{out}}$ is the power required for EER i, when there are many paths, it is challenging to solve Eq. (12) directly, and the nonlinear Eq. (12) needs to be solved iteratively, which is similar to the solution Eq. (8).

3) According to Eq. (13), power output constraint is checked for all production-type EERs, and those that do not meet the constraint are adjusted according to formula (14) and recalculated according to formula (12) until the constraint is met.

$$P_{pm}^{\max} - P_{pm} \geq 0, m = 1, 2, 3, \ldots, n_{\text{multi}} \tag{13}$$

$$P_{pm} = P_{pm}^{\max} \tag{14}$$

4) After meeting the power output constraint, avoid congestion for each line according to Sect. 3.2.2. The power supply volume and the optimal power transmission path of each production-type EER are selected according to the final results.
5) When key-type EER demand is fully met, the nearby power supply principle is adopted to distribute the surplus power to minimize the power loss and drive more loads to restore power supply as much as possible.

4 Simulation and Analysis

4.1 Key Data

Taking the energy internet in Fig. 1 as an example, the EER routing control and fault recovery strategy are simulated and analyzed. The transaction information of each EER is shown in Table 1, and the parameters of each line in e-WAN are shown in Table 2. The power conversion efficiency η_{in} is 0.99, and η_{out} is 0.98.

Table 1. Transaction information of each EER.

Type	No	Price ($/kW)	P (kW)	Type	No	Price ($/kW)	P (kW)
Production-type	1	0.100	145	Consumption-type	2	0.098	178
	3	0.104	106		4	0.088	120
	5	0.082	136		7	0.105	134
	6	0.086	149		10	0.086	108
	9	0.096	142	Key-type	8	0.110	122

Table 2. The parameters of each line in e-WAN.

Line	R (Ω)	U (V)	Line	R (Ω)	U (V)	Line	R (Ω)	U (V)
[1 → 2]	0.3182	DC800	[3 → 7]	0.3175	DC800	[6 → 7]	0.2582	DC800
[1 → 4]	0.2837	DC800	[4 → 5]	0.2205	DC800	[6 → 10]	0.3837	DC800
[2 → 3]	0.3201	DC800	[4 → 8]	0.2019	DC800	[7 → 10]	0.3201	DC800
[2 → 5]	0.3182	DC800	[5 → 6]	0.2182	DC800	[8 → 9]	0.2447	DC800
[2 → 6]	0.2837	DC800	[5 → 8]	0.3837	DC800	[9 → 10]	0.2345	DC800
[3 → 6]	0.2201	DC800	[5 → 9]	0.2231	DC800			

4.2 Simulation and Analysis of Routing Control Strategy

1) Simulation and analysis of MLP selection

Through centralized trading, the clearing price is 0.098 \$/kW, and the clearing power is 427 kW. The EER allowed being traded shown in italic font in Table 1. Two priority ranking methods are used to determine the trading pairs' priority, and then the loss of each transmission path between the two EERs is sorted. The MLP is selected as the optimal power transmission path. The final results are shown in Table 3. Taking EER5 → EER8 as an example, there are three alternative transmission paths: p_1: 5 → 4 → 8, p_2: 5 → 8, p_3: 5 → 9 → 8. By calculation we can get that $P_{p1.loss} = 18.8423$ kW, $P_{p2.loss} = 15.3121$ kW, $P_{p3.loss} = 19.6787$ kW, so p_2 is selected as the optimal path. It should be noted here that since the reverse power flow may change the direction of the previous power flow and related routes, the power reverse flow is not allowed in this paper. Once the line's power flow direction is determined, it will be prohibited as the transmission line of reverse power trading.

Priority will affect the final power loss. When there is no intersection between each transaction's optimal paths, the proposed priority ranking method will not affect the loss. However, if there is an intersection between paths, the loss of priority execution path will be reduced. As shown in Table 3, when using the ranking method of incentive transaction volume, the trading pair EER6 → EER7 will be ahead of the transaction pair EER5 → EER7, so the loss of EER6 → EER7 is reduced, that is a part of the loss of EER6 → EER7 is transferred to EER5 → EER7.

2) Simulation and analysis of minimum loss congestion avoidance.

In the above simulation, it was assumed that there is no capacity limitations for each line. Here, we considered the capacity of entire lines is limited to 120 kW. After inspection, it is found that the transmission power of the line [6 → 7] is 129.5410 kW, exceeding the capacity limit; congestion avoidance is required. The congestion avoidance results of the method in this paper and in reference [18] are shown in Table 4.

Due to the light congestion, the congestion can be avoided when n_{circ} is 1, and the bypass line is 6 → 3 → 7. From Table 4, it can be found that the minimum loss congestion avoidance method proposed in this paper can cause less power loss and improve user income compared with the other existing methods.

Table 3. Priority of each trading pair and optimal transmission path.

Priority ranking method	Priority of trading pairs	Trading power (kW)	Optimal path	Power loss (kW)
Incentive reasonable quotation	EER5 → EER8	122	5 → 8	15.3121
	EER5 → EER7	14	5 → 6 → 7	1.3412
	EER6 → EER7	120	6 → 7	13.5464
	EER6 → EER2	29	6 → 2	2.0431
	EER9 → EER2	142	9 → 5 → 2	26.3582
Incentive transaction volume	EER9 → EER2	142	9 → 5 → 2	26.3582
	EER5 → EER8	122	5 → 8	15.3121
	EER6 → EER7	120	6 → 7	12.3509 (↓)
	EER6 → EER2	29	6 → 2	2.0431
	EER5 → EER7	14	5 → 6 → 7	2.5367 (↑)

Table 4. The power distribution results obtained by different congestion avoidance methods.

Congestion circumvention method	Selected lines	Distributed power (kW)	Power loss (kW)	Total loss (kW)
Previous method introduced in ref [18]	6 → 7	120	9.2124	9.8430
	6 → 3 → 7	9.5410	0.6306	
Proposed method	6 → 7	98.1191	6.6922	9.2944
	6 → 3 → 7	31.4219	2.6022	

4.3 Simulation and Analysis of Fault Recovery Strategy Based on MCPS

The energy internet fault simulations consider the production-type EER 6 with the largest transaction volume is out of operation. At this time, it is necessary to rely on MCPS to restore power for key-type EER 8, and then restore power supply to other EERs nearby. The simulation results of using the fault recovery strategy using the routing control strategy are shown in Table 5.

It can be seen from Table 5 that the adoption of fault recovery strategy can reduce the power loss in the whole e-WAN after meeting the power consumption of key-type EER. In the case of insufficient output caused by a fault, the smaller the loss is, the more load can be driven to restore power supply, which clearly shows the superiority of the proposed fault recovery strategy applying fault conditions.

Table 5. The simulation results obtained by different strategies in the case of fault.

Strategy	Trading pairs	Trading power (kW)	Optimal path	Power loss (kW)	Total loss (kW)
Routing control strategy	EER5 → EER8	122.0000	5 → 8	15.3121	45.1744
	EER5 → EER7	14.0000	5 → 2 → 3 → 7	1.8367	
	EER9 → EER2	142.0000	9 → 5 → 2	28.0256	
Fault recovery strategy	EER5 → EER8	45.3310	5 → 8	3.7886	28.6291
	EER9 → EER8	71.0809	9 → 8	5.9375	
	EER5 → EER2	90.6690	5 → 2	9.0560	
	EER9 → EER7	70.9191	9 → 10 → 7	9.8470	

5   Conclusion

In this paper, we addressed the advent of emerging internet technology, energy internet, and routing control schemes to utilize RERs effectively. We also highlighted the limitations found in the literature regarding optimal path selection, congestion, and fault recovery and proposed new routing control methods taking into account these limitations. The feasibility and superiority of the proposed methods are verified via obtained simulation results and comparison; the proposed methods also show the following:

1) New transaction priority ranking methods for EER, namely MLP and MCPS using incentive reasonable quotation and incentive transaction volume, are proposed, which can be achieved through power loss transfer.
2) The proposed routing control strategy based on the MLP has enhanced path identification ability compare to other techniques with smaller power loss in congestion avoidance, which consequently improves the user's income.
3) The proposed fault recovery strategy based on MCPS can guarantee the key power consumption when fault occurs. Compared with the routing control strategy in stable operation, it can drive the normal load to resume power supply to a greater extent.

Acknowledgments. The authors are grateful for the financial support from the National Nature Science Foundation of China under Grant No. 51877089.

References

1. Elavarasan, R.M., et al.: A comprehensive review on renewable energy development, challenges, and policies of leading Indian states with an international perspective. IEEE Access **8**, 74432–74457 (2020)
2. Aftab, M.A., Hussain, S.M.S., Ali, I., Ustun, T.S.: Dynamic protection of power systems with high penetration of renewables: a review of the traveling wave based fault location techniques. Int. J. Electr. Power Energy Syst. **114**, 105410 (2020)

3. Telukunta, V., Pradhan, J., Agrawal, A., Singh, M., Srivani, S.G.: Protection challenges under bulk penetration of renewable energy resources in power systems: a review. CSEE J. Power Energy Syst. **3**, 365–379 (2017)
4. Ullah, Z., Wang, S., Radosavljević, J.: A novel method based on PPSO for optimal placement and sizing of distributed generation. IEEE J. Trans. Electr. Electron. Eng. **14**, 1754–1763 (2019)
5. Elkadeem, M.R., Elaziz, M.A., Ullah, Z., Wang, S., Sharshir, S.W.: Optimal planning of renewable energy-integrated distribution system considering uncertainties. IEEE Access **7**, 164887–164907 (2019)
6. Ullah, Z., Wang, S., Radosavljević, J.: A novel method based on PPSO for optimal placement and sizing of distributed generation. IEEJ Trans. Electr. Electron. Eng. **14**(12), 1754–1763 (2019)
7. Ullah, Z., Elkadeem, M.R., Wang, S., Akber, S.M.A.: Optimal planning of RDS considering PV uncertainty with different load models using artificial intelligence techniques. Int. J. Web Grid Serv. **16**(1), 63–80 (2020)
8. Ullah, Z., Elkadeem, M.R., Wang, S., Sharshir, S.W., Azam, M.: Planning optimization and stochastic analysis of RE-DGs for techno-economic benefit maximization in distribution networks. Internet Things **11**, 100210 (2020)
9. Ullah, Z., Wang, S., Radosavljevic, J., Lai, J.: A solution to the optimal power flow problem considering WT and PV generation. IEEE Access **7**, 46763–46772 (2019)
10. Hannan, M.A., et al.: A review of internet of energy based building energy management systems: Issues and recommendations. IEEE Access **6**, 38997–39014 (2018)
11. Xue, Y.: Energy internet or comprehensive energy network? J. Mod. Power Syst. Clean Energy **3**, 297–301 (2015)
12. Guo, H., Wang, F., Zhang, L., Luo, J.: A hierarchical optimization strategy of the energy router-based energy internet. IEEE Trans. Power Syst. **6**, 4177–4185 (2019)
13. Suhail Hussain, S.M., Aftab, M.A., Nadeem, F., Ali, I., Ustun, T.S.: Optimal energy routing in microgrids with IEC 61850 based energy routers. IEEE Trans. Ind. Electron. **67**, 5161–5169 (2020)
14. Li, P., Sheng, W., Duan, Q., Li, Z., Zhu, C., Zhang, X.: A Lyapunov optimization-based energy management strategy for energy hub with energy router. IEEE Trans. Smart Grid **11**(6), 4860–4870 (2020)
15. Wang, R., Wu, J., Qian, Z., Lin, Z., He, X.: A graph theory based energy routing algorithm in energy local area network. IEEE Trans. Ind. Informatics **13**, 3275–3285 (2017)
16. Ge, X., Pan, L., Li, Q., Mao, G., Tu, S.: Multipath cooperative communications networks for augmented and virtual reality transmission. IEEE Trans. Multimed. **19**, 2345–2358 (2017)
17. Guo, H., Wang, F., James, G., Zhang, L.: Luo, J: Graph theory based topology design and energy routing control of the energy internet. IET Gener. Transm. Distrib. **12**, 4507–4514 (2018)
18. Guo, H., Wang, F., Zhang, L., Feng, X., Luo, J.: Matchmaking tradeoff based minimum loss routing algorithm in energy internet. Dianli Xitong Zidonghua/Autom. Electr. Power Syst. **42**, 172–179 (2018)
19. Wang, S., Liu, Q., Ji, X.: A fast sensitivity method for determining line loss and node voltages in active distribution network. IEEE Trans. Power Syst. **33**, 1148–1150 (2018)
20. Sandgani, M.R., Sirouspour, S.: Priority-based microgrid energy management in a network environment. IEEE Trans. Sustain. Energy **9**, 980–990 (2018)
21. Wang, J., Huang, M., Fu, C., Li, H., Xu, S., Li, X.: A new recovery strategy of HVDC system during AC faults. IEEE Trans. Power Deliv. **34**, 486–495 (2019)

End-to-End Data Pipeline in Games for Real-Time Data Analytics

Noppon Wongta[✉] and Juggapong Natwichai

Department of Computer Engineering, Faculty of Engineering,
Chiang Mai University, Mueang Chiang Mai, Thailand
noppon.w@cmu.ac.th, juggapong@eng.cmu.ac.th

Abstract. Data pipeline architecture in game is a system that captures, organizes, and routes data, which enables reporting or data analytics. Such results can be utilized by game developers. A main challenge is that to allow huge data from the game clients to flow to the system, while the core game processing must not be interrupted. This paper reviews thorough concepts related to the game analytics. In addition, the observations of the data pipeline components in games are made and discussed to evaluate the challenges which are real-time data pipeline and in-memory process implementation.

1 Introduction

Data pipeline design generally composes of many manual processes that need to define what, where, and how data is collected. Besides, such processes need an automated flow of data for extracting, transforming, combining, validating, or loading data from one process to the next one. Usually, the operation of the data pipeline has to be ensured that the streamlined data has be processed and available 24/7 [5]. So, the data pipeline would be designed to divide each data stream processing into smaller chunks that can be processed in parallel, and thus it could be utilized the computing power more efficiently.

One of the most important differences for a data pipeline in games is that the implementation to obtain the data from the game client, while it should not interrupt the core game processing, i.e. object rendering or game playing. Meanwhile, game companies need the data for analysis to improve the gaming experience, game design, and reduce player churn [19]. The balance between such two objectives, gameplay, and analytics, must be addressed carefully.

In general, data pipeline in game analytics has been designed to meet various objectives. Such objectives define system design both for data collection and data processing. For example, if a game studio needs to know how many players play their game on-line. The system needs to design in order to collect player information from clients such as player name and time stamp. If the number of players is increasing, the system should be able to adapt its scale to support such data collection. Also, infrastructure defines components of the system and how to operate them. For instance, if an E-sport host would like to share game

L. Barolli et al. (Eds.): EIDWT 2021, LNDECT 65, pp. 269–275, 2021.
https://doi.org/10.1007/978-3-030-70639-5_25

replay to other players in game, the system could provide broadcast data to game clients in different regions around the world without network latency that could cause game interruption. Thus, the infrastructure must support broadcasting efficiently.

There are various attempts to address the issues. For example, in [13], an AI solutions service is proposed by Electronic Arts Inc. (EA) that provided analytic services for other studios under cooperating with EA. This work also explained a horizontal scaling strategy for delivering AI services, by combining a data warehouse for player history, machine learning, and the engine for the action layer. In [9], the authors proposed to collects real-time data in games for match result prediction by Game state integration (GSI)[1]. However, this research did not focus on designing data flow for real-time analytics, which might be needed in practices.

In this paper, the work related to game analytics are thoroughly reviewed. Not only the general analytics, but also the data pipelines in games analytics. Such concepts can help analysts to further design/redesign/response to the game engine more effective and efficient. In addition, we discuss two approaches to improve real-time data analytic process pipeline, i.e. 1) real-time data flow model processing to combine between historical and stream data 2) in-memory computing for end-to-end data pipeline.

The structure of this paper is organized as follows. First the related work, which composes of background of game analytics, data pipeline for game analytics, and real-time processing issues, is reviewed in Sect. 2. Then, the challenges of the game analytics which is the focused of this paper is presented in Sect. 3. Finally, we conclude the paper in Sect. 4.

2   Related Work

In this section, firstly we present attempts to address game analytics that the game industry usually need to do to know more on their customer's experiences or reactions from their own game. Secondly, we describe work related to the data pipeline for game analytics. Lastly, we explain the needs to process the data real-time for game analytics.

2.1   Game Analytics

In general, analytics in games [3] demands to understand the reasoning behind game engaging, monetization, fraud and player investigation, game performance, and error reporting. To gain more engagement in games, usually, developers take the data into development process to enhance game design to attract more players and increase engagement. Game monetization is the process of gain money or revenue from players which in general the process focuses on mobile games.

---

[1] https://developer.valvesoftware.com/wiki/Counter-Strike: Global Offensive Game State Integration and https://github.com/antonpup/Dota2GSI.

The developer could use game analytics to find the most effective strategy for players to make purchases in-game. Moreover, using player investigations analytics to identify player behavior in game could improve game environment and mechanics for player experience. In game playing data, there are several work attempts to research on win prediction in Esports [1,10,11,16,18]. Such type of work can use data of player performance, spatiotemporal data, and the correlation between players and game-character skills. Also, in [6], the authors present the classification on player roles in a MOBA (Multiplayer online battle arena) game, DotA2. In [20], MOBA-Slice which is based on a supervised learning model is proposed to demonstrate the idea of streaming processing data with game records. In terms of performance and error reporting, developers could generally analyze data in games to understand important engine metrics such as CPU and memory utilization, and measure latency-sensitive indicators.

2.2 Data Pipeline for Game Analytics

In general, a simple architecture data pipeline to collect the data and enable game analytics, has a game client as the front-end that could send data from player to game analytics services as the back-end. In [19], an efficient data flow for games analytics is proposed as follows. First, the flow allows game clients to send requests to dynamic difficulty adjustment system, then the service determines top five candidate random seeds back to the game client for selecting suitable difficulty level.

From the literature, data pipeline services for game analytics can be categorized into two groups 1) service for game data pipeline 2) custom microservices that run on cloud infrastructures.

For the first category, there are three main services for the game data pipeline in the industry. First, Game Spark[2] from Amazon Inc. that highlights features for the custom front-end to view data such as dashboard and harness testing system. Secondly, Playfab[3] by Microsoft which has the Full-stack LiveOps and Real-Time Control features for understanding and reacting to player behavior. Last, GameAnalytics[4] which has analytics tracking systems to empower game developers and publishers with free of charge fee.

The second category, the most popular custom microservice usually run on cloud infrastructure using Google Cloud and Amazon Web Services (AWS). Google Cloud[5] provides two main architecture patterns for analyzing mobile game events: Real-time processing for individual events using a streaming processing pattern and Bulk processing for aggregated events using a batch processing pattern. Amazon Web Services (AWS)[6] provides a scalable serverless data pipeline for game developers to stream ingest, store, and analyze telemetry data generated from games and services. These solutions lack of support for

---

[2] https://www.gamesparks.com/.
[3] https://playfab.com/.
[4] https://gameanalytics.com/.
[5] https://cloud.google.com/solutions/mobile/mobile-gaming-analysis-telemetry.
[6] https://aws.amazon.com/solutions/implementations/game-analytics-pipeline/.

the library on game clients or game engines, so the developers need to implement the analytic engine by themselves. Usually, REST API or low-level network applications are applied.

Aside from the mentioned two main approaches, some providers that specialize in data analytics present alternative solutions for adaptive AWS from their analytics system. For example, Doit[7] offers solutions that automatically launch and helps to configure many Amazon microservices to analyze player experiences, advertising effectiveness, and game mechanics. Also, Databrick[8] is a unified platform for data science and artificial intelligence, built on lake house architecture. This platform creates end-to-end streaming data pipeline for mobile games and addresses results of real-time analytics, real-time KPI's and real-time visualization.

2.3 Real-Time Processing Issues

In General, real-time data pipelines are data pipelines that handle millions of events at scale of real-time processing and response. The process usually operates from in-memory computing architecture and variation of in-memory databases such as query, and intermediate results [12].

A main components for creating a real-time data pipeline are database system that can process the workloads. When data is streamed to the system, database is first part of component which stores dataset from game client. The in-memory database systems for real-time workloads can be categorized into two types [5].

1. Online Transaction Processing (OLTP) focuses on a high volume of low-latency operations such as S-Store [17] which is an extension of H-Store that presents how can simultaneously accommodate OLTP and streaming applications.
2. Online Analytical Processing (OLAP) focuses on the capture of transactions such as Trill [2] which is a query processor for analytics, it derives real-time streaming queries, temporal queries on a historical log.

Both types can process in batches and timestamp-data for separate operators to support MapReduce [4], execute optional scheduler, and improve performance for real-time process.

On the other hand, when optimization is needed, the needs to combine relational and linear algebra operator with user define functions(UDFs) and control flow is addressed in [14]. This work proposes Lara, a domain-specific languages (DSLs) that combines collection processing and machine learning. It provides monadic view [8] and combinator view [7] on the intermediate representation(IR) [15] to perform diverse optimizations end-to-end data pipeline.

---

[7] https://www.doit-intl.com/services/gaming-analytics-pipeline-on-aws/.
[8] https://databricks.com/blog/2018/07/02/build-a-mobile-gaming-events-data-pipeline-with-databricks-delta.html.

3 Challenges in Game Analytics

We propose two approaches to improve real-time data analytic process. 1) real-time data flow model processing to combine between historical and stream data 2) in-memory computing for end-to-end data pipeline. After the related work has been reviewed in the previous section, this section discusses challenges in game analytics. First, we observe that the game analytics services usually do not consider end-to-end analytical process. One might considered from the data sources, ETL sub-systems, and basic analytics, meanwhile the other groups might consider only machine learning features. Though, considering overall process from the sources until the advanced analytics, e.g. classification, clustering, or prediction to understand deep information of the players' behavior, could be of importance.

Also, incorporating the real-time data flow model which operates on temporal dataset, could benefit analysts who could use further extended query model. It is able to diverse spectrum of analytics including: real-time, offline, temporal, relational, and progressive. The data can be stored as columnar data-batches in order to serializers encode arrays directly without any fine-grained encoding, and also memory pools help reuse data-batches. In addition, temporal operators can be associated with a data window (or interval). They can access dataset as a separate insertion and deletion on temporal datasets. Which have two timestamps (sync-time and other-time) for optimized operation. This idea demonstrates analytical processes by combining a historical and stream data process that could improve the accuracy of machine learning models.

In terms of in-memory computing for end-to-end pipelines. We consider that the program execution by type-based DSLs and IR should be implemented well to reflect on the completeness of analytics. Using quotation-based DSLs for collection processing operators, which have meta programming capabilities to access the Abstract syntax tree (AST) such as LINQ, LMS, and Squid, should be included. Also, IR representation on AST as a low-level intermediate representation (LIR) and two higher-level views such as monadic view which represent monad comprehensions and combinator view that represent high-level operators in an operator tree. For the execution, it would process on top of domain-specific languages and intermediate representation, that it could improve in-memory computing for analytics and increase processing time.

Both concepts provide an important contribution of an end-to-end pipeline and optimizations access to dataset for real-time game analytics.

4 Conclusion

In this paper, we have presented related work and challenges for game analytics. We have discussed that real-time data analytics is very important for game developers to analyze players' behaviour using basic analytics and machine learning. We have presented two approaches concerning concepts to combine data from with streams and to implement the in-memory computing operators, that could improve the data pipeline performance. The main concept can contribute to end-to-end data pipeline and real-time analytics, which lead to conducting a thorough implementation of game analytic modules.

For the future work, we will explore the efficiency issues of the individual components for creating real-time data pipeline. For the in-memory computing, concept of monad comprehensions will be a main candidate to be investigated.

References

1. Agarwala, A., Pearce, M.: Learning Dota 2 team compositions, 2–6 (2014). http://cs229.stanford.edu/proj2014/Atish Agarwala, Michael Pearce, Learning Dota 2 Team Compositions.pdf
2. Chandramouli, B., Goldstein, J., Barnett, M., DeLine, R., Fisher, D., Platt, J.C., Terwilliger, J.F., Wernsing, J.: Trill: a high-performance incremental query processor for diverse analytics. Proc. VLDB Endowment **8**(4), 401–412 (2014). https://doi.org/10.14778/2735496.2735503
3. Dawson, C., Dawson, C.: Game analytics (2019). https://doi.org/10.4324/9781351044677-21
4. Dean, J., Ghemawat, S.: MapReduce: simplified data processing on large clusters. JTN Weekly, 22 June September, 6 (1996). https://doi.org/10.1145/1327452.1327492
5. Doherty, C., Orenstein, G., Camiña, S., White, K.: Building Real-Time Data Pipelines. O'Reilly Media, United States (2015)
6. Eggert, C., Herrlich, M., Smeddinck, J., Malaka, R.: Classification of player roles in the team-based multi-player game dota 2. Lecture Notes in Computer Science (including subseries Lecture Notes in Artificial Intelligence and Lecture Notes in Bioinformatics), vol. 9353, 112–125 (2015). https://doi.org/10.1007/978-3-319-24589-809
7. Grust, T.: Comprehending Queries. August 2001 (2000). https://doi.org/10.1007/978-3-322-84823-9
8. Grust, T.: Monad comprehensions: a versatile representation for queries. In: The Functional Approach to Data Management, pp. 288–311 (2004). https://doi.org/10.1007/978-3-662-05372-012
9. Hodge, V., Devlin, S., Sephton, N., Block, F., Cowling, P., Drachen, A.: Win prediction in multi-player esports: live professional match prediction. IEEE Trans. Games, October 2019. https://doi.org/10.1109/tg.2019.2948469
10. Kalyanaraman, K.: To win or not to win? A prediction model to determine the outcome of a DotA2 match (2014). https://cseweb.ucsd.edu/~jmcauley/cse255/reports/wi15/Kaushik_Kalyanaraman.pdf
11. Kinkade, N., Jolla, L., Lim, K.: DOTA 2 Win Prediction. Univ. Calif., pp. 1–13 (2015)
12. Kleppmann, M.: Designing Data-Intensive Applications (2017)
13. Kolen, J.F.J., Sardari, M., Mattar, M., Peterson, N., Wu, M.: Horizontal scaling with a framework for providing AI solutions within a game company. In: 32nd AAAI Conference on Artificial Intelligence, AAAI 2018, pp. 7680–7687, February 2018
14. Kunft, A.: Optimizing end-to-end machine learning pipelines for model training, August 2019

15. Kunft, A., Katsifodimos, A., Schelter, S., Breß, S., Rabl, T., Markl, V.: An intermediate representation for optimizing machine learning pipelines. Proc. VLDB Endowment. **12**(11), 1553–1567 (2018). https://doi.org/10.14778/3342263.3342633
16. Makarov, I., Savostyanov, D., Litvyakov, B., Ignatov, D.I.: Predicting winning team and probabilistic ratings in "Dota 2" and "Counter-strike: Global offensive" video games. Lect. Notes Comput. Sci. (including Subser. Lect. Notes Artif. Intell. Lect. Notes Bioinformatics), 10716 LNCS, January 2018, pp. 183–196 (2018) https://doi.org/10.1007/978-3-319-73013-417
17. Meehan, J., Tatbul, N., Zdonik, S., Aslantas, C., Cetintemel, U., Du, J., Kraska, T., Madden, S., Maier, D., Pavlo, A., Stonebraker, M., Tufte, K., Wang, H.: S-store: streaming meets transaction processing. Proc. VLDB Endowment. **8**(13), 2134–2145 (2015). https://doi.org/10.14778/2831360.2831367
18. Pobiedina, N., Neidhardt, J., Moreno, M.D.C.C., Werthner, H.: Ranking factors of team success. In: WWW 2013 Companion - Proceedings of the 22nd International Conference on World Wide Web, pp. 1185–1193, May 2013. https://doi.org/10.1145/2487788.2488147
19. Xue, S., Wu, M., Kolen, J., Aghdaie, N., Zaman, K.A.: Dynamic difficulty adjustment for maximized engagement in digital games. In: 26th International World Wide Web Conference 2017, WWW 2017 Companion, pp. 465–471 (2019). https://doi.org/10.1145/3041021.3054170
20. Yu, L., Zhang, D., Chen, X., Xie, X.: MOBA-Slice: a time slice based evaluation framework of relative advantage between teams in MOBA games. Commun. Comput. Inf. Sci. **1017**, 23–40 (2019). https://doi.org/10.1007/978-3-030-24337-1_2

Decision Analysis of Winter Road Conditions by Crowd Sensing Platform

Yoshitaka Shibata[1]([⊠]), Akira Sakuraba[1], Yoshikazu Arai[2], Yoshiya Saito[2], and Jun Hakura[2]

[1] Regional Corporate Research Center, Iwate Prefectural University, Sugo, Takizawa 152-89, Iwate, Japan
{shibata,a_saku}@iwate-pu.ac.jp
[2] Faculty of Software and Information, Iwate Prefectural University, Sugo, Takizawa 152-52, Iwate, Japan
{arai,y-saito,hakura}@iwate-pu.ac.jp

Abstract. In order to realize autonomous driving system on ordinal roads, it is necessary for drivers to precisely predict and know the road state before running the objective road. In this paper, accuracy of states decision on actual winter road using various on-board sensors including 9-axis dynamic sensors, near infrared razer sensor, far infrared temperature sensor and humid sensor is analyzed.

1 Introduction

Autonomous driving is very attractive technology for various fields such as car industries, public and private transportation industries, logistics, tourist industries from economical point of view. Currently well developed countries such as the U. S, China, Japan and Europe countries have produced the practical autonomous cars which can run on the exclusive roads and highway roads at level 3 or 4. The roads conditions in those countries on which autonomous car run are ideal for autonomous car driving because the driving lanes are clear, the driving direction is the same and opposite driving car lanes are completely separated [1]. So far, the autonomous driving cars are autonomously running on the roads while sensing obstacles and deciding dangerous or safe condition by themselves using LiDAR and camera around the relatively narrow areas within hundred meter ahead.

However, most of the ordinal roads are not always well maintained compared with the exclusive roads and highway roads. In particular, the road conditions in local areas are so rough and dangerous due to luck of regular road maintenance in addition to falling obstacles from other vehicles or precipice from roadside.

On the other hand, in the cold or snow countries, such as Japan and Northern countries, most of the road surfaces are occupied with heavy snow and iced surface in winter as shown in Fig. 1 and many slip accidents occur even though the vehicles attach anti-snow slip tires. In fact more than 90% of traffic accidents in northern part of Japan is caused from slipping car on snowy or iced road [2].

In order to resolve those problems, particularly for winter season in snow countries, we introduce a new crowd sensing platform, particularly for winter. In road state sensing,

L. Barolli et al. (Eds.): EIDWT 2021, LNDECT 65, pp. 276–287, 2021.
https://doi.org/10.1007/978-3-030-70639-5_26

Fig. 1. Road conditions in snow country

the sensor data from various environmental sensors including accelerator, gyro sensor, infrared temperature sensor, quasi electrical static sensor, camera and GPS attached on vehicle are periodically sampled and integrated to precisely determine the various road conditions and identify the dangerous locations. This road information is transmitted to the neighbor vehicles and road side servers in realtime using V2X communication network. This road information is also collected in wide area as bigdata to the cloud computing server on Internet through the roadside servers and analyzed to predict the future road states and unobserved local roads by combining with open weather data and 3D terrain data along roads. Thus, wide area road state GIS information platform can be attained.

In the following, the related works with road state information by sensor technology are explained in section two. A general system and architecture of the proposed road surface state information platform are explained in section three. Next, crowd sensing system to collect various sensor data and to identify road states is introduced in section four. The temporal and geological prediction of road conditions using the observed sensor data and meteorological meshed data is explained in section five. Preliminary experiment by prototyped platform to evaluate our proposed system is precisely explained in section six. In final, conclusion and future works are summarized in section seven.

2   Related Works

With the road state sensing method, there are several related works so far. Particularly road surface temperature is essential to know the road state such as snowy or icy in winter season whether the road surface temperature is under minus 4 °C or below.

In the paper [3], the road surface temperature model by taking account of the effects of surrounding road environment to facilitate proper snow and ice control operations is introduced. In this research, the fixed sensor system along road is used to observe the precise temperature using the monitoring system with long-wave radiation. They build the road surface temperature model using heat balance method.

In the paper [4–6], cost effective and simple wide area road surface temperature prediction method while maintaining the prediction accuracy of current model is developed. Using the relation between the air temperature and the meshed road surface temperature, statistical thermal map data are calculated to improve the accuracy of the road surface temperature model. Although the predicted accuracy is high, the difference between the ice and snow states was not clearly resolved.

In the paper [7], a road state data collection system of roughness of urban area roads is introduced. In this system, mobile profilometer using the conventional accelerometers to measure realtime roughness and road state GIS is introduced. This system provides general and wide area road state monitoring facility in urban area, but snow and icy states are note considered.

In the paper [8], a measuring method of road surface longitudinal profile using build-in accelerometer and GPS of smartphone is introduced to easily calculate road flatness and International Road Index (IRI) in offline mode. Although this method provides easy installation and quantitative calculation results of road flatness for dry or wet states, it does not consider the snow or icy road states.

In the paper [9], a statistical model for estimating road surface state based on the values related to the slide friction coefficient is introduced. Based on the estimated the slide friction coefficient calculated from vehicle motion data and meteorological data, the road surface state is predicted for several hours in advance. However, this system does not consider the other factors such as road surface temperature and humidity.

In the paper [10], road surface temperature forecasting model based on heat balance, so called SAFF model is introduced to forecast the surface temperature distribution on dry road. Using the SAFF model, the calculation time is very short and its accuracy is higher than the conventional forecasting method. However, the cases of snow and icy road in winter are not considered.

In the paper [11], blizzard state in winter road is detected using on-board camera and AI technology. The consecutive ten images captured by commercial based camera with 1280×720 are averaged by pre-filtering function and then decided whether the averaged images are blizzard or not by Convolutional Neural Network. The accuracy of precision and F-score are high because the video images of objective road are captured only in the daytime and the contrast of those images are almost stable. It is required to test this method in all the time.

In the paper [12], road surface state analysis method is introduced for automobile tire sensing by using quasi-electric field technology. Using this quasi-electric field sensor, the changes of road state are precisely observed as the change of the electrostatic voltage between the tire and earth. In this experiment, dry and wet states can be identified.

In the paper [13], a road surface state decision system is introduced based on near infrared (NIR) sensor. In this system, three different wavelengths of NIR laser sensors is used to determine the qualitative paved road states such as dry, wet, icy, and snowy states as well as the quantitative friction coefficient. Although this system can provides realtime decision capability among those road states, decision between wet and icy states sometimes makes mistakes due to only use of NIR laser wavelength.

With all of the systems mentioned above, since only single sensor is used, the number of the road states are limited and cannot be shared with other vehicles in realtime. For those reasons, construction of communication infrastructure is essential to work out at challenged network environment in at inter-mountain areas. In the followings, a new road state information platform is proposed to overcome those problems.

3   Road Surface State Information Platform

In order to resolve those problems in previous session, we introduce a new generation wide area road surface state information platform based on crowd sensing and V2X technologies as shown in Fig. 2. The wide area road surface state information platform is organized mainly by mobile wireless nodes, so called Smart Mobile Box (SMB) and roadside wireless nodes, so called Smart Relay Shelters (SRS). Each SMB is furthermore organized by a sensor server unit and wireless communication unit, and installed in a vehicle. The sensor server unit includes various sensor devices such as semi-electrostatic field sensor, an acceleration sensor, gyro sensor, temperature sensor, humidity sensor, infrared sensor and sensor server. Using those sensor devices, various road surface states such as dry, rough, wet, snowy and icy roads can be qualitatively and quantitatively decided.

On the other hand, the wireless communication unit in SMB and SRS includes multiple wireless network devices with different N-wavelength (different frequency bands) wireless networks such as IEEE802.11b/g/n (2.4 GHz), 11ac (5.6 GHz), 11ad (28 GHz), 11ah (920 MHz) and organizes a cognitive wireless node. The network node can selects the best link among the cognitive wireless network depending on the observed network quality, such as RSSI, bitrate, error rate by Software Defined Network (SDN). If none of link connection is possible, those sensor data are locally and temporally stored in the database unit as internal storage until the vehicle approaches to the region where a link connection to another mobile node or roadside node is possible. When the vehicle enters the region, it starts to connect a link to other vehicle or roadside node and transmits sensor data by DTN Protocol. Thus, data communication can be attained even though the network infrastructure is not existed in challenged network environment such as mountain areas or just after large scale disaster areas.

Thus, in our system, SRS and SMB organize a large scale information infrastructure without cellular network. The collected sensor data and road state information on a SMB are directly exchanged to near SMB by V2V communication. Those received data from near SMB are displayed on the digital map of the viewer system while alerting the dangerous locations such as snowy and icy road locations. Thus, the vehicle can know and pay attention to the ahead road state before passing through.

On the other hand, when the SMB on own vehicle approaches to the SRS, the SMB make a connection to the SRS, delivers the sensor data and the decided road state information in the other SMBs through the N-wavelength wireless network. At the same time, those collected sensor data are sent to the cloud computing server in data center through the cloudlet in SRS. By combining those sensor data form SRS on various locations with public mesh weather data such as temperature, pressure, amount of rain and snow as open data, a wide area road states can be geographically and temporally predicted and provided as road state Geographic Information System (GIS) information service through public network not only for current running drivers but for even ordinal users.

Thus, the proposed system can provide safer and more reliable driving environment even though winter season in snow countries. This network not only performs various road sensor data collection and transmission functions, but also provide Internet access network functions to transmit the various data, such as sightseeing information, disaster

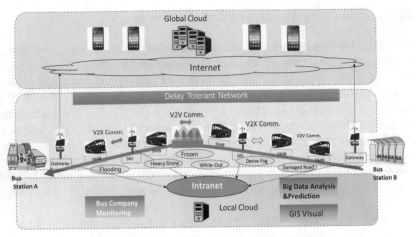

Fig. 2. Road surface state information platform

prevention information and shopping and so on as ordinal public wide area network for residents. Therefore, many applications and services can be realized.

4   Road State Sensor System

In order to detect the precise road surface states, such as dry, wet, dumpy, showy, frozen roads, various sensing devices including accelerator, gyro sensor, infrared temperature sensor, humidity sensor, quasi electrical static sensor, camera and GPS are integrated to precisely and quantitatively detect the various road surface states and determine the dangerous locations on GIS in sensor server as shown in Fig. 3. The 9 axis dynamic sensors including accelerator, gyro sensor and electromagnetic sensors can measure vertical amplitude of roughness along the road. The infrared temperature sensor observes the road surface temperature to know whether the road surface is snowy or icy without touching the road surface. The quasi electrical static sensor detects the snow and icy states by observing the quasi electrical static field intensity. Camera can detect the obstacles on the road. The far-infrared laser sensor precisely measures the friction rate of snow and icy states. The sensor server periodically samples those sensor signals and performs AD conversion and signal filtering in Pre-filtering module, analyzes the sensor data in Analyzing module to quantitatively determine the road surface state and learning from the sensor data in AI module to classify the road surface state as shown in Fig. 4. As result, the correct road surface state can be quantitatively and qualitatively decided. The decision data with road surface state in SMB are temporally stored in Regional Road State Data module and mutually exchanged when the SMB on one vehicle approaches to other SMB. Thus the both SMBs can mutually obtain the most recent road surface state data with just forward road. By the same way, the SMB can also mutually exchange and obtain the forward road surface data from roadside SRS.

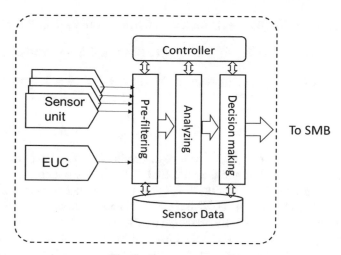

Fig. 3. Sensor server unit

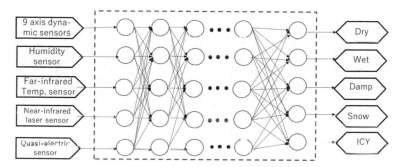

Fig. 4. Road state decision system by AI

5 Temporal and Geological Prediction of Road Conditions

In order to predict the road conditions in future time, the current observed sensor data and predicted meteorological data from meteorological agency must be integrated as bigdata and analyzed. To formulate the prediction process, the whole area R is divided into the region Rn for n = 1, N as shown in Fig. 5. In Rn, we define a road position as lk for 1 to d and time ti for 1 to e. Here each temperature is defined as Observed road surface temperature TMn(lk, ti) for k = 1, d, Observed air temperature TSn(lk, ti) for k = 1, d, the average air temperature TSn(lkav, ti) where lkav is expressed as an average of whole road position. Then, TSn(lkav, ti) can be expressed as:

$$TSn(lkav, \ ti) \ = \ a \ * \ TMn(lkav, \ ti) \ + \ b \qquad (1)$$

We can obtain the meteorological mesh data at the region Rn from meteorological agency, the future temperature TLn(lkav, ti) for i = 0, d. Therefore, the future TSn(lkav, ti + 1) can be predicted using TLn(lkav, ti + 1) as follows.

$$TSn(lkav, \ ti + 1) \ = \ a' \ * \ TLn(lkav, \ ti + 1) \ + \ b' \qquad (2)$$

Therefore, the future TSn(lk, ti + 1) for k = 1, d can be estimated as follows:

$$TSn(lk, \, ti + 1) = TSn(lk, \, ti) + a' * \{TLn(lkav, \, ti + 1) - TSn(lkav, \, ti)\} \quad (3)$$

Finally, the future road surface temperature TMn(lk, ti + 1) for k = 1, d can be estimated as the follows:

$$TMn(lk, \, ti + 1) = TMn(lk, \, ti) + a'' * \{TSn(lk, \, ti + 1) - TSn(lk, \, ti)\} \text{ for } k = 1, \, d \quad (4)$$

The parameters, a, a′, a″, b, b′, b″, can be determined by statistical process between the observed temperature data and the predicted meteorological data. As the same process, the future air humidity HMn(lk, ti + 1) for k = 1, d can be also derived. Based on TMn(lk, ti + 1), TSn(lk, ti + 1), HMn(lk, ti + 1) and the other sensor data, the road condition for i = 1, d can be finally predicted by Road State Decision System using AI model which is shown in Sect. 4. For the other Rn for n = 1, N, the same process can be applied and predicted the road condition for i = 1, e. Eventually, the road condition in regions R can be predicted.

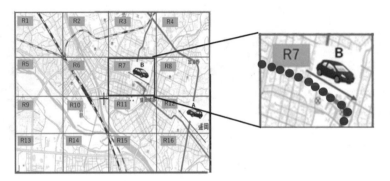

Fig. 5. Objective road condition region

6   Prototype System and Social Experiment

In order to verify the effects and usefulness of the proposed system, a prototype system is constructed and those function and performance are evaluated. The prototype system with sensor server system and communication server system of both SRS and SMB for two-wavelength communication is shown in Fig. 6. As communication components of cognitive wireless network, OiNET-923 of Oi Electric Co., Ltd. For 920 MHz band, WI-U2-300D of Buffalo Corporation for of 2.4 GHz band and T300 of Ruckus for 5.6 GHz band are used. OiNET-923 is used as control data communication link to exchange UUID, security key, password, authentication, IP address, TCP port number, socket No. On the other hand, both WI-U2-300D and T300 are used for sensor data transmission links. Raspberry Pi3 Model B+ is used for N-wavelength cognitive communication server unit to perform cognitive controller and SDN function. Intel NUC Core i7 is used for sensor database registration and data analysis by AI based road state decision.

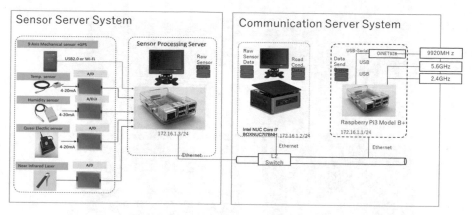

Fig. 6. Crowed road surface sensing system

On the other hand, in sensor server system, several sensors including BL-02 of Biglobe as 9 axis dynamic sensor and GPS, CS-TAC-40 of Optex as far-infrared temperature sensor (FIR sensor), HTY7843 of azbil as air temperature and humidity sensor and RoadEye of RIS system as near-infrared laser sensor and quasi electrical static field sensor for road surface state are used. Those sensor data are synchronously sampled every 0.1 s and averaged every 1 s to reduce sensor noise by another Raspberry Pi3 Model B+ as sensor server. Then those data are sent to Intel NUC Core i7 which is used for sensor data storage as sensor database and data analysis by AI based road state decision. Both sensor and communication servers are connected to Ethernet switch.

In order to evaluate the sensing function and accuracy of sensing road surface states, both sensor and communication servers, and various sensors are set to the vehicle as shown in Fig. 7. We ran this vehicle about 4 h to evaluate decision accuracy in realtime on the winter road with various road states such as dry, wet, snowy, damp and icy state around our campus in winter as shown in Fig. 8. In order to evaluate the road surface decision function, the video camera is also used to visually compare the decided road surface state and the actual road surface state.

As evaluation, accuracy and response time of realtime monitoring of air temperature, road surface temperature and humidity are evaluated. The Fig. 9 shows a typical output form those sensors while the vehicle is running and periodically sampling their sensor data. Through the test running and comparing the measured air temperature data and open data at fixed roadside temperature and humidity sensors, our system could identify the real winter road states more than about 80% as accuracy.

Also, from this typical output, the measurement data of road surface temperature quickly responds to the change of locations compared with air temperature and humidity due to near infrared sensor in realtime with 0.1 s interval. Therefore, using this far infrared sensor, the dangerous locations such as snowy and icy locations can be correctly identified. On the hand, from the air temperature and humidity, wide area weather conditions can be temporally and spatially measured. Those data together with public open weather data can be used to predict the future wide area road states in cloud computer server.

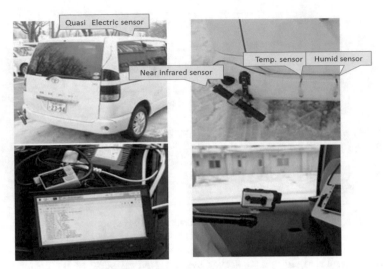

Fig. 7. Sensing vehicle in winter road

Fig. 8. Screen shot of field test

Finally, the sensing function and accuracy of decision for all of road surface states are evaluated using near infra-red laser sensor and the other sensors. In order to evaluate the road surface decision function by near infra-red laser sensor, the video camera is also used to visually compare the decided road state and the actual road state.

Figure 10 shows the changes of various sensor data and the determined road state, estimated friction coefficient, air and road temperature. Basically, the road surface with moisture freezes under zero degree Celsius. When the road surface freezes, the friction coefficient rapidly decreases. If not, the friction maintains at higher values.

We analyzed the relation between the estimated friction coefficient and road temperature with Pearson's correlation coefficient. As the result, the correlation coefficient of road surface between them was 0.6151 and t-static was 23.831, and p-value was less than 0.22. From this statistic values, there is positive correlation road temperature and the estimated road friction coefficient. Through this analysis, the locations of the road

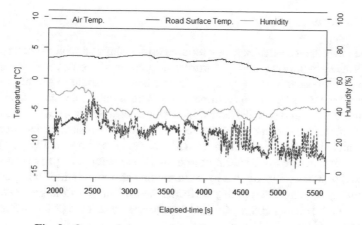

Fig. 9. Output of air temp., humidity and road surface sensor

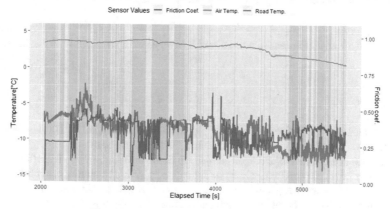

Fig. 10. Road states vs. various sensor data

where both the friction coefficient and road temperature very low are snowy or icy. Thus our proposed system has capability of detecting the critical road locations for drivers to be alerted when he/she passes through.

However, in case of the other days experiment, we experienced mis-determination of the road state as icy or snow even though both air and road surface temperature were beyond plus Celsius degree. This mis-determination is estimated due to the operational principle of NIR sensor based on the ratio of the reflection energies of NIR laser for different wavelengths on the road surface regardless of the temperature on the pavement. Therefore the system should compensate this mis-determination by considering multiple sensor values and organizing AI learning system with a lot of training data. Unfortunately, in our experiment, since there are quite few snow or icy days in this winter, enough training data could not obtained. It is reminded that more enough training data on different weather days should be collected and to converge the AI model and to attain higher accurate decision making for road states.

7   Conclusions

In this paper, in order to realize safe and reliable driving environment in challenged snow countries, a new generation road state information platform based on IoT crowed sensing technology is proposed. Various sensors are attached on vehicle to sample sensor data and to determine the road state in realtime. Those sensor data and road state information are exchanged and shared with the near vehicles and road side server using V2X communication network. Driver can receive the sensor data and road surface state information from the vehicle in opposite direction or road side server and eventually pay attentions to his/her driving before encountering the dangerous location. A prototype system of this proposed system is constructed to evaluate the various functions and performance as a preliminary system. Though this performance evaluation of the prototype system, we could verify the effects and capability of the proposed method of identifying various road states in realtime.

As future works, we are constructing the practical road state information platform by cooperating with the bus operating company in the snow country in Japan.

Acknowledgement. The research was supported by JSPS KAKENHI Grant Numbers JP 20K11773 and Strategic Research Project Grant by Iwate Prefectural University in 2020.

References

1. SAE International: Taxonomy and Definitions for Terms Related to Driving Automation Systems, for On-Road Motor Vehicles. J3016_201806, June 2018
2. Police department in Hokkaido: The Actual State of Winter Typed Traffic Accidents, November 2018. https://www.police.pref.hokkaido.lg.jp/info/koutuu/fuyumichi/blizzard.pdf
3. Takahashi, N., Tokunaga, R.A., Sato, T., Ishikawa, N.: Road surface temperature model accounting for the effects of surrounding environment. J. Jpn. Soc. Snow Ice **72**(6), 377–390 (2010)
4. Fujimoto, A., Nakajima, T., Sato, K., Tokunaga, R., Takahashi, N., Ishida, T.: Route-based forecasting of road surface temperature by using meshed air temperature data. In: JSSI & JSSE Joint Conference, P1–34, September 2018
5. Saida, A., Sato, K., Nakajima, T., Tokunaga, R., Sato, G.: A study of route based forecast of road surface condition by melting and feezing mass estamation method using weather mesh data. In: JSSI & JSSE Joint Conference, P2–57, September 2018
6. Hoshi, T., Saida, A., Nakajima, T., Tokunaga, R., Sato, M., Sato, K.: Basic consideration of wide-scale road surface snowy and icy conditions using weather mesh data. Monthly report of Civil Engineering Research Institute for Cold Region, No. 800, pp. 28–34, January 2020
7. Fujita, S., Tomiyama, K., Abliz, N., Kawamura, A.: Development of a roughness data collection system for urban roads by use of a mobile profilometer and GIS. J. Jpn. Soc. Civ. Eng. **69**(2), I_90–I_97 (2013)
8. Yagi, K.: A measuring method of road surface longitudinal profile from sprung acceleration and verification with road profiler. J. Jpn. Soc. Civ. Eng. **69**(3), I_1–I_7 (2013)
9. Mizuno, H., Nakatsuji, T., Shirakawa, T., Kawamura, A.: A statistical model for estimating road surface conditions in winter. In: The Society of Civil Engineers, Proceedings of Infrastructure Planning (CD-ROM), December 2006

10. Saida, A., Fujimoto, A., Fukuhara, T.: Forecasting model of road surface temperature along a road network by heat balance method. J. Civ. Eng. Jpn. **69**(1), 1–11 (2013)
11. Okubo, K., Takahashi, J., Takechi, H., Sakurai, T., Kokubu, T.: Possibility of blizzard detection by on-board camera and AI technology. Monthly report of Civil Engineering Research Institute for Cold Region, No. 798, pp. 32–37, November 2019
12. Takiguchi, K., Suda, Y., Kono, K., Mizuno, S., Yamabe, S., Masaki, N., Hayashi, T.: Trial of quasi-electrical field technology to automobile tire sensing. In: Annual conference on Automobile Technology Association, 417–20145406, May 2014
13. Casselgren, J., Rosendahl, S., Eliasson, J.: Road surface information system. In: Proceedings of the 16th SIRWEC Conference, Helsinki, 23–25 May 2012. https://sirwec.org/wp-content/uploads/Papers/2012-Helsinki/66.pdf
14. Ito, K., Shibata, Y.: V2X communication system for sharing road alert information using cognitive network. In: Proceedings of 8th International Conference on Awareness Science and Technology, Taichung, Taiwan, pp. 533–538, September 2017
15. Ito, K., Shibata, Y.: Experimentation of V2X communication in real environment for road alert information sharing system. In: Proceedings on IEEE International Conference on Advanced Information Networking and Applications (AINA2015), Gwangju, Korea, pp. 711–716, March 2015
16. Ito, K., Shibata, Y.: Estimation of communication range using Wi-Fi for V2X communication environment. In: The 10th International Conference on Complex, Intelligent, and Software Intensive Systems (CISIS2016), Fukuoka, Japan, pp. 278–283, July 2016

Analysis for Real-Time Contactless Road Roughness Estimation System with Onboard Dynamics Sensor

Akira Sakuraba[1](✉), Yoshia Saito[2], Jun Hakura[2], Yoshikazu Arai[2], and Yoshitaka Shibata[1]

[1] Regional Cooperative Research Division, Iwate Prefectural University, Takizawa, Iwate, Japan
{a_saku,shibata}@iwate-pu.ac.jp
[2] Faculty of Software and Information Science, Iwate Prefectural University, Takizawa, Iwate, Japan
{y-saito,hakura,arai}@iwate-pu.ac.jp

Abstract. In the cold districts, there is a large portion of traffic collision which is caused by skidding on icy road in entire traffic collisions. Some method in the latest research can detect exact road surface weather condition with vehicle onboard sensor. In effort to realize high-level autonomous driving vehicle, understanding of road surface condition or road state is an essential knowledge. This paper introduces an estimation method for road surface state contactless sensors. This method utilizes 9-axis dynamics sensors to obtain vehicle behavior on the road. In particular, obtaining roughness of the road weather state could be a key feature to estimate road characteristics. The designing proposed system considers real-time measurement for roughness of the road. This allows to calculate roughness to estimate characteristics of road surface condition while vehicle is moving, and exchanges road state information with other moving vehicle with vehicle-to-everything (V2X) communication. This paper also reports the result of evaluation for our method on actual road to analyze with the prototype onboard system. The result shows that proposed approach can evaluate road surface roughness in real-time while vehicle is moving, whereas it is suggested that relationship among other onboard sensors and roughness value is independent from across road surface weather state.

1 Introduction

In the cold districts, there is a large portion of traffic collision which is caused by skidding on snowy and/or icy road due to snowfall and low air temperature. According to statistics of traffic collision subjected to Hokkaido prefecture, which is located in the northern part of Japan, skidding by icy road was amount for 83.9% of entire traffic collision [1]. To secure road traffic safety, there is a demand to detect the particular points of critical weather condition on the road before the vehicle approaching those points, and warn it for to the driver on the vehicle. The critical road weather information could be essential knowledge for the driver on conventional manual operated vehicles.

On the other hand, automotive industries introduce high-level autonomous vehicle prototypes which is SAE 4 or 5 as defined in SAE definition [2]. These type of vehicles obtain surrounding situation which includes location and movement of obstacles, other vehicles, pedestrian etc. using onboard sensor such 2D/3D light detection and ranging (LiDAR), radar, RGB camera, or far-infrared camera. There is technical challenge to implement autonomous vehicle which has availability in cold district. For instance, false detection caused by noise on LiDAR image due to snowfall or splash or path planning on snow-covered roads, and path planning on snow-covered roads [3].

In addition, the authors concern other technical issue for path planning algorithm on the vehicle which depends on slippery road condition. In effort to avoid vehicle skidding or spinning, it is important to assess road surface weather condition and utilize those knowledge for realizing autonomous vehicle in very low temperature region in winter season.

This paper introduces a system for evaluation of road surface roughness level with onboard sensor. The system installs several types dynamics sensor such acceleration, gyro, and magnetic field sensors on the vehicle and estimates roughness on the road pavement. This system processes real-time evaluation from obtain behavior of the vehicle with dynamics sensor to associate roughness and geolocation while vehicle is moving. Designing of this real-time calculation allows to immediate exchange for ahead road condition via vehicle-to-everything (V2X) communication.

This paper also reports relationship among road roughness value and road surface weather state which is determined by the other onboard sensor.

2 Related Works

Some researchers introduced onboard sensor based sensing method which is subject to by behavior of moving vehicle.

Du et al. presented onboard accelerometers to measure road roughness of pavement [4]. Their method calculates international roughness index (IRI) with 3-axis acceleration sensor. Their system has the distributed architecture which consists of processing system located at fixed location and car mounted terminal on the probe vehicle. They evaluated and analyzed that relative error was about up to 7% thus proposed algorithm meets the required accuracy of IRI measurement and offers a highly efficient.

Yagi reported a practical method to evaluate characteristics of road pavement with generic smartphone which is installed on dashboard on the vehicle [5]. Their method BumpRecorder is very simple system configuration composed of only single smartphone which equips acceleration sensor, and considered easy-to-install on the vehicle. The system measures roughness and IRI. They compared measured roughness with physical measured value obtained from road profiling device and concluded as high correlation between them. Although estimation value was 0.55 to 0.7 times less value than the truth value.

Nishimura et al. introduced extended application [6] which intends to realize real-time processing for Yagi's method. Their proposed system can perform concurrent process to obtain acceleration and roughness which is a post-process on Yagi's one. They also use acceleration sensor on smart device which has 50 Hz refresh rate and analyzing node which is connected the smart device. They evaluated that processing time was around 100 ms to calculate a roughness value which is required to solve every second. They concluded that the measured value was almost reasonable however there is not enough discussion.

3   Algorithm for Road Roughness Estimation System

This section describes our proposed system which designed to estimate road pavement roughness using dynamics sensors with highly real-time. The following algorithm is based on Yagi's [5] and Nishimura et al.'s [6] methods.

In the following method, X-axis describes right and left of the vehicle from driver's view, Y-axis describes vertical direction, and Z-axis describes travelling direction of the vehicle as described in Fig. 1.

Fig. 1. The coordinate system corresponds to the vehicle.

3.1   Obtain Values from the Sensor and Correction on Y-axis Acceleration

Firstly, the onboard system fetches 3-axis acceleration values from dynamics sensor. In this phase, the system buffers raw acceleration values which are generated in the last 4 s in order to reduce drift using moving average for 3 s.

Basically, while vehicle is moving, onboard acceleration sensor can observe vertical acceleration A_Y which results from not only movement of vertical direction. In the case of sensor installation was not perpendicular angle for all of three-axis, the sensor observes false vertical acceleration to be attributable to change of vehicle speed or turning of the vehicle.

Thus proposed application corrects vertical acceleration after obtaining raw value from the sensor. Our proposed system solves corrected vertical acceleration A_{YY} with least square. In formula (1), N_{buf} is the data buffer length.

$$A_{YY} = A_Y - A_X \frac{N_{buf} \sum_{i=1}^{N_{buf}} A_X A_Y - \sum_{i=1}^{N_{buf}} A_X \sum_{i=1}^{N_{buf}} A_Y}{N_{buf} \sum_{i=1}^{N_{buf}} A_x^2 - \sum_{i=1}^{N_{buf}} A_x^2}$$
$$- A_z \frac{N_{buf} \sum_{i=1}^{N_{buf}} A_z A_Y - \sum_{i=1}^{N_{buf}} A_z \sum_{i=1}^{N_{buf}} A_Y}{N_{buf} \sum_{i=1}^{N_{buf}} A_z^2 - \sum_{i=1}^{N_{buf}} A_z^2} \tag{1}$$

3.2 Estimate Vertical Unsprung Movement from Corrected Acceleration

Secondly, system calculates dynamic component of vertical acceleration dY in order to remove components which do not attribute vertical movement such gravitational acceleration, the thermal drift on the sensor etc. Formula (2) describes i^{th} dynamic component of vertical acceleration with sampling rate of dynamics sensor is N_{samp} Hz. To realize real-time processing, this process regards average of vertical acceleration in the last 2 s as static vertical acceleration.

The process solves vertical movement velocity V_Y by adding dynamic vertical acceleration in formula (3).

$$dY(i) = A_{YY}(i) - \frac{\sum_{j=i+1}^{i+N_{samp}} A_Y(i)}{2N_{samp}} \tag{2}$$

$$V_Y(i) = V_Y(i-1) - \frac{\sum_{j=i+1}^{i+N_{samp}} dy(i)}{N_{samp}} \tag{3}$$

Next, this algorithm attempts to solve dynamic component of vertical movement velocity dV_Y to reduce vertical velocity component which is resulting from climbing a gradient using formula (4). This process also subtracts static component of static components on vertical axis by the average of velocity in the last 2 s. Finally, the system can obtain the displacement amount on the vertical axis L_Y with sum of vertical movement velocity with formula (5). Incidentally, $L_Y(i)$ is equivalent to sprung displacement amount of movement.

$$dV_Y(i) = V_Y(i-1) - \frac{\sum_{j=i+1}^{i+N_{samp}} V_Y(i)}{2N_{samp}} \tag{4}$$

$$L_Y(i) = L_Y(i-1) + \frac{\sum_{j=i+1}^{i+N_{samp}} dV_Y(i)}{N_{samp}} \tag{5}$$

3.3 Assumed Suspension Model of the Vehicle

Suspension system on modern vehicle consists of complicated structure. Although in this paper, the proposed system regards suspension model on the vehicle as single-degree-of-free (SDOF) system which is composed of a spring and a shock absorber described as Fig. 2. To estimate unsprung displacement amount of movement u from sprung movement L_z solved above. Our proposed system will solve the equation of motion which is described in Eq. (6) and (7) where ω is equivalent to natural frequency of suspension spring and h is equivalent to attenuation ratio of shock absorber. This method solves for u, and then roughness estimation system can estimate vertical movement at particular time interval in a second.

$$dY + 2h\omega(dV_Y - \dot{u}(i)) + \omega^2(L_Y(i) - u(i)) = 0 \tag{6}$$

where,

$$u(i) = u(i-1) + \frac{\dot{u}(i) + \dot{u}(i-1)}{2N_{samp}} \tag{7}$$

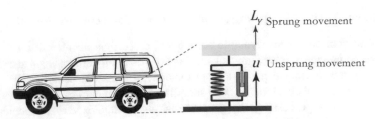

Fig. 2. SDOF system describes suspension on the vehicle.

Finally, system can solve the roughness value with formula (8). The roughness value can be described as variance of the unsprung displacement amount of movement. In the formula, $\bar{u}$ is the average of unsprung displacement amount of movement in the past a second. This system deals the roughness as the representative value which is vertical movement of the vehicle every second while the vehicle is moving.

$$roughness(i) = \sqrt{\frac{\sum(u(i)-\bar{u})}{N_{samp}}} \tag{8}$$

3.4 Associated Roughness Values with Geolocation

Table 1. Data schema on RSI database which is implemented by SQLite

Field	Data class of SQLite	Description
ID	INTEGER	Unique ID of records
timestamp	TEXT, NOT NULL	Timestamp of this record has been created
latitude	REAL, NOT NULL	Latitude of the position
longitude	REAL, NOT NULL	Longitude of the position
state	TEXT	Determined road surface weather state *dry|damp|wet|slash|snow|ice|unknown*
friction	REAL	Coefficient of friction
air_temp	REAL, NOT NULL	Outside air temperature
road_temp	REAL, NOT NULL	Road surface temperature
humidity	REAL, NOT NULL	Outside air humidity
roughness	REAL	Roughness value calculated with vehicle dynamics analyzing

As requirement to representation of road surface weather condition, it is important to associate estimated roughness value with geolocation.

The proposal system considers to be a subsystem which is a part of our previous road surface weather sensing system [7]. The sensing system records road state information (RSI) with 1 Hz update ratio which consists of data in Table 1. This is an example for implementation on SQLite lightweight database.

The roughness estimation system determines to update estimated roughness value into RSI database with timestamp. Due to calculation is delayed and algorithm is based on moving average of current and past acceleration values. To determine which existing record to update, the system identifies intermediate timestamp of whole data subjected to calculation of roughness value. Assuming that roughness estimation process requires the last 2 s data to evaluate and a sensor which has sampling rate in 100 Hz, the system regards that the roughness value was generated at the intermediate time between 100th and 101st data was recorded on sensor device.

After determination of timestamp, system updates a RSI record which includes timestamp of roughness on RSI database on the field roughness. Each RSI records contain geolocation of the vehicle where the data was observed, therefore, it is possible to associate the roughness value and geolocation.

4 Estimation System and Configuration

4.1 Configuration

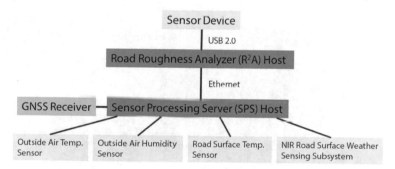

Fig. 3. System configuration

We have built a prototype system Road Roughness Analyzer (R^2A) which is shown in Fig. 3.

As described above, design of R^2A considers as subsystem of previous road surface sensing system, we installed the R^2A system and the Sensor Process Server (SPS) host on the vehicle. SPS is an onboard sensor unit which observes, records and accumulates sensors value in efforts to analyze and determine road surface weather state. SPS can obtain and record various types of sensor includes air temperature, road temperature, outside air humidity, near infrared (NIR) road surface weather sensing subsystem, and Global Navigation Satellite System (GNSS) for positioning of the vehicle.

4.2 Architecture of Application

System architecture of the application is illustrated as Fig. 4.

In the proposed implementation, sensor device caches own internal dynamics sensors which contain acceleration, gyro, Earth's magnetic field, geolocation, and timestamp when the record was recorded. These values were cached on the sensor device as CSV formatted text.

Next, Sensor Data Fetching Module on R^{2}A device obtains partial cache data on the sensor device as much as required to calculate roughness. This module extracts timestamp and 3-axis acceleration fields, and pass these data to Road Roughness Estimation Module.

Then, in response to receive passed data, Road Roughness Estimation Module calculates roughness value which is described in Sect. 3. This module also determines regarding timestamp in roughness value. This module should be written by native code to secure highly real-time ability.

Visualization Module plots roughness value to present changing over timeline in real-time for sensor vehicle's crew as R^{2}A Real-time Monitor. Due to buffering process in Road Roughness Estimation Module, there is constant delay to render the plot.

On the other hand, RSI Database Updater also obtained roughness data in effort to update RSI database on SPS. To update the database remotely, this module sends timestamp and roughness value over HTTP POST request. On SPS, RSI Database Writer deals to receive roughness data and updating existing particular RSI record depending on timestamp.

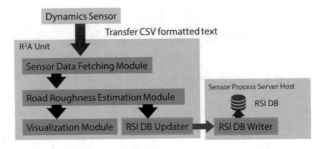

Fig. 4. System architecture in current implementation

4.3 Implementation

R^{2}A host is implemented on a Linux based single-board computer which runs Raspberry Pi OS (Kernel 5.4.79) on Raspberry Pi 3B+ hardware. Application is written by Python 3.8 and C++ on GCC 8.3.0 in the native code part. Visualization of roughness value is based on Seaborn 0.11.0 which is a statistical data visualization library of Python.

Acceleration sensor is the internal 9-axis sensor on Android 6.0 based device OPTiM BL-02 which has capability of 100 Hz refresh rate. Fetching sensor record cache from Android device was via Android Debug Bridge (ADB) and Android application which is written by Android SDK and Java.

SPS is also based on Raspberry Pi based system and installed some Python based applications. Web interfaces are implemented with Nginx 1.14.2, Flask 1.0.2, and uWSGI 2.0.19.1.

5 Experiment

This section describes and reports the experiment to investigate relationship among roughness values and existing road surface weather related data.

5.1 Onboard Data

To analyze relationship between road roughness and other onboard sensor element, we attempted to obtain these sensor output on actual public road. The onboard system collects output values of dynamics or environment sensors while the vehicle is moving.

During the trial, system obtained 5,553 records on RSI database in 1.85 h and associated the 2,741 records roughness value with adequate record on the database. As described in Table 1, each records consist of timestamp, latitude and longitude coordinate, determined road state, friction coefficient, road surface temperature, outside air temperature, outside air humidity, and roughness value.

Fig. 5. Travelling path of the vehicle on the test track

5.2 Test Track

We set a testing track which extends over Morioka-city and Takizawa-city located at northern part of the mainland Japan, as illustrated in Fig. 5. The length of the test track was about 36 km. The route includes a national highway route, residential roads in urban area, and farming area which located at the foot of a mountain.

5.3 Vehicle for Experiment and Profile

We setup the evaluation system on Toyota TA-AZR60G minivan class passenger car with about 102,000 km mileage. The vehicle is installed NIR road surface weather sensor, road temperature sensor, outside air temperature/humidity sensor, and GNSS receiver. Except smart device which is connected to R^2A host, any other sensor probe exposed to the outside of the vehicle.

The system determines vehicle's profile related suspension system which is composed of natural frequency of suspension spring and the attenuation ratio of the shock absorber using fast Fourier transform (FFT) from corrected vertical acceleration which is obtained while test drive. The process estimates that natural frequency of suspension spring set to 1.58 Hz and the attenuation ratio of the shock absorber was determined as 0.1.

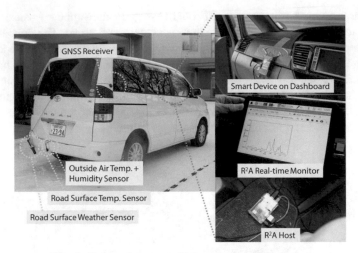

Fig. 6. Onboard sensor vehicle and sensor installation

We installed a set of SPS unit with an Optex CS-30TAC infrared temperature sensor to measure road surface temperature, an Azbil HTY7843 temperature and humidity sensor to measure outside air temperature and humidity, an Optical Sensors RoadEye Model SD road surface weather sensor, a dashboard camera on the vehicle. Installation on the vehicle is described in Fig. 6.

5.4   Result and Analysis

First, we compared road surface weather with roughness values. Figure 7 is a partial plot of changing of both values by time series while the vehicle is moving.

As analysis of data which is obtained entire trial, the system measured 0.014 m roughness (standard deviation was 0.014) on the average. The most frequently section of roughness value was 0 to 0.010 m in any road states. As observing of the result, there is no significant change of roughness among states of road surface weather.

Surprisingly, regardless of the state of road surface weather, system measured large instant values of roughness at multiple times and the value does not correspond to result of concavity and convexity on actual road surface. Most of this instant values were indicated beyond 0.05 m roughness. System also observed a roughness value which was 0.131 m at the maximum on entire test track. It is obviously a false value by comparison of road roughness value and the analysis of dashboard camera video.

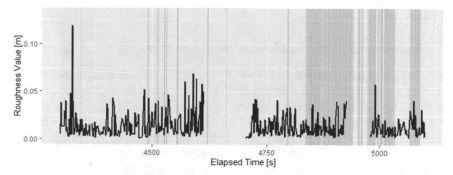

Fig. 7. Elapsed time from trial begins vs. varying of roughness value. Background colors are filled with the determined road surface state and corresponding to gray: dry, green: damp, yellow: snow, and red: ice.

Next, we have compared roughness value and geolocation. We have visualized RSI database to digital map which is illustrated as Fig. 8. We can find large roughness which is rendered as bigger and reddish circles in various point on the map. According to observe that, there is very large instant values is located at the points on the road near intersections, tight corners, or varying point of longitudinal gradient. At these points, vehicle is required to change own speed including start moving or halt, climbing/descending of the hill, or steering. Therefore it appears that at the points, system estimated false roughness value due to varying of behavior of the vehicle.

Then, we focused on variance of roughness value in a similar road surface weather state. Figure 8(2) describes the roughness value which is limited to show road state was *dry* and *damp*. Road section A on the map contains better pavement and indicates lesser road roughness value. Opposite to road section B on the map presents larger road roughness and pavement was poor relatively on actual particular road section. This result can be explained that this system can examine road roughness while the vehicle is moving.

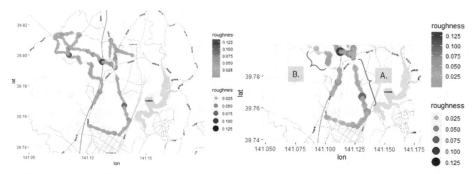

(1)All records visualization on RSI DB   (2) Filtered to limit 'dry' or 'damp' state

Fig. 8. Visualized RSI database which is recorded while the vehicle is moving on the designated test track

6 Discussion

6.1 Dependency and Independency of Roughness Value

Firstly, we found notable result of the roughness evaluation experiment that there is a dependency on vehicle behavior by controlling. Nishimura et al. [6] discussed removal of static composition in vertical movement in their algorithm. Although their algorithm still remains the influence of the composition which is ascribed to pitching behavior by climbing/descending of gradient or changing of vehicle speed, and rolling and yawing by steering etc. This result suggests that it is required to utilize other dynamics sensors to correct the false large instant roughness values such gyro sensor.

Besides, the result suggested that there is possibility of no significant relationship among road roughness value and onboard environment sensors. It is also clear that roughness was poorly correlated to any other onboard quantitative sensors with Pearson correlation heatmap which is represented in Fig. 9. Hence, we believe that roughness value does not describe and be depended on road surface weather condition directly at this time.

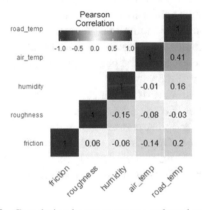

Fig. 9. Correlation heatmap among onboard sensors

6.2 Possibility of Quantitative Expression in Qualitative Field

Road surface weather state sensor determines several qualitative categories such *dry*, *damp*, *snow*, *ice*, etc. This state field does not describe road quantitatively, namely, current implementation can categorize road surface weather however it does not represent how much the particular road point is accumulating snow. Through the evaluation described above, this approach associates road roughness and state with geolocation where the roughness value is describing.

Therefore it has a possibility of quantitative expression by quantitative sensor data by associating road roughness and road surface weather state. This quantitative information could be an essential part for technology to determinate moving path of vehicle in implementation of high-level autonomous vehicle in cold districts.

6.3 Difficulty of Determination for Road Surface Weather Using Microscale Data Set

Thirdly, as this paper described above, through the field experiment the system remained at less data amount acquisition which has less than 2 h duration and test track was also just located into 8 km square area. Microscale road state such as this data set is easy to exchange among inter-vehicle or vehicle-to-road (V2R) wireless communication.

Despite, it is insufficient to analyze road state statistically due to noise which is resulting from instant large values. To reduce effect of the noise, it would be effective to acquire huge amount of macroscale RSI database in time and space. The system should collect and record RSI database in any season, it is possible to observe reference roughness value in non-snow accumulated season, not only statistical reason.

7 Conclusion

This paper presented a road roughness estimation system which considers to evaluate road surface weather condition. Designing of this system deliberates to realize real-time estimation while vehicle which installs the system is moving. The algorithm of this approach is based on calculation for unsprung displacement amount of movement from raw value output from vertical 3-axis acceleration sensor value. After calculation of the displacement amount, the system associates it with geolocation as road roughness value.

We had experiment to investigate relationship among the roughness value and other existing environment sensor value such as outside air and road surface temperature, outside air humidity etc. The result indicated that there is no significant relationship in this time. Another analysis compares qualitative road surface weather state, this result also presented no specific different across each weather state.

In the future work, we are planning to obtain macroscale road state information database in time and space for a long time to analyze relationship among dynamics sensor and other environment sensor.

Acknowledgments. This work was supported by JSPS KAKENHI JP20K19826 and Strategic Information, Communications R&D Promotion Program (SCOPE) No. 181502003, Ministry of Internal Affairs and Communications, Japan, and Strategic Research Project Grant, Iwate Prefectural University, Japan.

References

1. Takada, T., Tokunaga, R., Takahashi, N.: Categories of traffic accidents in winter season on national routes in Hokkaido. Hokkaido No Seppyo, no. 29, pp. 69–72 (2010). (in Japanese)
2. Taxonomy and Definitions for Terms Related to Driving Automation Systems for On-Road Motor Vehicles J3016_201806. SAE International (2018)
3. Yoneda, K., Sugamura, N., Yanase, R., Aldibaja, M.: Automated driving recognition technologies for adverse weather conditions. Int. Assoc. Traffic Saf. Sci. Res. **43**, 253–262 (2019)
4. Du, Y., Liu, C., Wu, D., Li, S.: Application of vehicle mounted accelerometers to measure pavement roughness. Int'l. J. Distrib. Sens. Netw. **2016**, 1–8 (2016)

5. Yagi, K.: A measuring method of road surface longitudinal profile from sprung acceleration, and verification with road profiler. J. JSCE **69**(3), I_1–I_7 (2013). (in Japanese)
6. Nishimura, H., et al.: Real time detection of rode conditions using in-vehicle force sensor. In: The 79th National Convention of IPSJ, 6U-04 (2020). (in Japanese)
7. Sakuraba, A., Shibata, Y., Sato, G., Uchida, N.: Evaluation of end-to-end performance on N-wavelength V2X cognitive wireless system designed for exchanging road state information. In: Advances in Internet, Data and Web Technologies, pp. 595–604 (2020)

Automatic Classification and Rating of Videogames Based on Dialogues Transcript Files

Alessandro Maisto[✉], Giandomenico Martorelli, Antonietta Paone,
and Serena Pelosi

University of Salerno, via Giovanni Paolo II, Fisciano, SA, Italy
{amaisto,gmartorelli,spelosi}@unisa.it, antonietta.paone@outlook.it

Abstract. Video games industry represents one of the most profitable activity connected with entertainment and visual arts. Even more than movie industry, video game industry involves a great number of different professionals who work together to create products expected to reach people in many countries. A substantial part of these people are teenagers, strongly attracted and influenced by video games. For these reasons, various systems of labels have been created. They indicate the recommended age ranges for each product. These systems are based on different criteria, but they have in common the presence of descriptors, or labels, which identify the type of contents in the game. One of them is PEGI (Pan-European Game Information), and we will mainly take this system in consideration for the purposes of our study. The rating procedure includes questionnaire enquires compiled by the publisher for the automatic attribution of the label and a large process of manual control of each submitted game. In order to help this large and demanding process, we propose a system of video games rating based on automatic classification of the products performed over the "transcript" or script, files that display the full transcription of dialogues in a video game. The proposed automatic classification algorithm is based on large, specialized dictionaries. Such as the dictionary of offensive language. This is based on semantic vector spaces and on sentiment analysis, and is able to provide an age rating and a genre classification of video games. It works in a more efficient way in games with a consistent amount of dialogues. The experimentation of the proposed algorithm is returning encouraging results.

1 Introduction

The necessity of a regulatory system to suggest the minimum age for a video game user started in the early nineties. At the time the importance of this medium was rapidly increasing, and technology was allowing the creation of

Alessandro Maisto edited Sect. 3, 4, 5, and 6; Giandomenico Martorelli edited Sect. 1 and 2; Antonietta Paone worked on data collection and dictionaries; Serena Pelosi edited Sect. 4.1.1.

more advanced graphics. At the same time, some video game with very violent contents were introduced in the market, triggering a series of controversies due to inappropriate contents. Regulatory laws on minimum age for video games came into effect for the first time at the end of 1992, but only by 2005 they achieved a legal status permitting the application of fines and penalties. Before that date, their nature was essentially non-legislative [9]. The American and Canadian regulatory system, named ESRB, differs from other system mainly for the great number of descriptors. The descriptors represent the basics of the evaluation of a video game. After the verification of the descriptors, a label will be applied to the cover of the product. The European evaluation system, called PEGI, is simplified and includes the following descriptors:

- bad language: with a series of sub-descriptors referring to the type of bad language used.
- fear: the contents of the game may cause fear on children and teens.
- sex/nudity: in the game there are scenes with nudity or sexual contents, more or less explicit. Violence: the game may contain representations of violent acts or blood.
- drugs: during the game there is use of drugs, or alcohol/tobacco.
- discrimination: in the game there are racial stereotypes or discrimination against certain gender, religions or populations.
- gambling: the game depicts gambling acts, may resulting in an encouragement towards this behaviour.
- online: it is possible to play the game with other real people on the Internet.
- In-game purchases: this descriptor has been added recently (since 2018) to indicate the presence of paying contents in the video game (micro-transactions).

With 30 descriptors, the American system is more intricate, even if the categories are very similar (use of substances, bad language, nudity etc.) When PEGI was launched, the producers had various aims in order to make it real, and at least three main interests:

1. The producers could show to the public opinion and the customers their interest towards the morality of their product.
2. The European system would be clearer, and the producers did not have to worry about multiple systems and legislation anymore.
3. The system was not so harsh in case of infraction, so the worse that could happen was a half-million dollars fine and the suspension of the product from the PEGI system.

These were great incentives that lead the producers to support this evaluative method. The presence of a rating system are a way that corporation use to prove to the general public their commitment towards their younger users [10].

The greatest part of labels includes violence and bad language descriptors. Other descriptors, such as "discrimination" and "gambling" are less used. The one which notifies the inclusion of in-game purchases is more and more used due to the presence of many video games during recent years which feature payments, micro-transactions and loot boxes. Sometimes these methods of in-game purchases can be actually considered gambling.

This paper will face the problem of automatic video game rating through a linguistic approach by exploiting lexical resources such as electronic dictionaries and Distributional Semantics matrices. The basic idea is to use high-specialized lexicons to extract the linguistic indicators of the main features used to rate video games by the most important regulatory systems from video game Dialogue Transcripts. In addition, we will analyze texts in order to propose an unsupervised classification of the products. The Semantic Similarity scores extracted from a large DS matrix will be enveloped in two processes: first, we use semantic neighbors to expand lexicons, then we generate a vector description of Transcripts that helps the automatic classification.

In Sect. 3 we will describe more thoroughly the problems of video game rating and the classification task. In Sect. 4 we will describe the adopted methodologies; Sect. 5 will show the results of the application of these methodologies over a small group of dialogue transcripts. In Sect. 6 we will present our considerations about the project.

2   How the Current System Works

The current regulatory system procedure, for both ESRB and PEGI is divided in the following phases:

- The video game developer creates a list of descriptors for the product, that will be examined by a cloud software.
- Parts of the game are presented to a diversified audience who works for the department. They are even shown some dialogue transcripts. This second phase does not happen with every product.
- The department verifies the data obtained from the cloud software and the audience.
- One or more label(s) are applied to the video game cover.

During the first phase, the developers fill out an online form which contains 37 questions to verify the presence of a negative content in the game. The cloud software is able to recognize inappropriate contents and it proposes one or more label(s). The final part of the process is supervised by VSC and NICAM, the teams which manage the PEGI system. They will verify every information and decide the final label for the game [10].

2.1   Critics to the Rating Systems

It is conventional wisdom that video games rating only returns partial or sometimes incorrect results. This is due to a variety of factors. The main issue resides in the fact that only some parts of the entire product are supervised. This means only some graphics imagery, reference pictures, only some extracts of the text transcription and few other game components. Furthermore, it has been proved that the outcome can be influenced from the developer and the departments.

Even if ESRB received the approval by the Congress, it has often been subject to controversies. One of the main debates about the efficiency of the department was its reluctance to apply ad Adult-Only rating. At the date of 2008, on 13.000 video games rated by the department, only 23 received an AO label. For this reason it has been believed that the video games rating systems are too industry-friendly and their interest is more influenced from business reason instead of returning a public service [8].

Other inaccuracies of the system rely on the importance of graphics imagery instead of the deep meaning of the product. Nowadays, video games are often very complex, with long storylines, plenty of characters and their evolution. There are cases of games which feature a shallow and apparently innocent aesthetics, but their contents are effectively harsh, violent or hard to be understood by a teenager. In these cases, a complete analysis of the dialogue transcription would be recommended, whereas the evaluation criteria are mainly based on the graphical impact of the game.

Studies have shown parents perspective of the current entertainment mediums are not satisfactory in providing adequate information, adding to the primary focus of age-based ratings, and content descriptors do not represent the context of the medium in full [17].

It has been suggested from various authors who studied the topic [9–11,13], that the evaluation criteria should become more detailed and dedicated. Sometimes the content descriptors are considered misleading and parents have to make an effort to understand the real mood and purpose of the video game, and often these people do not have the minimum base acquaintance and knowledge to fully understand the medium. So if the label does not cover the full spectrum of the game content, the risk is that the very people whose the system is addressed are not provided with the right tools in order to fully take advantage of. The possible ways to improve the systems could be the major focus on text transcription and the consideration of the positive contents instead of focusing only on the negative ones.

3   State of the Art

Classification algorithms can be classified into supervised, unsupervised and semi-supervised according to specific learning rules [22]. The former requires a human intervention for assigning labels to a set of training documents and then apply a learning algorithm to the classifier building. The most used supervised parametric algorithms are logistic regression adopted to predict binary classes and the Naïve Bayes [15] which assume that all attributes of the examples are independent of given category [18]. Among the non-parametric classifiers, the mostly applied are: support vector machines, k-nearest neighbor, decision trees and neural networks [23].

The unsupervised algorithms allow to identify the structure of unlabeled data and to explore text patterns perhaps understudied or unknown [12]. Classification training of the unsupervised approaches is usually based on the inferences that are carried out by clustering data into different clusters without labeled responses [22].

A combined use of both supervised and unsupervised techniques is provided by semi-supervised approaches which use both labeled data and unlabeled data to improve the classification accuracy [1].

Building networks of similarity between texts could allow the use of network classification algorithms to classify texts [16] or extract a set of objects from the network [2]. The use of specialized lexicon to give a structure to unstructured texts with classification purpose could improve the classification task [7]. In particular, the use of specific keywords allows the identification of domain objects, roles or events [4]. Moreover, a lexicon enriched with semantic information, gives to the text a semantic structure that could be used for future analysis [3].

4   Methodology and Resources

The methodology we present in this paper focuses on two tasks: the automatic identification of linguistic indices and the semantic classification of video games transcripts.

As we saw in previous sections, the video games rating process involves many features related to different aspects of the product, such as the language used in dialogues, representation of violence, drugs, gambling, etc. Although different companies evaluate video games, some of these features are shared among the different rating processes. Moreover, some of these features are strictly related to the visual dimension and can not be assessed by analyzing the video game's linguistic sphere.

In our methodology, we take as input data the video games transcripts, which mainly contains the dialogues and, in some cases, description or menu texts. Since we analyze only the linguistic feature of a video game, we concentrate our analysis on three indicators that can be automatically calculated with a basic text analysis. We analyze the presence of *Bad Language*, *Violence*, and *Drugs*. Once extracted the five indices, we perform a semantic amplification of each transcript's frequent words to classify the products automatically.

4.1   Extraction of Video Games Indices

The extraction of indices of *Violence*, *Bad Language*, presence of *Drugs* inside texts is performed by searching for specific lexical indicators inside the texts. These indicators were stored in electronic dictionaries that were manually created and automatically expanded in a second step.

For our analysis, we rely on a set of five dictionaries, as showed by Table 1.

Table 1. List of electronic dictionaries

Name	Dimension
Dictionary of Slang	5913
Dictionary of Violence-related terms	1173
Dictionary of Drugs	546
Dictionary of Discriminative-terms	635
NRC Affect Intensity Lexicon	
Anger	1515
Fear	1789
Disgust	1108

We collected the words of the dictionaries by examining the resources that have been published on different web-pages such as `urbandictionary.com`, `wikipedia.com`, and `theonlineslangdictionary.com`. The features of the four dictionaries are the following:

- Dictionary of Slang: it includes slang vulgar terms associated with bad language in general. Includes non-ambiguous insults and foul terms related to discrimination or sex and sexism.
- Dictionary of Violence: it includes terms that can be related to violence, such as weapons or war-related terms, among others.
- Dictionary of Drugs: it contains generic terms, commercial or scientific drug names, slang for common drugs.
- Dictionary of Discriminative Terms: it contains terms related to sexual, ethnical, regional, and religious discrimination.

To enrich the four resources, we expand the number of words by extracting the 10 most similar words for each word in a dictionary, using a Distributional Semantic Matrix generated by the algorithm COALS [21] applied to the BNC [14]. After this enrichment, the dictionaries have been manually verified to delete common words and other ambiguous terms that can produce falsified results.

4.1.1 Affect Intensity Lexicon

Emotions have been processed in this research through the exploitation of the *NRC Affect Intensity Lexicon* [19,20][1]. The original list of words, available for both the English and the Italian language, has been lemmatized and inflected, then integrated through the information contained into the lexical resources of the English and Italian module of Nooj[2]. This enriched lexical database now contains more than 20,000 entries (9,921 English and 10,334 Italian lemmas) endowed with labels that specify which one among the Plutchik's eight basic

---

[1] https://saifmohammad.com/WebPages/AffectIntensity.htm.
[2] http://www.nooj-association.org/.

emotions is connected with each lemma and specifies its degree. The emotions which have been taken into account are *Anger, Fear, Anticipation, Trust, Surprise, Sadness, Joy, Disgust.*

The emotions degrees range from 0, when the word conveys the lowest amount of emotion, to 1, when the word conveys the highest amount of emotion.

From the NRC Lexicon, we use a limited set of indicators that include *Fear, Anger,* and *Disgust.* These indicators, combined with the other dictionaries, can produce a value that can be used as an index of the presence/absence of specific features inside the texts.

4.2   Semantic Classification of Video Games Transcripts

The second task we faced in our project focuses on the automatic classification of video games based on semantic analysis of transcripts. We employ a Distributional Semantic Matrix to extract similarity values between words and search for the nearest neighbors of a single transcript's most frequent terms.

The distributional semantic Algorithm we use to build the matrix is COALS, trained on the BNC corpus and with vectors of 14.000 dimensions. The matrix contains the 18.000 most frequent terms of the BNC.

The Matrix has been used to perform a semantic expansion of the transcripts' more significant words: we extracted 100 terms with the higher frequency values for each test and searched for the 10 nearest neighbors of each of these terms. The result is a vectored text in which the texts are represented as the list of high-frequency terms included in it, more their semantic similar words associated with the value of frequency and similarity respectively.

After obtained the vectored texts, we perform a classification task. The first step is to generate values of similarity between text vectors by applying the Cosine Similarity algorithm. Once compared all texts and generated a large network of text-per-text edges, we represent this network graphically, assigning the similarity value as weight and deleting nodes with similarity zero. We use Gephi [5] to represent the network.

Once we build a network of video game similarity, we perform the Modularity Class [6] over the network in order to generate a classification.

5   Experiment

This paper aims to test the possibility of helping the rating process of video games. To test the preliminary resources that we create with this purpose, we start developing a corpus of video games transcripts. There are some on-line resources about video game dialogues, but they are not verified and belong to User Generated Content. We found the transcript we use for the experiment from the web-pages `transcripts.fandom.com` and `game-scripts.fandom.com`. Table 2 presents the title we select for the experiment, associated with the year of production and the punctuation of the two main rating system, where included.

Table 2. Description of the corpus

Title	Year	PEGI	ESRB	Genre
Assassin's Creed	2007	18	M	Action adventure stealth
Assassin's Creed: Bloodlines	2009	16	M	Action adventure stealth
Batman: Arkham Origins	2013	16	T	Adventure stealth
Bioshock	2007	18	M	FPS story-driven
Dark Souls II: Scholar of the First Sin	2015	16	T	Action RPG
Death Stranding	2019	18	M	Action adventure
Dungeon Siege III	2011	16	T	Action RPG
God of War	2018	18	M	Action adventure
Horizon: Zero Dawn	2017	16	T	Action RPG
Infamous	2009	16	T	Action adventure
Kane & Lynch: Dead Men	2007	18	M	Third person shooter
Lego the Incredibles	2018	7	E 10+	Action adventure
Mafia II	2010	18	M	Action adventure
Marvel's Spider-Man	2018	16	T	Action adventure
MediEvil	1998	12	T	Hack and slash
Metal Gear Rising: Revengeance	2013	18	M	Hack and slash
Middle-Earth: Shadow of Mordor	2014	18	M	Action adventure
Mortal Kombat vs DC universe	2008	16	T	Beat 'em up
Ninja Gaiden	1988	-	-	Hack and slash
Prince of Persia: The Sands of Time	2003	7	T	Action platform
Resident Evil 2	1998	18	M	Survival horror
Resident Evil 7 Biohazard	2017	18	M	Survival horror
Soulcalibur VI	2018	16	T	Beat 'em up
South Park: The Stick of Truth	2014	18	M	Role-playing
Spider-Man	2000	12	E	Action adventure
Star wars: Knights of the Old Republic 2	2004	12	T	Role playing
Tenchu 2	2000	18	M	Action adventure stealth
The Last Of Us	2013	18	M	Survival horror
The Evil Within 2	2017	18	M	Survival horror
The Witcher	2007	18	M	Action RPG
Tomb Raider	2013	18	M	Action adventure
Uncharted 4: A Thief's End	2016	16	T	Action adventure
Yakuza 0	2015	18	M	Action adventure

We adopt different criteria in choosing these titles such as temporal variety, genre variation, different target players, and dimension of the transcript.

Once collected the corpus, we start with the rating phase. In this step, we apply dictionaries and extract the frequency of each feature inside a single text. The indices for each feature has been calculated as follow:

- Bad Language value is calculated as the sum of the total frequency of the Bad Language and the Discrimination dictionaries words.
- Violence includes the frequency of Violence, Anger, fear, and Disgust dictionaries.
- The Drugs indicator only contains the total frequency of extracted words from the Drugs Dictionary.

The following step consists in comparing the results of our indicators with indicators of Bad Language, Violence, and Drugs given to a video game from the Rating System. To compare the results, we extract the description of the video game rating of PEGI and ESRB and define a set of rule of transformation:

- if the video game rating description includes the word "Violence" or "bad Language", we set the correspondent feature level as 1.
- If an adjective is present as "insane/intense/strong violence/language", we set the level as 2.
- if there is no mention of a particular feature, the level is 0.

Table 3 shows the results of this comparison.

Table 3. Comparison of the level of indicators

	Bad language	Violence	Drugs
Exact rating	0.6	0.58	0.63
Presence/absence	0.73	0.97	0.79

In Table 3 we present two different precision calculation: in the first one we consider correct only rating score with an exact correspondence, in the latter we consider correct all the rating in which the presence/absence of an indicator has been well calculated. As obvious, the latter precision is higher then the first. In the exact rating precision, the best score is achieved by the drugs indicator, but the three values are very similar. As we expect, the violence indicator obtains a slightly lower value due to the strong visual component of this feature. However, in the presence/absence precision, the highest value of precision is related specifically to the violence feature.

The second experiment we perform regards the classification of video games. As we said in Sect. 4, we use Gephi to create the network and calculate the classes. We perform a Modularity Class algorithm over the video game network with a resolution of 1. The result is showed in Fig. 1.

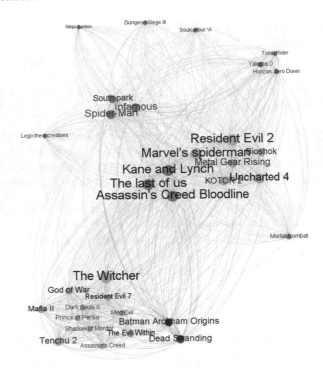

Fig. 1. Video games similarity network and classification

6   Conclusion

In this paper, we present a new methodology for the rating and classification of video games based on linguistic features. The corpus on which we perform our analysis includes Dialogue Transcripts files of 33 all-time video games. The methodology takes advantage of a set of lexical and semantic resources: for the automatic rating of video games, we select 3 features, such as Bad Language, Violence, and Drugs, and base the analysis on 6 electronic dictionaries. The unsupervised classification has been performed through the use of a Distributional Semantics Matrix calculated with the algorithm COALS: first, we extract the words with a higher frequency of each text, then we use the matrix in order to extract the semantic nearest neighbors of each word and calculate the similarity between texts. These values have been represented as a network on which the Modularity Class algorithm generates n different classes.

The results we obtained are encouraging. We must consider that video games' rating includes many features that must be better analyzed by visualizing scenes of the game. The linguistic features related to violence, for example, are often inadequate to describe the presence of this feature in a game. Nevertheless, the results we achieve could be considered a good starting point for future analysis. In particular, this kind of methodology could be advantageous to automatically extract keywords that can be considered a trusty index of some specific feature.

In the future, we plan to expand our dictionaries and refine the methodology to achieve better results, in particular with the Bad Language feature, which is the one more directly related to language.

References

1. Altmel, B., Ganiz, M.C.: A new hybrid semi-supervised algorithm for text classification with class-based semantics. Knowl.-Based Syst. **108**, 50–64 (2016)
2. Amato, F., Castiglione, A., Mercorio, F., Mezzanzanica, M., Moscato, V., Picariello, A., Sperlì, G.: Multimedia story creation on social networks. Future Gener. Comput. Syst. **86**, 412–420 (2018). Cited by 13
3. Amato, F., Cozzolino, G., Moscato, V., Moscato, F.: Analyse digital forensic evidences through a semantic-based methodology and NLP techniques. Future Gener. Comput. Syst. **98**, 297–307 (2019)
4. Amato, F., Moscato, V., Picariello, A., Sperli'ì, G.: Extreme events management using multimedia social networks. Future Gener. Comput. Syst. **94**, 444–452 (2019). Cited by 19
5. Bastian, M., Heymann, S., Jacomy, M., et al.: Gephi: an open source software for exploring and manipulating networks. In: ICWSM 2009, vol. 8, pp. 361–362 (2009)
6. Blondel, V.D., Guillaume, J.-L., Lambiotte, R., Lefebvre, E.: Fast unfolding of communities in large networks. J. Stat. Mech: Theory Exp. **2008**(10), P10008 (2008)
7. Catone, M.C., Falco, M., Maisto, A., Pelosi, S., Siano, A.: Automatic text classification through point of cultural interest digital identifiers. In: International Conference on P2P, Parallel, Grid, Cloud and Internet Computing, pp. 211–220. Springer (2019)
8. Chalk, A.: Inappropriate content: a brief history of videogame ratings and the ESRB. The Escapist (2007)
9. Dogruel, L., Joeckel, S.: Video game rating systems in the US and Europe: comparing their outcomes. Int. Commun. Gaz. **75**(7), 672–692 (2013)
10. Felini, D.: Beyond today's video game rating systems: a critical approach to PEGI and ESRB, and proposed improvements. Games Cult. **10**(1), 106–122 (2015)
11. Gentile, D.A., Humphrey, J., Walsh, D.A.: Media ratings for movies, music, video games, and television: a review of the research and recommendations for improvements. Adolesc. Med. Clin. **16**(2), 427–446 (2005)
12. Grimmer, J., Stewart, B.M.: Text as data: the promise and pitfalls of automatic content analysis methods for political texts. Polit. Anal. **21**(3), 267–297 (2013)
13. Humphreys, A., Jen-Hui Wang, R.: Automated text analysis for consumer research. J. Consum. Res. **44**(6), 1274–1306 (2017)
14. Leech, G.N.: 100 million words of English: the British national corpus (BNC) (1992)
15. Lewis, D.D.: Naive (Bayes) at forty: the independence assumption in information retrieval. In: European Conference on Machine Learning, pp. 4–15. Springer (1998)
16. Maisto, A., Pelosi, S., Stingo, M., Guarasci, R.: A hybrid method for the extraction and classification of product features from user generated contents. Lingue e Linguaggi **22** (2017)
17. Marston, H.R., Smith, S.T.: Understanding the digital game classification system: a review of the current classification system and its implications for use within games for health. In: International Conference on Human Factors in Computing and Informatics, pp. 314–331. Springer (2013)

18. McCallum, A., Nigam, K., et al.: A comparison of event models for Naive Bayes text classification. In: AAAI-1998 Workshop on Learning for Text Categorization, vol. 752, pp. 41–48. Citeseer (1998)
19. Mohammad, S.M.: Word affect intensities. arXiv preprint arXiv:1704.08798 (2017)
20. Mohammad, S.M., Turney, P.D.: Crowdsourcing a word–emotion association lexicon. Comput. Intell. **29**(3), 436–465 (2013)
21. Rohde, D.L.T., Gonnerman, L.M., Plaut, D.C.: An improved model of semantic similarity based on lexical co-occurrence. Commun. ACM **8**(627–633), 116 (2006)
22. Thangaraj, M., Sivakami, M.: Text classification techniques: a literature review. Interdisc. J. Inf. Knowl. Manag. **13**, 117–135 (2018)
23. Vasa, K.: Text classification through statistical and machine learning methods: a survey. Int. J. Eng. Dev. Res. **4**, 655–658 (2016)

Requirements and Technical Design for Online Patient Referral System

Vivatchai Kaveeta[1]([⊠]), Supaksiri Suwiwattana[1], Juggapong Natwichai[2], and Krit Khwanngern[1]

[1] Princess Sirindhorn IT Foundation Craniofacial Center, Chiang Mai University, Chiang Mai, Thailand
{vivatchai.k,supaksiri.s,krit.khwanngern}@cmu.ac.th
[2] Center of Data Analytics and Knowledge Synthesis for Healthcare, Chiang Mai University, Chiang Mai, Thailand
juggapong@eng.cmu.ac.th

Abstract. Upon patients being referred between healthcare facilities, the medical historical data transfer becomes a crucial process. The requirement for the transfer including correctness, completeness, and security. In Thailand, traditionally patients are required to physically carry their printed-out medical documents when they are referred to another hospital. This process is troublesome and prone to human error. Electronic medical record (EMR) transfer is a direct answer to this problem. However, the biggest challenge of developing the referral data transfer system comes from the fact that hospitals are using different EMR systems. Another cause is hospitals in Thailand operate under different authorities such as the Ministry of Health or Ministry of Higher Education. Each has its data protection and data transfer policies. In this paper to address these issues, we propose a technical design for an online patient referral system. The platform can manage a list of hospitals and their treatment capacities. When physicians need to refer a patient, they can search and locate nearby hospitals with the patient's specific needs. Required modules for the system will be presented. We present the possibility of implementing the system based on blockchain technology for secure data transfer across providers.

1 Introduction

Hospitals in Thailand can be classified into healthcare levels as primary, secondary and tertiary care. Like most countries, patients usually start by visiting a hospital near their neighborhood. When the patient required more sophisticated procedures that depending on equipment and specialists beyond local hospitals, they can be referred to other larger hospitals. Upon the referring, their medical records need to be transferred between the source and the target hospitals. Traditionally, patients carried their printed-out medical record papers themselves to the target hospital. This process is prone to human error. This is crucial as the missing or incomplete data can be life-threatening. Additionally, printed out documents are very high privacy risk.

The electronic medical record transfer is the direct answer to this problem. Thai hospital had been migrating to electronic health record systems for many years. Many attempts such as ThaiRefer [4] and Thai Care Cloud tried to create online patient referral systems but they were faced with many difficulties. Most importantly the relatively limited number of adoption by hospitals. The lack of communication between the hospital in primary care level and specialty hospitals [6]. And the absence of traceable data transfer history and auditing process [5]. In the following sections, we dissect the root causes of this problem. And eventually, we propose a new design for a referral system that tries to mitigate these issues.

We adopt mobile applications as one of the user interface in our patient referral system. This is based on works in the field of mHealth which is an adaptation of mobile devices in healthcare objectives. Health mobile applications are successfully utilized in many developing countries as examples shown in [1–3]. In general, mobile applications have low requirements for operation. The devices are readily available, users are likely to already own smart mobile devices.

This paper is organized as follows. Section 2, we start by looking into the challenges the patient referral system. We derive them into requirements which are the base for the system structure in Sect. 3. The process of patient referral on the proposed system are explained in Sect. 4. We discuss some important aspects of the system and how they relate to the previously mentioned challenges. And finally, we conclude and give a future direction for this work.

2   Challenges

To address the challenges and requirements for the referral system, the previously mentioned referral platforms are examined. Also, we discuss with the users in the referral process including hospital managers, healthcare workers, data administrators, and patients. We conclude that the existing systems have some drawbacks. In the following section, we categorize and examine these challenges in detail.

2.1   System Separation

We found that the existing systems are self-containing platforms. They operate with their own set of interfaces, data formats and databases. Therefore, they cannot be connected with any hospital's EMR systems. To start a new referring request, users need to manually copy all patient medical information from the internal system into the referral system. This additional work became a huge struggle for many hospital personnel and prone to human input error. Also, it raises serious security concerns as the data were through the external transfer system. Any weakness in the system's security can result in massive data leakage. In contrast, our system will act as the bridge between the internal hospital's EMR systems. It relied on the existed data in the hospital databases, eliminate the need for manual data copying. This reduces personnel workload and privacy concerns.

2.2   Hospital Authorities

Most patient referral systems were developed to only connecting hospitals within the same authorization. For example, "Thai Refer" system only available for hospitals under the Ministry of Health. As the result, it cannot be used for patients who are referred from and to university hospitals. The reason behind this usage limitation may come from the different policies, requirements and data structures regulated by authorities. Our system will not change their proven internal systems, instead, it piggybacks them as the standardized data transfer route across hospitals in any authorizations.

2.3   Hospital Capabilities

Current systems lack any up to date record of hospital capabilities. The hospital capability can help inform the physician to appropriate facility, equipment and specialist to perform specific medical procedures. Instead, they rely on existing knowledge or external communication to obtain information. Complications of this problem are including delayed referral process, patient travel longer distance than necessary, and in the worst-case patient cannot receive the required treatment. In our system, we maintain a list of hospitals in our network along with their capabilities. Doctors can search for the closest hospital for the patient's specific needs.

2.4   Data Transfer History

All previous systems are centralized systems that are managed by a single entity. The patient private information is directly stored within the system. Any vulnerability of the system can be disastrous. In our design, the referral system only keeps the history of referral requests, referral responses and data transfer. No actual patient's medical information is transport through our system. As the result, we can eliminate the data privacy concern but still able to maintain the data exchange transparency. The ability to keep a full history of user interactions and data transfers is very advantageous. Our system will keep the log begin with new referral requests, referral confirmations or denials, and the results of data transfer between hospital nodes.

2.5   Patient Involvement

Previous systems lack any patient involvement in the data transfer process. To comply with personal data protection regulation, patients need to be able to control their medical data and must give explicit approval for each transfer. Our system introduces a patient application as the interface for patients to create a new referral request and approve the data transfer.

3   System Structure

In this section, we introduce the structure of our proposed patient referral system. We explain each system's components in detail and how they act together to form a complete referral system. Figure 1 shows an overview of our referral system. On top, each box represents an application that operates inside hospital infrastructure. A hospital can join the referral network anytime without affecting the hospitals in the network. We introduce data formatter to translate each hospital's proprietary data format into standardized data for transferring.

3.1   Hospital Electronic Medical Record System

The first component is the existing hospital EMR systems. Each hospital likely operates on a different system tuned to fit their specific needs. Like we mentioned, we are not make any change to the existing system but introduce a bridge between them to form the transfer network. However, we rely on these internal systems and their databases as the source of patient medical history. The data in the proprietary format will be translated into the standard format by data formatter.

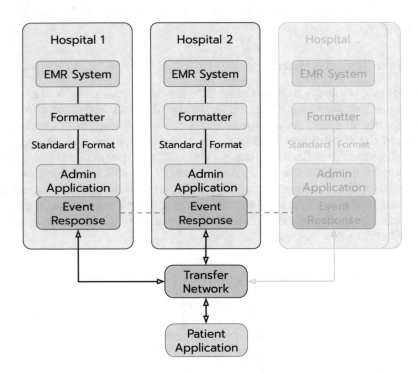

Fig. 1. Structure of patient referral system

3.2   Standard Data Format

To transfer the patient medical data across hospitals with a different EMR system, we need to introduce a universal format. Our standard data format consists of five basic data structures.

1. Patient personal information
   Basic patient information will be used to identify the patient in the internal EMR system. The list including name, national identification number, home address, hospitals along with hospital number, contact information and guardian list.
2. Diagnosis
   Date of diagnosis and doctor's prognoses.
3. History
   Medical record on hospital visits including body measurements, chief complaint, problem, treatment.
4. Gallery
   Including disease photography, medical scan images, scanned document, video. This is crucial for some systems as they still keep old records as scanned images. Therefore, we can transfer this legacy data as well as data in the standard format.
5. Appointment
   Appointment date, hospital, room, doctor and procedure.

3.3   Data Formatter

Data formatter is a translation layer between the hospital's proprietary data format and standard format. Formatter has one job but in two roles as follows.

1. Data Retrieve Interface
   Retrieve patient information for the hospital EMR system and translate it into the standard format.
2. Data Updater Interface
   - Receive patient information from admin application in the standard format and translate them into hospital EMR format.
   - Validate the data is in the correct format and compatible with the hospital's internal system.
   - Insert new records into a hospital database or update the existing records with new data.

3.4   Event Response

Event response is headless software that detects real-time events in the transfer network. The events are status change for referral request, data transfer request. If the event is related to its controlled hospital, it will forward the data to the administrator application or data formatter. Two main functions of the event response are.

1. Event request handler
   - Monitor and detect event from transfer network
   - Request and receive data from data formatter
   - Send data to data formatter
2. Data receiver interface
   - Receive data from another hospital's event response
   - Validate data in correct format
   - Validate data compatible with hospital internal format
   - Send data to another hospital's event response
   - Update transfer status to transfer network

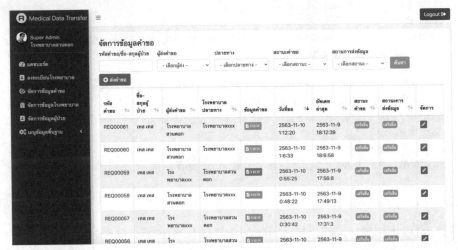

(a) Patient referral and data transfer management.

(b) Form for create referral request.

Fig. 2. Administrator application

3.5 Administrator Application

This component is a web application designed for administrators (healthcare workers, hospital managers). This acts as the back office application for hospitals to manage the referral request. The main functions of this application are as follows.

1. Patient management
2. Create referral request
3. Approve incoming referral request
4. User management

3.6 Transfer Network

This is the main network that kept the history of all transfer requests and responses between facilities. It acts as the shared event database. We need to emphasize that it only the event history, no actual medical data are passing through this network. The medical data only directly transferred between event response nodes. The referral request needs to get approved by the target hospital and patient. There are two types of interactions with the transfer network as follows.

1. Receive request from patient application (mobile application).
2. Receive status change from event response in hospital.

As you can see from the structure of the network, this is very similar to how the blockchain network operates. We can apply the blockchain technology and distribute the transfer network into multiple nodes in each network hospitals. This will further strengthen the data integrity of the whole network.

3.7 Patient Application

This component is a mobile application designed for patients. Each patient needs to register with the application by entering their national identification number and hospital number, Fig. 3. It will connect them with our referral transfer network. The main functions of this application are as follows.

1. Register and login
2. View enrollment
3. Create referral request
4. Approve data transfer
5. View request history
6. View Appointment calendar
7. Edit personal information

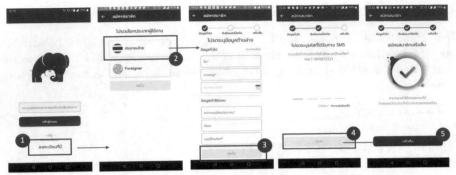

(a) 1. Register button, 2. Nationality, 3. Basic information   (b) 4. One-Time-Password, 5. Finish

Fig. 3. Patient application - registration module

4   Process

4.1   Patient Registration

Patients need to register into the referral system before their data can be transferred by the system. Figure 3 shows the registration process on patient mobile application. This process is preferably done on the first hospital visit. Patients need to register once then they can be referred to any hospital in the networks.

4.2   Hospital Search

The administrator application (Fig. 2) maintain an up-to-date hospital list along with their capabilities. It contains a list of medical procedures that can be performed at each hospital. This information informs users about each hospital's facility, equipment and personnel readiness for the procedures. Healthcare workers can enter the procedure name and search for the appropriate facility.

4.3   Referral Request

4.3.1   Create Request

Users need to select the source and target hospital to which data will be transferred. Our system allows both patients and health workers to create a new referral request.

- In the case of health workers, they can create them on administrator application shown in Fig. 2b. To comply with the private data policy, patients will need to give their approval again in the mobile application for each request.
- In the case of patients, they can create a new request on the patient mobile application shown in Fig. 4b.

(a) 1. Patient login, 2. OTP   (b) 3. New request, 4. Detail, 5. Referral History

Fig. 4. Patient application - request module

4.3.2   Target Hospital Approval

After a worker or a patient creates a new referral request, the request is submitted to the transfer network. Administrator application receives the new request and notifies appropriate users at the target hospital. User can decide either to approve the request and confirm the appointment date or to deny and send back a reason. Finally, the source hospital and patient get notifications for the request results.

4.4   Data Transfer

Upon patient approval, the transfer network record the event in the shared storage. The data transfer process is as shown in Fig. 5.

4.4.1   Source Hospital
1. Event response: acknowledge the request
2. Event response: send data request to data formatter
3. Data formatter: retrieve data from the hospital database
4. Data formatter: translate the data into standard format
5. Data formatter: send standardized data to Event response
6. Event response: send the data to target hospital event response

4.4.2   Target Hospital
1. Event response: receive the data
2. Event response: validate the data format
3. Event response: send data to data formatter
4. Data formatter: receive the data
5. Data formatter: insert the data into hospital database
6. Event response: send the transfer result

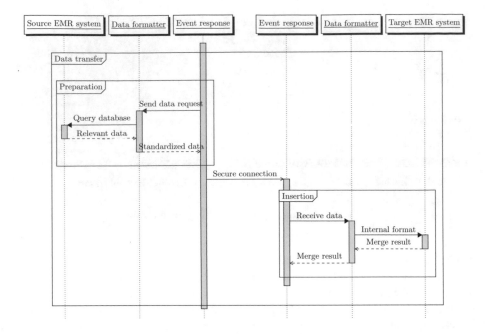

Fig. 5. Data transfer

5   Conclusion

In this paper, we introduce a design for an electronic patient referral system. We start by examining the existing platforms and point out their shortcomings. Based on this information, we design the structure and processes of our referral system. The most unique aspect of this system is how it connects with the existing hospital's electronic health record systems. To connect with a new hospital system, we only need to develop a new simple data formatter to translate the hospital's internal data structure into the standard format. This eliminates the manual data copying in the previous referral system, data are being retrieved and stored directly in the hospital's EMR systems instead.

We show in detail the processes of how users interact with the system. Help reduce the workload of healthcare wakers. An administrator application and patient application were introduced as the interfaces between the system and user groups. Patient directly involves in the transfer process as an approval step which is missing in any previous systems. We discuss the advantages of our design in the security aspect. We emphasize that patient medical information is never sent through the transfer network. Instead, the network only keeps the history of all referral requests and approvals. The advantage of this is the transparency of the transfers toward all users including physicians and patients. We propose the possibility of utilizing blockchain technology to store this transfer history.

In future work, we will deploy and evaluate the system in both technical and practical aspects. We expect some changes will be needed to satisfy broader user targets. Eventually, we hope that it becomes a baseline system for the future national patient referral system.

Acknowledgements. This research was partially supported by Chiang Mai University and Center of Data Analytics and Knowledge Synthesis for Healthcare, Chiang Mai University.

References

1. Agarwal, S., Perry, H.B., Long, L.A., Labrique, A.B.: Evidence on feasibility and effective use of mHealth strategies by frontline health workers in developing countries: systematic review. Trop. Med. Int. Health **20**(8), 1003–1014 (2015)
2. Chib, A.: The promise and peril of mHealth in developing countries. Mob. Media Commun. **1**(1), 69–75 (2013)
3. Chib, A., van Velthoven, M.H., Car, J.: mHealth adoption in low-resource environments: a review of the use of mobile healthcare in developing countries. J. Health Commun. **20**(1), 4–34 (2015)
4. Ketvichit, T., Mungsing, S.: The development of information management system model for patient referral using Thairefer program. J. Thai Med. Inform. Assoc. **1**(1) (2015)
5. Shephard, E., Stockdale, C., May, F., Brown, A., Lewis, H., Jabri, S., Robertson, D., Moss, V., Bethune, R.: E-referrals: improving the routine interspecialty inpatient referral system. BMJ Open Qual. **7**(3), e000249 (2018)
6. Zuchowski, J.L., Rose, D.E., Hamilton, A.B., Stockdale, S.E., Meredith, L.S., Yano, E.M, Rubenstein, L.V., Cordasco, K.M.: Challenges in referral communication between VHA primary care and specialty care. J. Gen. Intern. Mod. **30**(3), 305–311 (2015)

Dynamic Pricing Method for One-Way Car Sharing Service to Meet Demand and to Maximize Profit Under Given Utility Function

Ryuta Kikuchi and Hiroyoshi Miwa[✉]

Graduate School of Science and Technology, Kwansei Gakuin University,
2-1 Gakuen, Sanda-shi, Hyogo, Japan
{kikuchi.ryuta,miwa}@kwansei.ac.jp

Abstract. A car sharing service is a model of car rental where people rent cars for short periods of time. A round-trip car sharing service, in which a user must return the car to the same station where the user picked up the cars is generally used for shopping, leisure and occasional trips; on the other hand, one-way car sharing service such that a user can return a car at an arbitrary station is used for all purposes and more convenient for users than a round-trip car sharing service. However, due to imbalance of demand between users' departures and destinations, when a user want to use a car at an arbitrary station, no cars may be returned yet till the time at the station and the user cannot always use a car. Some strategies have been proposed to solve this problem, one of which is a dynamic pricing. Charging higher price decreases demand, and charging lower price increase demand; thus, we can make demand match supply of available cars by using the strategy of dynamic pricing. This strategy is effective; however, too high price decreases demand too much, so that the total profit may decrease. On the other hand, too low price increase demand too much, so that the imbalance of demand cannot be sufficiently balanced. Therefore, a control method of the dynamic pricing to maximize profit with matching demand to supply is needed. In this paper, we formulate the optimization problem to determine price between any two station at each time slot. This problem is formulated as a network flow problem on a time and space network. We evaluate this control method by numerical experiments. The result shows that the proposed method achieves more profit than a fixed pricing.

1 Introduction

A car sharing service is a model of car rental where people rent cars for short periods of time. The service gains in popularity today as an alternative to private car use and public transport. The users of the service can have the benefit of more flexible car use than traditional public transportation such as bus and train without the expense of owning private cars. Furthermore, the spread of the car sharing service reduces the amount of pollutant emissions. Thus, the car sharing service is now spreading all over the world.

There are two types of operation of the car sharing service. One is a round-trip car sharing service, in which a user must return the car to the same station where the user picked up the cars; the other is a one-way car sharing service, in which a user can return a car at an arbitrary station. A round-trip car sharing service is generally used for shopping, leisure and occasional trips; while a one-way car sharing service is used for all purposes, which is more flexible and convenient for users than a round-trip car sharing service. However, a one-way car sharing service has a drawback; for example, car stocks at a station often become imbalanced, because, since demand for cars varies for pairs of origin-destination stations during a day, cars tends to be accumulated at stations where they are not needed and it causes a shortage in stations where there is needed.

Some strategies have been proposed to solve the drawback, one of which is a dynamic pricing. Charging higher price decreases demand, and charging lower price increase demand; thus, we can make demand match supply of available cars by using the strategy of dynamic pricing (Fig. 1).

This strategy is effective; however, too high price decreases demand too much, so that the total profit may decrease. On the other hand, too low price increase demand too much, so that the imbalance of demand cannot be sufficiently balanced. Therefore, a control method of the dynamic pricing to maximize profit with matching demand to supply is needed.

In this paper, we address a method for determining prices in one-way car sharing service to maximize profit with matching demand to supply. For that purpose, we formulate the optimization problem to determine price between any two stations at each time slot. This problem is formulated as a network flow problem on a time and space network. When the size of a problem instance is small, we can solve the problem by solving the mixed non-linear integer programming; on the other hand, when the size is large, we can solve the problem heuristically by using a network flow algorithm after relaxation of integer variables to continuous variables Then, we have an approximation solution by the rounding of the solution. We evaluate this control method by numerical experiments. The result shows that the proposed method achieves more profit than a fixed pricing.

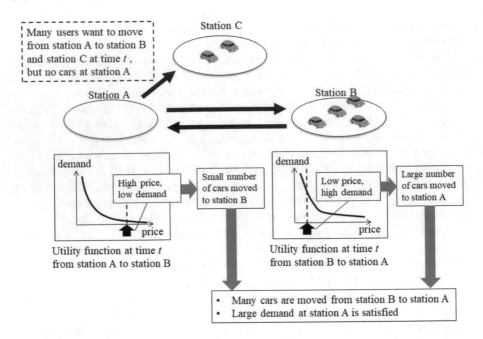

Fig. 1. Dynamic pricing and one-way car sharing service

2   Related Research

Several approaches have been proposed regarding dynamic pricing in one-way car sharing service.

The reference [1] defines the optimization problem to determine price and proposes a method based on a meta-heuristic algorithm. Furthermore, it shows by a case study that the profit of 1161 Euro is obtained when the proposed dynamic pricing is applied, while the deficit of 2068 Euro is obtained when the proposed dynamic pricing is not applied. This means that the dynamic pricing is effective from the viewpoint of profit for a service provider. This research takes a specified utility function into consideration. The utility function defines the relationship between price and demand. In [1], the utility function is the same for all pairs of two stations, but, in our research, we generalize so as to deal with different utility functions among different pairs of two stations. Moreover, transportation of cars by staff is not considered in [1], but, in our research, we take this into consideration.

The reference [2] deals with the problem of determining the small number of transportation of cars by staff so as to satisfy demand and proposed two methods based on the optimization problem and the simulation. It shows by a case study that the profit of 854 Euro is obtained by the simulation method and the profit of 3865.7 Euro is obtained by the method based on the optimization problem, while the deficit of 1160 Euro is obtained when no methods are applied. This research does not deal with dynamic pricing, but focus on reduction of transportation of cars by staff.

The reference [3] defines the problem of determining the location of stations so as to avoid the imbalance of demand. It shows by a case study that, if demand near to a station must be matched by supply at the station, the charge of 4 or more Euro per 10 min is necessary to avoid the deficit; however, it competes with a taxi service.

The reference [4] proposes a method of sharing a car with other users so as to reduce the transportation of cars by staff. The result by the simulation shows the proposed method can reduce 42% of the transportation of cars by staff.

The reference [5] proposes a method using reinforcement learning, and it shows by the simulation that the imbalance of demand can be avoid by the proposed dynamic pricing.

3   Dynamic Pricing for One-Way Car Sharing Service

In this section, we address a method for determining prices in one-way car sharing service to maximize profit with matching demand to supply. We take different utility functions among different pairs of two stations and transportation of cars by staff into consideration.

First, we formulate the optimization problem for dynamic pricing.

We show the notation used to formulation in the following Table 1 and Table 2.

Table 1. Input

Notation	Description
S	Set of stations
T	Set of time slots
$D_{t,i,j}$	Trip time of a car from origin station $i(\in S)$ departing at time $t(\in T)$ to destination station $j(\in S)$. When $D_{t,i,j} = 0$, the trip from i to j is impossible
$A_{t,i,j}$	A parameter of the utility function (1) corresponding the relationship between price and demand. ($t \in T, i, j \in S$)
$B_{t,i,j}$	The other parameter of the utility function (1) corresponding the relationship between price and demand. ($t \in T, i, j \in S$)
$C_{\min}$	The lower bound of price
$C_{\max}$	The upper bound of price
PSC	Cost of parking space for a car
CC	Cost of a car
RC	Cost of transportation of a car by staff per a time slot

Table 2. Variables

Notation	Description
PS_i	Parking space of station $i(\in S)$
NC_i	Number of cars at station $i(\in S)$ at time slot 0
$R_{t,i,j}$	Number of cars transported by staff departing from origin station $i(\in S)$ at time $t(\in T)$ to destination station $j(\in S)$
$P_{t,i,j}$	Price of a car departing from origin station $i(\in S)$ at time $t(\in T)$ to destination station $j(\in S)$
$X_{t,i,j}$	Number of cars used by a user from origin station $i(\in S)$ departing at time $t(\in T)$ to destination station $j(\in S)$
$Y_{t,i,j}$	Auxiliary variable. $(t \in T, i, j \in S)$
$Z_{t,i,j}$	Auxiliary variable. $(t \in T, i, j \in S)$

$$X_{t,i,j} = A_{t,i,j} P_{t,i,j} + B_{t,i,j} \quad (A_{t,i,j} < 0, B_{t,i,j} \geq 0) \tag{1}$$

We formulate the optimization problem to determine price between any two stations at each time slot as follows.

$$\text{maximize} \sum_{t\in T} \sum_{i\in S} \sum_{j\in S} (Y_{t,i,j} - D_{t,i,j} R_{t,i,j} \text{RC}) - \sum_{s\in S} (\text{PS}_s \text{PSC} + \text{NC}_s \text{CC}) \tag{2}$$

$$\text{subject to} \sum_{t\in T} \sum_{i\in S, t+D_{t,i,s}=t'} (X_{t,i,s} + R_{t,i,s}) - \sum_{j\in S} (X_{t',s,j} + R_{t',s,j}) = 0, \quad s \in S, t' \in T, t' \neq 0, t' \neq |T| \tag{3}$$

$$\text{NC}_s - \sum_{i\in S} (X_{0,s,i} + R_{0,s,i}) = 0, \quad s \in S \tag{4}$$

$$\sum_{t\in T} \sum_{i\in S, t+D_{t,i,s}=t'} (X_{t,i,s} + R_{t,i,s}) \leq \text{PS}_s, \quad s \in S, t' \in T, t' \neq 0 \tag{5}$$

$$\text{NC}_s \leq \text{PS}_s, \quad s \in S \tag{6}$$

$$X_{t,i,j} \leq B_{t,i,j}, \quad t \in T, i, j \in S, i \neq j, D_{t,i,j} \neq 0 \tag{7}$$

$$X_{t,i,j} \leq (A_{t,i,j} C_{\min} + B_{t,i,j}) Z_{t,i,j} \quad t \in T, i, j \subset S, i \neq j, D_{t,i,j} \neq 0 \tag{8}$$

$$X_{t,i,j} \geq A_{t,i,j} C_{\max} + B_{t,i,j} \quad t \in T, i, j \in S, i \neq j, D_{t,i,j} \neq 0 \tag{9}$$

$$X_{t,i,j} = 0, \quad t \in T, i, j \in S, D_{t,i,j} = 0 \tag{10}$$

$$Y_{t,i,j} \leq D_{t,i,j} X_{t,i,j} \frac{X_{t,i,j} - B_{t,i,j}}{A_{t,i,j}} \quad t \in T, i, j \in S, i \neq j \tag{11}$$

$$Y_{t,s,s} = 0, \quad t \in T, s \in S \tag{12}$$

$$R_{t,i,j} = 0, \quad t \in T, i, j \in S, D_{t,i,j} = 0 \tag{13}$$

$$X_{t,i,j}, Y_{t,i,j}, R_{t,i,j}, \text{PS}_i, \text{NC}_i \in \mathbb{N}, Z_{t,i,j} \in \{0,1\} \quad t \in T, i, j \in S \tag{14}$$

The objective function (2) is the profit considering income, cost of station, and cost of transportation by staff; the constraints (3) and (4) are the flow conservation in the time-space network; the constraints (5) and (6) are the capacity constraints in the time-space network; the constraints (7), (8), and (9) are the upper and lower bounds of the number of cars used by users; the constraints

(10) and (13) indicate that cars do not move from station i to j when the move is not defined; $Y_{t,i,j}$ of the constraints (11) and (12) are the auxiliary variables to make the objective function a linear function; the constraints (14) indicate the number of cars is the non-negative integers and so on.

Next, we address the case that the price between two stations is a constant regardless of time. We formulate the optimization problem to determine a constant price between any two stations as follows. Variable C indicates the constant price per a time slot.

$$\text{maximize} \sum_{t \in T} \sum_{i \in S} \sum_{j \in S} (Y_{t,i,j} - D_{t,i,j} R_{t,i,j} \text{RC}) - \sum_{s \in S} (\text{PS}_s \text{PSC} + \text{NC}_s \text{CC}) \tag{15}$$

$$\text{subject to} \sum_{t \in T} \sum_{i \in S, t+D_{t,i,s}=t'} (X_{t,i,s} + R_{t,i,s}) - \sum_{j \in S} (X_{t',s,j} + R_{t',s,j}) = 0, \quad s \in S, t' \in T, t' \neq 0, t' \neq |T| \tag{16}$$

$$\text{NC}_s - \sum_{i \in S} (X_{0,s,i} + R_{0,s,i}) = 0, \quad s \in S \tag{17}$$

$$\sum_{t \in T} \sum_{i \in S, t+D_{t,i,s}=t'} (X_{t,i,s} + R_{t,i,s}) \leq \text{PS}_s, \quad s \in S, t' \in T, t' \neq 0 \tag{18}$$

$$\text{NC}_s \leq \text{PS}_s, \quad s \in S \tag{19}$$

$$X_{t,i,j} \leq B_{t,i,j}, \quad t \in T, i, j \in S, i \neq j, D_{t,i,j} \neq 0 \tag{20}$$

$$X_{t,i,j} \leq (A_{t,i,j}C + B_{t,i,j} + 0.5)Z_{t,i,j} \quad t \in T, i, j \in S, i \neq j, D_{t,i,j} \neq 0 \tag{21}$$

$$X_{t,i,j} \geq A_{t,i,j}C + B_{t,i,j} - 0.5 \quad t \in T, i, j \in S, i \neq j, D_{t,i,j} \neq 0 \tag{22}$$

$$X_{t,i,j} = 0, \quad t \in T, i, j \in S, D_{t,i,j} = 0 \tag{23}$$

$$Y_{t,i,j} < D_{t,i,j} X_{t,i,j} \frac{X_{t,i,j} - B_{t,i,j}}{A_{t,i,j}} \quad t \in T, i, j \in S, i \neq j \tag{24}$$

$$Y_{t,s,s} = 0, \quad t \in T, s \in S \tag{25}$$

$$R_{t,i,j} = 0, \quad t \in T, i, j \in S, D_{tij} = 0 \tag{26}$$

$$C_{\min} \leq C \leq C_{\max} \tag{27}$$

$$X_{t,i,j}, Y_{t,i,j}, R_{t,i,j}, \text{PS}_i, \text{NC}_i \in \mathbb{N}, Z_{t,i,j} \in \{0,1\} \quad t \in T, i, j \in S \tag{28}$$

The constraints (21) and (22) means that the number of cars of the demand determined by the utility function is an integer. The constraint (27) means the lower and the upper bounds of price. Here, we replaced the constraints that $Y_{t,i,j} \leq D_{t,i,j} X_{t,i,j} C$ by the constraints (24) to solve rapidly.

4   Numerical Experiments

In this section, we evaluate the performance of the dynamic pricing determined by the proposed method. For that purpose, we compare the profit of the case using the dynamic pricing and the profit of the case using a constant price. Too high price decreases demand too much, so that the total profit may decrease; on the other hand, too low price increase demand too much, so that the imbalance of demand cannot be sufficiently balanced. Therefore, the control of price for dynamic pricing is not easy. Our proposed method determines the prices between all pairs of two stations at every time slot by solving the optimization problem maximizing profit with matching demand to supply. For comparison with the proposed method, we use the method determining the constant price by solving the optimization problem maximizing profit with matching demand to supply. We compare the profit by the proposed dynamic pricing with the profit of the case using the constant price.

The problem instances are randomly generated. We assume that the number of the stations is five and that the number of the time slots is 49 which means a unit time slot is one hour and a simulation of successive two days. We show the parameters as follows:

$$|S| = 5, \ |T| = 49$$
$$D_{t,i,j} \in \{1, 2, 3\} \qquad\qquad t \in T, i, j \in S$$
$$A_{t,i,j} = -0.01 \qquad\qquad t \in T, i, j \in S$$
$$0 \leq B_{t,i,j} \leq 15 \qquad\qquad t \in T, i, j \in S$$
$$C_{\min} = 400, C_{\max} = 800$$
$$PSC = 1000, CC = 3000, RC = 500$$

We used the solver SCIP [6] for solving the mixed integer non-linear programming.

Table 3 shows the average of the results for 50 instances. The total computation time is about 90 min on a PC using Windows10 with Intel Core i5-7200U 2.50 GHz and 8 GB RAM.

The availability ratio is the ratio of the sum of $\sum_{t \in T} \sum_{i \in S} \sum_{j \in S, j \neq i} X_{t,i,j} D_{t,i,j}$ for all instances to the sum of $\sum_{s \in S} NC_s$ for all instances. Intuitively, the availability ratio is the ratio of the number of cars used by users for a unit time slot.

Table 3. Average of results for 50 instances

	Dynamic pricing	Constant price
Profit	2569185.46	2162235.7
Availability ratio	0.736	0.606
Number of users	2615.96	2633.86
Ratio of transportation by staff	0.011	0.029
Number of cars	147.86	179.76
Parking spaces	156.54	213.3

The ratio of transportation by staff is the ratio of the sum of $\sum_{t \in T} \sum_{i \in S} \sum_{j \in S, j \neq i} R_{t,i,j} D_{t,i,j}$ for all instances to the sum of $\sum_{s \in S} NC_s$ for all instances. Intuitively, the ratio of transportation by staff is the ratio of the number of cars transported by staff for a unit time slot.

The profit by the dynamic pricing is about 1.19 times compared to that by the method using the constant price; in addition, the availability ratio by the dynamic pricing is larger than that by the method using the constant price.

On the other hand, the numbers of users in both methods are almost same.

From the above examination, the dynamic pricing whose objective is the increase of profit realizes efficient one-way car sharing service for a service provider; however, it does not always increase users.

5   Conclusions

In this paper, we addressed the control method of the dynamic pricing for one-way car sharing service. The objective of the proposed method is to maximize profit with matching demand to supply is needed. For that purpose, we formulated the optimization problem to determine price between any two station at each time slot.

We evaluated this control method by numerical experiments. Especially, we compared the profit of the case using the dynamic pricing and the profit of the case using a constant price. The profit by the dynamic pricing is about 1.19 times compared to that by the method using the constant price; in addition, the availability ratio by the dynamic pricing is larger than that by the method using the constant price. On the other hand, the numbers of users in both methods are almost same. The result shows that the proposed method achieves more profit than a fixed pricing; therefore, our proposed method realizes efficient one-way car sharing service for a service provider.

In this research, we assumed that demand is predictable and that demand changes according to utility functions. However, in a actual environment, these assumptions do not always hold. As the future work, we develop a method that works well under environment that it is difficult to predict and control demand.

Acknowledgements. This work was partially supported by the Japan Society for the Promotion of Science through Grants-in-Aid for Scientific Research (B) (17H01742) and JST CREST JPMJCR1402.

References

1. Jorge, D., Molnar, G., de Almeida Correia, G.H.: Trip pricing of one-way station-based carsharing networks with zone and time of day price variations. Transp. Res. Part B Methodol. Part 2 **81**, 461–482 (2015)
2. Jorge, D., de Almeida Correia, G.H., Barnhart, C.: Comparing optimal relocation operations with simulated relocation policies in one-way carsharing systems. IEEE Trans. Intell. Transp. Syst. **15**(4), 1667–1675 (2014)
3. de Almeida Correia, G.H., Antunes, A.P.: Optimization approach to depot location and trip selection in one-way carsharing systems. Transp. Res. Part E Logist. Transp. Rev. **48**(1), 233–247 (2012)
4. Barth, M., Todd, M., Xue, L.: User-based vehicle relocation techniques for multiple-station shared-use vehicle systems transportation research board. In: 80th Annual Meeting (2004)

5. Kamatani, T., Nakata, Y., Arai, S.: Dynamic pricing method to maximize utilization of one-way car sharing service. In: Proceedings of 2019 IEEE International Conference on Agents (ICA), Jinan, China, pp. 65–68 (2019)
6. https://www.scipopt.org/

A Proposition of Physician Scheduling Method for Improving Work-Life Balance

Yusuke Gotoh[1](✉), Naoki Iwamoto[2], Koji Sakai[3], Jun Tazoe[3], Yu Ohara[3], Akira Uchiyama[4], and Yoshinari Nomura[1]

[1] Graduate School of Natural Science and Technology, Okayama University, Okayama, Japan
gotoh@cs.okayama-u.ac.jp
[2] Department of Information Technology, Okayama University, Okayama, Japan
[3] Department of Radiology, Kyoto Prefectural University of Medicine, Kyoto, Japan
[4] Graduate School of Information Science and Technology, Osaka University, Osaka, Japan

Abstract. Due to the recent efforts to reform ways of working, it is important to expand employment opportunities and solve such current issues as the decrease in the working-age population and the balance between childcare and nursing care due to the diversification of working styles. To realize such an environment for work-style reform, the mechanism proposed in Society 5.0 that aggregates and analyzes real-space sensor data and feeds it back has been attracting attention. This mechanism is used in many fields, such as the transportation sector, construction, and the IT sector. However, in the medical field, it is necessary to satisfy the requirements for such actual medical practices such as the number of working physicians, the balance between their research and medical activities, and the educational activities by skilled physicians for junior doctors. This complicates creating system software to manage physician work assignments in an integrated manner. In this paper, we propose a physician-scheduling method to improve the work-life balance of physicians to reform ways of working in the medical field. Our proposed method reduces the workload on physicians by creating a duty roster based on the working environments of physicians according to an actual medical field. As a result of a simulation evaluation using our proposed method, we confirmed it significantly reduced the processing time to create a three-month duty roster for 51 physicians at a university hospital compared to a conventional method.

1 Introduction

Various issues are being tackled in Japan to achieve a smart society in which an information technology mechanism called Society 5.0 [1] is incorporated nationwide. The challenges in Japanese society include a decline in the working-age population and the balance between childcare and nursing care due to the diversification of working styles. To solve these issues, we must create an information infrastructure based on Society 5.0.

© The Author(s), under exclusive license to Springer Nature Switzerland AG 2021
L. Barolli et al. (Eds.): EIDWT 2021, LNDECT 65, pp. 333–343, 2021.
https://doi.org/10.1007/978-3-030-70639-5_31

In March 2017 [2], a Japanese government plan addressed working style reform. To achieve this plan, the Ministry of Health, Labor and Welfare (MHLW) is addressing many issues, including reducing working hours, promoting diversity, raising wages, and increasing productivity [3].

For such work-style reform, using information technology is very effective. Practical use is progressing in many fields, such as transportation, construction, and IT.

On the other hand, in the medical field, the duty rosters of physicians are managed manually because no system software exists. In addition to daily hospital care, physicians are burdened by a variety of tasks: maintaining specialist certification, teaching younger physicians, and daily research activities. Therefore, to reform how physicians work, their burden must be reduced by creating duty rosters that shrink their workload and shorten their working hours.

For a radiologist in a typical university hospital, the scheduling manager (who is also a physician) with a certain amount of medical experience creates the duty roster. He/She prepares it for the next three months based on pre-interviewed travel schedules, individual skills, and family situations for every physician. However, last minute changes due to vacations and business trips can severely burden the scheduling manager.

In this paper, we propose a physician-scheduling method to improve the work-life balance of physicians. Our proposed method reduces their burden by creating an algorithm for duty rosters based on the actual working environments of doctors.

The remainder of the paper is organized as follows. Society 5.0 is introduced in Sect. 2. In Sect. 3, we explain the details of work-style reform. Related works are explained in Sect. 4. In Sect. 5, we explain the details of our proposed method and evaluate its performance in Sect. 6. Finally, we conclude in Sect. 7.

2   Society 5.0

2.1   Basic Idea

Society 5.0 is a mechanism for constructing a human-centered society that achieves both economic development and social equality through a system that highly integrates virtual cyber space and actual physical space. Society 5.0 is defined as a new society, following previous societies: Society 1.0 (hunting), Society 2.0 (agrarian), Society 3.0 (industrial) and Society 4.0 (information). The Japanese Cabinet Office proposed Society 5.0 as a future model that the country should aim for in the 5th Science and Technology Basic Plan [4].

2.2   Social Structure in Society 5.0

In the previous information society, Society 4.0, users connected to cloud services in cyberspace by the internet to obtain and analyze information. In Society 5.0, users will acquire big data from sensors in physical space and store them in cyberspace, where artificial intelligence (AI) analyzes such big data and returns the results to users in physical space.

The key technologies required by Society 5.0 are described below.

2.2.1 IoT

In Society 4.0, one problem was that this knowledge was not shared among users because each user acquired it. In Society 5.0, the Internet of Things (IoT) technology will connect all users and objects. Users can also create new value by sharing various pieces of knowledge.

2.2.2 AI

In Society 4.0, users struggled to quickly find and analyze required information from a huge amount of information. Society 5.0 uses AI technology to allow users to extract and analyze information in a shorter amount of time.

2.2.3 Automation Technology

In Society 4.0, since users need to perform many deeds independently, their actions might be limited by physical limitations. In Society 5.0, robots and self-driving cars will expand the users' range of motion by assisting them.

3 Work-Style Reform

3.1 Basic Idea

In Japan, the goal of work-style reform [3] is to allows workers to choose a diverse and flexible work style based on their individual circumstances. To solve such tasks as the decrease of the working-age population due to the declining birthrate, aging population, and diversified demands on workers, investment and innovation will make them more productive. By creating a society in which workers can choose a variety of ways to work, employment opportunities might be increased and their own abilities improved.

3.2 Work-Style Reform in Medical Fields

Recently, there has been an increasing demand for improving the working styles of physicians. In particular, their working hours are increasing due to more patients who require long-term treatment in hyper-aged societies. In addition, the physical and mental burden on physicians will inevitably increase to maintain the quality of medical care, ensure communication with patients and their families, and achieve regional cooperation with other medical facilities. In such a work environment, the health of physicians will be severely impacted.

Previous studies have shown that the long working hours by physicians lead to a decline in performance and more medical accidents. For example, by the

Ministry of Internal Affairs and Communications in 2012, the average percentage of employees in Japan who work 60 h or more per week was 14.0% in all occupations; the percentage of physicians is the highest: 41.8% [5]. Therefore, it is important to improve the duty roster of physicians to reduce working hours and provide high quality care while maintaining health care.

3.3 Challenges in Current Physician Scheduling Method

In university hospitals, a scheduling manager who is also a physician typically manages the duty roster for the next three months by interviewing physicians about their travel schedules, particular skills, and childcare status. In addition, the scheduling manager must revise the duty roster each time schedules change due to leave or conference travel. In many medical settings, the task of scheduling physicians impacts normal workloads and increases the burden on scheduling managers.

4 Related Works

4.1 Nurse-Scheduling Problem

As a study related to physician scheduling, we describe the nurse-scheduling problem (NSP) [6]. When the scheduling manager creates a working schedule based on each skill and the work styles of nurses, the time required to find a combination that satisfies all nurses is lengthened.

Deriving combinatorial solutions using general-purpose solvers addresses these problems. There are two main types of derivation methods. The first is an exact solution method that seeks the optimal solution even if the time is lengthened [7]. In such a solution method, a general-purpose solver based on Mixed Integer Linear Programming (MILP) and Satisfiability Problem (SAT) has been developed. The second is an approximate solution method [8] that finds a better feasible solution in a more realistic computation time. In the approximation method, a generic solver based on the Weighted Constraint Satisfaction Problem (WCSP) has been developed.

Although several scheduling methods have been proposed for nurses, few have been proposed for physicians. Proposing a scheduling method that improves the work-life balance of physicians will significantly affect work-style reform in the medical field.

4.2 Matching of Children to Daycare Centers Using Game Theory

In other fields, a maching method has eliminated waiting lists for children [9]. When determining the admission of children in each family based only on the order of preference for kindergartens and nursery schools, higher priority cannot be given if siblings wish to enter the same kindergarten. By introducing an algorithm based on a depth-first search, this method can complete the selection of thousands of children in a few seconds [10]. This process used to take about 1,000 h. This matching method has been systematized and introduced into many municipalities.

5   Proposed Method

5.1   Research Purpose

In this paper, we propose a scheduling method to improve the work-life balance of physicians. Our proposed method reduces the burden on physicians by constructing an algorithm that makes duty rosters considering working environments based on actual medical practices. We also create low-cost work assignments with a database manually created and accumulated by the scheduling manager.

5.2   Interviews

We propose a physician-scheduling method based on the working conditions at the Department of Diagnostic and Therapeutic Radiology, Kyoto Prefectural University of Medicine. By interviewing the scheduling manager, we identified the following initial information for radiologists:

1. Positions
   a. Staff (professor, associate professor, assistant professor)
   b. Graduate student (qualified specialist)
   c. Graduate student (not yet qualified as a specialist)
   d. Second-year major physician (qualified specialist)
   e. Late major physician (not yet qualified as a specialist)
   f. First-year major physician
   g. Resident
2. Responsible department: Primary responsibility for one of the following (a) to (d) and sub-responsibilities for multiple departments other than the main one.
   a. Treatments
   b. Interventional radiology (IVR)
   c. Diagnosis
   d. Radio isotope (RI)
3. Days and times
   a. Monday-Friday
   b. Arrangements of time (AM, PM, night)
4. Ability ranks of physicians based on their leadership authority
   a. Authority level A
   b. Authority level B
   c. Training
5. Main staff (Diagnosis)
6. Main staff (RI)

5.3   Procedure for Making Physician Schedules

Similar to the conventional manual scheduling of physicians, the proposed method assumes that a shift list with internal and external workdays for all physicians is provided in advance to the scheduling manager, who creates a three-month duty roster. For example, the scheduling manager prepares duty rosters for July, August, and September after the work schedule outside the hospital for physicians is provided in June.

We describe below the procedure for creating a duty roster based on interviews with the scheduling manager.

5.3.1   Making a Three-Month Duty Roster

1. Based on the initial information and the outside roster described in Sect. 5.2, we created a duty roster for each month and each week for three months.
2. Referring to the duty roster for each week, it is updated by deleting the personnel who fall under (a) through (e) below from the candidate list of days of the week (Monday, $\cdots$, Friday) and hours (AM, PM, night).
   a. Working in hospitals
   b. Student practice
   c. Working with other hospitals
   d. Image diagnostic services
   e. Other (study abroad, research)
3. Assignment of diagnostic department:
   The duty roster can be updated to include as many diagnostic sections as possible as the primary contact of a scheduling manager with authority level A.
4. Departmental personnel coordination of RI section:
   Update the duty roster to ensure a sufficient number of main staff (RI) for each period on each day.
5. Coordination of personnel in Diagnostics Department:
   Update the duty roster so that the following (a) through (e) are available for each period on each day:
   a. Remote response of jurisdictional hospitals
   b. Research activities
   c. Medical photography
   d. Specialists with a authority level A or B
   e. Image diagnostic services

5.3.2   Updating Duty Roster (Two Weeks to Three Months)

Two weeks prior to the start of use, the list of personnel who will be removed from the duty roster due to business trips or vacations are removed. Then it is updated based on the physician scheduling described in Subsect. 5.3.1.

5.3.3   Last Minute Changes to Duty Roster (Three Days to Two Weeks)

Whenever a person is removed due to illness, the physician scheduling described in Subsect. 5.3.1 is carried out, and the duty roster is updated.

5.4   Implementation

In this paper, we implemented a physician scheduling based on the outside roster of next three months.

Table 1. Example of initial information (20 physicians, CSV format)

```
1 ,1 ,3 ,3, 2, 3, 1, 6, 6, 5, 6, 4, 6, 6, 5, 6, 4, 0, 0, 0, 0, 0, 0, 0, 0, 0, 0, 0
2 ,1, 3, 2, 3, 2, 3, 6, 5, 6, 5, 6, 6, 5, 6, 5, 6, 0, 0, 0, 0, 0, 0, 0, 0, 0, 0, 0
3 ,1, 3, 3, 0, 3, 3, 6, 6, 0, 6, 6, 6, 6, 0, 6, 6, 0, 0, 0, 0 ,0, 0, 0, 0 ,0, 0, 0
4 ,1, 3, 3, 3, 1, 1, 6, 6, 6, 4, 4, 6, 6, 6, 4, 4, 0, 0, 0, 0, 0, 0, 0, 0, 0, 0, 0
5 ,1, 3, 2, 3, 1, 3, 6, 5, 6, 4, 6, 6, 5, 6, 4, 6, 0, 0, 0, 0, 0, 0, 0 ,0, 0, 0, 0, 0
6 ,1, 3, 1, 3, 3, 2, 6, 4, 6, 6, 5, 6, 5, 6, 6, 5, 0, 0, 0, 0, 0, 0, 0, 0, 0, 0, 0
7, 1, 1, 3, 0, 3, 1, 4, 6, 0, 6, 4, 4, 6, 0, 6, 4 ,0 ,0 ,0 ,0, 0, 0, 0, 0, 0, 0, 0
8, 1, 2, 3, 3, 2, 3, 5, 6, 6, 5, 6, 5, 6, 6, 5, 6, 0, 0, 0, 0, 0, 0, 0, 0, 0, 0, 0
9, 1, 3, 3, 3, 3, 0, 6, 6, 6, 6, 0, 6, 6, 6, 6, 0, 0, 0, 0, 0, 0, 0, 0, 0, 0, 0, 0
10, 1, 0, 3, 3, 0, 1, 0, 6, 6, 0, 4, 0, 6, 6, 0, 4, 0, 0, 0, 0, 0, 100, 0, 0, 0, 0, 0
11, 1, 3, 3, 0, 3, 0, 6, 6, 0, 6, 0, 6, 6, 0, 6, 0, 0, 0, 0, 0, 0, 0, 0, 0, 0, 0, 0
12, 1, 0, 3, 3, 0, 3, 0, 6, 6, 0, 6, 0, 6, 6, 0, 6, 0, 0, 0, 0, 0, 0, 0, 0, 0, 0, 0
13, 1, 0, 2, 3, 0, 0, 5, 6, 6, 0, 0, 5, 6, 6, 0, 0, 0, 0, 0, 0, 0, 0, 0, 0, 0, 0, 0
14, 6, 3, 1, 3, 0, 3, 6, 4, 6, 0, 6, 6, 4, 6, 0, 6, 0, 0, 0 ,0 ,0, 0, 0, 0 ,0, 0, 0
15, 7, 3, 3, 3, 3, 3, 6, 6, 6, 6, 6, 6, 6, 6, 6, 6, 0, 0, 0, 0, 0, 0, 0, 0, 0, 0, 0
16, 1, 6, 6, 6, 6, 0, 3, 3, 3, 3, 0, 6, 6, 6, 6, 0, 0, 0, 0, 0, 0, 0, 100, 0, 0, 0, 0, 0
17, 1, 6, 0, 6, 6, 6, 3, 0, 3, 3, 3, 6, 0, 6, 6, 6, 0, 0, 0, 0, 0, 0, 0, 0, 0, 0, 0
18, 1, 6, 6, 0, 6, 6, 3, 3, 0, 3, 3, 6, 6, 0, 6, 6, 0, 0, 0, 0, 0, 100, 0, 0, 0, 0, 0
19, 1, 6, 0, 6, 0, 6, 3, 0, 3, 0, 3, 6, 0, 6, 0, 6, 0, 0, 0, 0, 0, 100, 0, 0, 0, 0, 0
20, 1, 6, 0, 0, 0, 6, 3, 0, 0, 0, 3, 6, 0, 0, 0, 6, 0, 0, 0, 0, 0, 100, 0, 0, 0, 0, 0
```

Table 2. Initial information for each physician in CSV format

Items	Content
1	ID number
2	Position
3–22	Assignment information (day, time)
23	Competence based on leadership authority of scheduling manager
24	Main staff (Diagnosis)
25	Main staff (RI)
26	Diagnostic imaging services
27	Remote response of jurisdictional hospitals
28	Full-time imaging services

An example of the initial information for 20 physicians entered in the CSV format is shown in Table 1. This initial information, which is stored in the CSV format, consists of 28 items, the contents of which are shown in Table 2. The item numbers in Table 2 correspond to the order in which they are counted from left to right.

First, we create a CSV file based on the initial information of all the physicians mentioned in Sect. 5.2. Next, we update the values that compose the CSV file based on the procedure described in Sect. 5.3.1, output the CSV file that reflects the process results, and create a duty roster.

Table 3. Processing time for creating duty roster

Term	Proposed	Conventional
1 week	30.2 min	60.0 min
3 months	32.0 min	720 min

6   Simulation Results

6.1   Evaluation Environment

We evaluated our method for scheduling physicians that improves the work-life balance of scheduling managers who actually created and evaluated physician scheduling using our proposed method. The evaluator is a physician in the Department of Diagnostic and Therapeutic Radiology, Kyoto Prefectural University of Medicine. The evaluator actually used the system with our proposed method and answered a questionnaire about its usability in creating a duty roster and its completeness based on interviews.

The evaluator judged the following items:

1. Processing time for making a duty roster:
   We compared the processing time to make a three-month duty roster in the future with our proposed and manual methods (conventional method).
2. Operation and integrity of system:
   We used a questionnaire to evaluate the usability of our proposed method for making duty rosters and its completeness based on interviews.

6.2   Processing Time for Making Duty Roster

The processing times for making a three month duty roster with the proposed and conventional methods are shown in Table 3.

In Table 3, the processing time was about 30.2 min with the proposed method and about 60.0 min with the conventional method, reducing the production time by almost 50.0%. In addition, the processing time for making a three-month duty roster was reduced to 32.0 min by the proposed method and approximately 720 min by the conventional method: an improvement of 95.6%.

In the proposed method, the processing time for making the CSV file based on the initial information took about 30 min. This processing time occupies most of the time required to make a duty roster. On the other hand, since the same CSV file can be used again to create a duty roster for three months, the processing time of the system is about 10 s. As the length of the term of the duty roster increases, the rate of the proposed method improves more than the conventional method.

6.3 Operability and Integrity

6.3.1 Evaluation Items

The evaluation items in the questionnaire on operability are listed below. In items 2 to 5, we evaluate the proposed method on a scale from 1 to 5.

Item 1: Was the duty roster created by the proposed method easier to use than the conventional schedule?

Item 2: Were all of the initial settings created based on the hearings satisfied?

Item 3: Was the duty roster created by the proposed method actually operational?

Item 4: Would you like to use the proposed method in the actual creation of duty rosters?

Item 5: Please describe what you liked about the proposed method that created a duty roster.

Item 6: Describe what must be improved about the duty roster created by the proposed method.

6.3.2 Evaluation Results

The results of the questionnaire evaluation are shown below.

Item 1: Value: 4 (good)
    Comment: Working physicians can use the duty roster if they understand the contents of each item. For the scheduling manager to quickly understand the duty roster, the reviewer requests an improvement in the output results.

Item 2: Value: 3 (average)
    Comment: For setting up the conditions of the initial information, the algorithm should be improved based on additional conditions in the interviews.

Item 3: Value: 4 (good)
    Comment: Many items are readily available. The reviewer wants to automate the input process of the initial information as well as more flexibility for inputting items.

Item 4: Value: 5 (excellent)
    Comment: The processing time to create the duty roster has been greatly reduced.

Item 5: In making a weekly duty roster, the system's processing time was less than 1 s

Item 6: The evaluator identified the following areas for improvement:
1. Inputting an assistance function of the initial information on a CSV file
2. Adding a display method for outputting the duty roster
3. Modifying the duty roster

In Subsect. 6.3, the reviewer positively rated the convenience and practicality of the system. On the other hand, in item 6, the reviewer listed the issues to be addressed in practice. For items 6 (1) and (2), we need to improve the user interface. For item 6 (3), we must make a duty roster using the general-purpose solver described in Subsect. 4.1.

7 Conclusion

We proposed a physician-scheduling method to improve the work-life balance of physicians to achieve a work-style reform in the medical field. Our proposed method reduces the burden on scheduling managers by making a duty roster based on the working environment of an actual medical field.

As a result of a the simulation evaluation, we confirmed that our proposed method significantly reduced the processing time to make a three-month working schedule for 51 physicians in the Department of Diagnostic and Therapeutic Radiology of Kyoto Prefectural University of Medicine compared to the conventional manual method. A reviewer confirmed that the duty roster made by our proposed method is operational.

In the future, we will improve its user interface and create a duty roster using a generic solver.

Acknowledgements. This paper was supported by Innovation Platform for Society 5.0 from Japan Ministry of Education, Culture, Sports, Science and Technology.

References

1. Cabinet Office, Government of Japan (2020). https://www8.cao.go.jp/cstp/english/society5_0/index.html
2. Government Decides "Action Plan for the Realization of Work Style Reform" Japan Labor Issues, vol. 1, no. 1 (2017)
3. Ministry of Health, Labour and Welfare: Outline of the "Act on the Arrangement of Related Acts to Promote Work Style Reform." https://www.mhlw.go.jp/english/policy/employ-labour/labour-standards/dl/201904kizyun.pdf
4. Cabinet Office, Government of Japan: Report on The 5th Science and Technology Basic Plan. https://www8.cao.go.jp/cstp/kihonkeikaku/5basicplan_en.pdf
5. Ministry of Health, Labour and Welfare: Survey on Employment Structure. https://www.mhlw.go.jp/english/database/db-slms/dl/slms-11.pdf
6. Cheang, B., Li, H., Lim, A., Rodrigues, B.: Nurse rostering problems - a bibliographic survey. Eur. J. Oper. Res. **151**, 447–460 (2003)

7. Thornton, J.R., Sattar, A.: Nurse rostering and integer programming revisited. In: International Conference on Computational Intelligence and Multimedia Applications (ICCIMA 1997), pp. 49–58 (2003)
8. Nonobe, K., Ibaraki, T.: A tabu search approach for the constraint satisfaction problem as a general problem solver. Eur. J. Oper. Res. **106**, 599–623 (1998)
9. Fujitsu Limited, Fujitsu AI Ideally Matches Children to Daycare Centers. https://www.fujitsu.com/global/about/resources/news/press-releases/2017/0830-01.html
10. Iwashita, H., Takagi, T., Suzuki, H., Goto, K., Ohori, K., Arimura, H.: Efficient constrained pattern mining using dynamic item ordering for explainable classification (2020). https://arxiv.org/pdf/2004.08015.pdf

Data Lake Architecture

David Taniar[1][✉] and Wenny Rahayu[2]

[1] Faculty of Information Technology, Monash University, Melbourne, Australia
david.taniar@monash.edu
[2] School of Engineering and Mathematical Sciences, La Trobe University,
Melbourne, Australia
w.rahayu@latrobe.edu.au

Abstract. This paper discusses the primary motivation for using data lakes – a new wave of database technology that is underpinned by the need to deal with data volume and variety of big data storage. The architecture will be described, together with prominent features of data lakes. To understand data lakes more deeply, we need to know the differences between data lakes and other systems, such as federated database systems, data warehouses, and big data systems. Finally, a number of challenges of data lakes will be described.

1 Introduction

A data lake is a centralized data repository to store data, which can be of any format, like structured, semi-structured, and unstructured data, in its native or raw format. The incoming data is not needed to be structured in a particular format; rather, it allows fast data ingestion in its current format into the data lake.

There are two particular historical reasons why data lakes are seen as favorable. Currently, most database systems are structured systems, where the structure or the schema of the data is predefined at the design stage. This works particularly well since the database or data warehouse is designed for a particular purpose, and therefore the data structure has been pre-defined to fit into the database or data warehouse design requirements. The functionalities surrounding the database systems, such as reporting, querying, integration, transformation, and so on, are well supported because they follow a suitable data structure that is already in place.

On the other hand, there are a number of growing practices that require the data to be handled differently. Using a train transportation industry as an example, an engineer who is in-charge of tamping the track might store data in a csv file for every tamping that is done to the track. From a mechanical engineering perspective, track tamping requires a lot of calculations and parameter adjustments, and these variables and parameters are stored in the tamping files. Another engineer might be in-charge in sensors which have been installed in the trains. These sensors capture various measurements related to the movement of the trains. These measurements are also critical in identifying passenger comfort during the trip, as well as giving indications of the health of the tracks, which will lead to off-scheduled maintenance. This information might be stored in a database for a reporting purpose to train management. Other measurements might also be done by other sections of the company that measure different aspects of train transportation. Traditionally, each of these is a silo. When an incident occurs, the engineers might recall similar incidents that had happened in the past. However, finding the data in various data storage is a challenging task. With data lakes, it is expected that this kind of problem would be minimized, as the data is centrally stored. The data will be stored in its native form, without the need to be adjusted with a predefined structure. Therefore, each silo system may still work individually, but when a cross over is needed, the data lake provides an environment that supports access to other datasets.

In other words, data lakes have features that allow silos and collaboration to take place simultaneously. This is due to a central data repository, data stored in its raw form, data is not required to comply to a certain standard or schema, and all of these are needed to support efficient and effective data analysis. This will help data scientists and engineers to explore data, to create hypotheses, to find anomalies, to correlate with previous cases of failures, and so on.

This paper presents a data warehouse architecture, which illustrates the components of a data lake and how they interact with each other. Additionally, we will also describe prominent data lake features, compare them with existing systems, as well as discuss some challenges.

2 Data Lake Architecture

Figure 1 shows a typical architecture for data lakes. The input data can come from various sources, providing various data format, from structured, semi-structured, to unstructured data [1]. The data is stored in its native or raw format. Internally, the data may go through some transformation process, including data cleaning and data integration. Raw data will co-exist with transformed data and enriched data with adequate catalogs. The data at any particular stage in the data lakes can be used by machine learning algorithms to perform data exploration, classification, and prediction. The results from the machine learning module are analysis models that can be used as hypotheses for further machine learning activities.

The enriched data may be transformed into enterprise data warehouses, which will then produce BI reports and visualization. The enriched data may also be stored in other database systems for use in the enterprise.

At any stage of the architecture, cataloging, curation, and version management could be applied. This is an essential element of the architecture that will shape the data from its native form to the transformed and finally to the enriched form. Security and governance also become the foundation of the architecture.

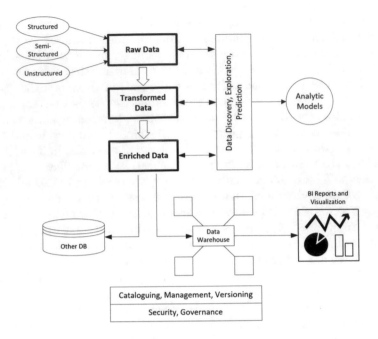

Fig. 1. A data lake architecture

3   Features of Data Lakes

Two categories of data lake distinctive features will be discussed in this section: one is related to the data types, and the other is related to the schema.

3.1   Types of Data and Structures in Data Lakes

There are a number of different kinds of data types that can be ingested to a data lake.

1. **Structured data**
   Data lakes can absorb data from the traditional database or data warehouse. Structured data is data that has a rigid structure, which is usually imposed by a database system. The data follows a table structure when the data is inserted into the database. Data lakes can accept structured data from an operational or transactional database or a data warehouse, and the data type can be the entire table from an existing database management system.

2. **Semi-structured data**
   Semi-structured data is un-normalized data, as it may contain nested data, and hence it has a nested structure. Additionally, the data may not necessarily uniform since new records may have additional features or attributes, or certain features or attributes can be absent in the new records. An example of semi-structured data is XML or JSON data. Multiple JSON records may be stored as different records in a data lake.

3. **Unstructured data**

Unstructured data literally means data that has no structure. This includes free text, paragraphs, notes, articles, etc. Multimedia data, which is also categorized as unstructured data, will be discussed separately. An example of a free text is a note written by a physician on a patient card. This free text may be addressed arbitrarily on the patient card, and hence it might be challenging to model free text as a text attribute in the context of a relational table.

4. **Opaque-structured data**

An opaque-structured data contains data that appears to have a certain structure or format pattern, but the structure or the schema is not defined. Hence, the structure is opaque. An example of opaque-structured data is data produced by sensors installed in the trains to measure certain mechanical aspects of the moving train. Although the data has specific attributes, such as time, location, and various measurements; however, the measurements may change from time to time due to the availability of the sensors as well as the changing of the environment that may affect the sensors in producing specific measurements. Although the data may look like following a particular structure or schema, the structure or the schema is not defined. The data may come in many csv files, where the columns may not be uniform from one sensor report to another.

5. **Multimedia data**

Multimedia data includes a number of data types, such as image, video, audio, and printed information (e.g. pdf files, leaflets, etc.). They may be stored as a BLOB (Binary Large Object). Machine learning and deep learning algorithms may use this kind of data for analysis and prediction.

These kinds of data are assumed to be dynamic whereby the underlying pattern or schema, if exists, is frequently changed. This means that the next batch of data can differ in terms of their formats from the previous batch of data. The changes can be as frequent as at a record level, rather than at a batch of records level.

3.2   Schema-on-Read, Instead of Schema-on-Write

Data lakes are said to have a feature of *Schema-on-Read*, instead of *Schema-on-Write* – the latter is commonly used in the traditional databases and data warehouses. Schema-on-Write means that the schema must be predefined before data is inserted into the database or data storage. This is very typical in database systems, where we need to create the table first before inserting the data. By creating a table, we mean that the table specifications (e.g. attribute names, data types, constraints, relationships) must be defined. Once the table is created, the incoming data must match with the table specification for the data to be inserted to the database; otherwise, the data will be declined. So the schema (or the specification) acts as an entrance guard to the database system. In other words, we cannot upload the data before the table is created, and we cannot create the table before we understand the schema of the data. Therefore, understanding the data (as well as future data) is critical in database design. Because the schema must be defined before the data is written to the database, the schema is called Schema-on-Write.

Schema-on-Write is needed in database systems because of various reasons. One is data integrity; the data in the database must be correct, not only in the context of itself but also in the relationships with other data. Another reason is concurrency management, where the database accessed simultaneously by many users must guarantee its correctness, both in the query results but also in the database itself. Databases and data warehouses are built for specific purposes, and therefore defining the constraints in advance is part of the business process, which can be done relatively easily. Having a predefined schema also helps data cleaning and data transformation. Data anomalies can also be minimized.

However, imposing schema-on-write has negative consequences too. Schema evolution will be a big challenge. If new information is suddenly present, or if the schema is changed, managing different datasets not previously defined by the schema and updating the exiting database is costly. The entire database needs to be reorganized, and this is not a small job.

In short, if the data is static, schema-on-write is ideal. It supports the originally planned operations, manages data integrity very well. However, if the data is dynamic, or even when the data cannot be predicted in advanced, schema-on-write may not be suitable.

Hence, data lakes offer Schema-on-Read. The schema is not predefined when the data is written to the data lake. Only when the data is processed, the schema (for that particular piece of data) is defined.

With growing data volume, the data upload is often required immediately when the data arrives, without any transformation. Fast data ingestion is needed. Therefore, the data might not follow any internal schema. Data ingestion simply copy and move the data to the data lake. Data handling becomes flexible. Additionally, various data types and formats can be accommodated simply because the schema is not imposed on data upload.

So the challenge is on data analysis. It needs to understand the structure of the data. However, the structure of the data is not defined upon data insertion, and the code for data analysis needs to understand the structure of the data. As a result, the code is adjusted to the data, rather than the opposite. When the data is read, the schema is identified; and then the analysis algorithm will be based on the schema of the data that is being read while processing it.

Data analysis still needs to understand the structure of the data and to figure out when the data is read. Ironically, data in the data lakes can have a lot of missing data, invalid data, duplicates, and many other problems that can lead to inaccurate analysis results. Nevertheless, understanding some level of schema design is inevitable.

4   Comparing Data Lakes with Other Systems

A data lake is basically a collection of a variety of data, is intended for data analysis, and is relevant to the context of big data. In this section, we compare data lakes with federated database systems, data warehouses, and big data systems.

4.1   Data Lakes vs. Federated Database Systems

Can *Federated Databases* be used, instead of Data Lakes, to address the need to store data from various sources? A federated database system is a database system that maps multiple autonomous database systems into a single federated database [2]. So, it is a meta-database management system whereby the elements are individual database systems that are interconnected via a computer network and may be geographically distributed. Each database system is entirely self-sustained and functional.

The benefit offered by a federated database system is that the application program may view this as a single entity, and hence the application program may invoke a query, and the federated database system may pass it on to the relevant database systems [3,4]. A federated database system may be seen as database virtualization that makes several databases appear as one [5]. Figure 2 illustrates a federated database architecture. The meta-database system binds autonomous individual database management systems and serves as an interface with external systems, such as application development, query requests, etc.

The same issue that faces relational database systems also faces federated database systems; simply because a federated database system is predominantly also a relational database system. The problem is that it requires schema-on-write. That means the incoming data must comply with the schema before it can

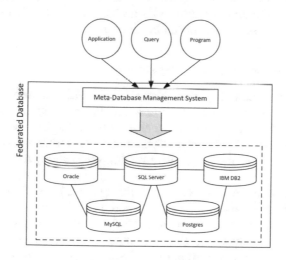

Fig. 2. A federated database system

be ingested into the database system. On the other hand, the need for a data lake is that incoming data should not be restricted by the schema for various reasons, such as the high velocity of incoming data, frequent schema changes of the data, non-uniform schema of the data, and other reasons that make data lake is the solution; and federated database systems cannot fulfil this requirement.

4.2   Data Lakes vs. Data Warehouses

A data lake is often conveniently compared with a data warehouse because a data warehouse is a data repository specifically designed for data analysis and business intelligence. A data lake is also meant to be a data repository for data analysis and discovery. Therefore, it is important to highlight the differences between these two data repositories.

A data warehouse is normally modeled using a star schema, where the fact containing fact measures is viewed from various angles or dimensions. A star schema is then a multidimensional database. The fact measure is a numerical value referenced by a cross product of all dimensions. Because of this structure, data warehousing is very much suitable to support decision making, as users are able to explore and navigate the data by drilling down data at different levels of granularity.

Table 1 summarizes the main differences between data lakes and data warehouses. In terms of the data, there are differences in data format, data granularity, data sources, and schema. Although there are some work on XML data warehouses, traditional data warehousing very much uses structured data [6–12]. There are also differences in purpose and usage; hence the user groups are also different. Data warehousing is very much used for Business Intelligence and visualization, whereas data lake is often used for machine learning related activities; although there is some previous work on machine learning activities using data warehousing [13]. Finally, in terms of its ability to change, there are significant differences between data warehouse and data lake.

4.3   Data Lakes vs. Big Data Systems

Big Data is often known by the 3Vs: Volume, Variety, and Velocity [14,15] (although some have suggested adding a few more Vs to it). There are some similarities between Big Data systems and Data Lakes. In data lakes, raw data is ingested at a high-velocity rate. These data come from non-traditional data producers, such as sensors, logs, social media, etc. Hence, data lakes also have Velocity.

Another feature of data lakes is the ability to absorb a variety of data formats – not only relational data, which is structured data, but also semi-structured, opaque-structured, and even unstructured data. This variety of data formats make data lake a rich data collection, and it should provide a more rounded data support for more comprehensive data analysis, machine learning, deep learning, and predictive analysis. Therefore, data lakes also feature Variety.

Table 1. Data warehouses vs. data lakes

	Data warehouses	Data lakes
Data format	Cleaned, filtered, pre-processed and transformed data; Structured data format	Raw data; Variety of data formats (e.g. structured, semi-structured, opaque-structured, unstructured)
Data granularity	Data at a summary or aggregated level of detail	Data at a low level of detail or granularity
Data source	Operational or transactional databases	Non-traditional data sources: Sensors, logs, texts, images Social networks, videos, etc.
Schema	Schema-on-Write; Schema is defined when data is written to data lake	Schema-on-Read Schema is defined when data is read or retrieved from data lake
Purpose	Built for a specific and definite purpose	Purpose not defined in advanced (some have a definite purpose, others just have the data handy)
Usage	For BI, visualization, and batch reporting, etc.	For machine learning, discovery, prediction, profiling, etc.
User	Business manager, operational user, business analyst	Data scientist, developer, engineer
Changes in data	Data is static; Update is difficult, not common, and costly	Data is dynamic; Continuous data ingestion; Frequent updates
Agility	Less agile process; Fixed configuration and structure	Highly agile process; Configuration dynamically adjusted when needed

As data lakes do not require building a schema in advance, which then allows them to absorb raw data (at a high speed), naturally, the volume of data in data lakes can quickly grow exponentially. This big data volume then requires not only more storage space but also more processing power, and in particular parallel processing, in order to speed up query processing, data analysis, and other machine learning activities, which normally require high processing power. So indirectly, data lakes also feature Volume.

Due to these similarities, very often, data lakes are associated with big data concept. However, it is important to highlight that data lakes are different from big data systems such as Hadoop or Hadoop-stack technologies. Hadoop is not a data centre repository like data lake, rather it is a processing framework for large data sets with the purpose of supporting big data analytics. Although Hadoop is used in some data lakes, it does not reflect the data lake architecture, as it is an enabling technology.

Hive is often associated with data lakes. Hive is a data warehousing framework that overlays a data infrastructure on top of Hadoop [16], so that data can be easily queried using an SQL-like language, called the Hive Query Language (or HiveQL). HiveQL performs queries on large datasets [17]. Comparing with traditional databases, traditional RBDMS is designed for interactive queries, whereas Hive uses batch processing to work across a very large distributed database. Hive transforms HiveQL queries into Map Reduce jobs that run on Hadoop distributed job scheduling framework [18]. As a result, Hive query performance on a large dataset is very fast, and is very much suitable for big data [19]. However, data lakes should provide richer capabilities than Hive. Data lakes can take data of various formats; it does not impose any predefined schema.

5 Challenges

Despite the promising features and benefits that data lakes offer, there are significant challenges that data lakes face. These must be carefully addressed so that they will not hinder a wider acceptance of data lakes in practice. Some challenges and trends have been listed in [20, 21]. However, in this paper, we will look at the challenges from a slightly different point of view.

5.1 Data Swamp

Data lakes could quickly become data swamps if not designed and organized properly. One of the main features of data lakes is high-velocity data ingestion [22]. But it could also become a drawback of data lakes. We need to have a proper strategy on the primary purpose of the data lake. This means that we should not just ingest anything into the data lakes. With a sound strategy and plan, data lakes will become data rich rather than data swamp.

Even with data lakes, if we have no knowledge on the data in the lake, finding, processing, and analysing data can become challenging [23–25]. Without Schema-on-Write, we often assume that we do not need to understand the data being ingested to the lake. This could have severe consequences since having the data does not automatically mean that we have the ability to process and analyze them, especially if we don't have enough knowledge about what kind of data we have in the lake. Although data lakes may not impose some schemas or metadata, having some understanding of the data is crucial in any data processing and analytics, and data lakes are not an exception [26].

Another feature of data lakes is the variety of data: from structured, semi-structured, opaque-structured to unstructured data. This feature, especially the unstructured data, could easily become a drawback, as it is very often that unstructured data become unusable data.

Data swamp occurs when data cannot be found, data cannot be processed, and data cannot be analyzed. We cannot find the data because we do not know where they are; We cannot process the data because we do not know what they are; We cannot analyze the data because they are not ready for analysis.

5.2 Internal Data Processing

Once the data is ingested to a data lake, it does not mean that the data is never further processed or transformed and then stored back into the data lake. Data in the data lake often undergoes internal data processing, such as data cleansing, data transformation, data integration, etc., and hence the data is a mixture between raw data and semi-processed or (to some degree) processed data.

Data cleaning or data cleansing is an essential element in the data processing life-cycle. It is even more critical in the context of data lakes because of the nature of high-velocity data ingestion and the variety of data formats allowed in the data lakes. In Relational Database Systems, data cleaning is very much assisted by the schema, such as data types, entity integrity, referential integrity, etc. In data lakes, since the notion of data schema may not exist, data cleaning becomes even more challenging. The schema in data lakes is often pushed to the application layer, as opposed to the data layer as in the traditional database systems; Consequently, data cleaning at the data layer imposes serious challenges. On the other hand, if the data cleaning phase is also pushed to the application layer, there will be lots of consequences as the mistakes in the data have been propagated to many layers, which may make the application unusable.

Data in the data lakes may go through some degrees of transformation. Initially, data ingestion input raw data into a data lake. It then transforms raw data into a pre-defined data model. Hence, a data lake is not totally free from data models or schemas; it is just that schema-less data is allowed to be ingested into the data lake. Managing raw data, in combination with not-so-raw data, raises issues, such as data duplication and data versioning. Data duplication issue will be part of the data cleaning, in which the data cleaning process not only cleans the data from mistakes but also clears the data from duplicates.

Data lakes are dynamic. It is possible that there are many versions of the same data, either through data ingestion or internal data processing. Maintaining all versions of data can be too costly, not so much in the context of storage, rather in processing them. Data analysis must then first search the correct version of data to work with, and this adds more complexity to the data analysis. A version manager must be part of the data lake architecture [27].

5.3   Metadata, Catalogs, and Schemas

Finding the right data in the data lakes can be problematic in the absence of a global schema. Having the data in the data lake does not guarantee that the right data can be found smoothly. Often, raw data is processed on a trial and error basis, as we do not know the data we need. More efficient data navigation and exploration are needed in data lakes, especially when metadata, catalogs, or schemas are incomplete [28].

Metadata or catalogs or schemas are essential, especially in the context of on-demand data exploration and discovery. Metadata or catalogs may be extracted from the raw data in the case where the raw data is not accompanied by adequate schema or metadata. In data lakes, data may be abundant, but metadata is scarce. Understanding the importance of metadata or catalogs in data processing and analysis is crucial to the success of data lake adoption.

5.4   Design Methodology

For data lakes to be adopted widely by practitioners, there must be a design methodology in-place. Schema-on-read and storing raw data into data lakes are features that make data lakes attractive compared to the existing paradigm, such as enterprise data warehousing, operational or transactional systems. However, without a robust design methodology, these features will become the brick wall of data lakes.

One of the reasons why existing technologies such as Relational DBMS and data warehousing are widely used is due to its underlying design methodology. Relational databases use Entity-Relationship (E/R design) as well as the under-pinning relational theory. Data warehousing uses star schema modeling, which is a form of dimensional modeling. However, data lakes lack a design model and are at the risk of collapsing.

Data lakes have the features of storing raw data; and often the same data or aggregation of the same data (this is even a harder problem), which results in data replication. Interesting data and semi preprocessed data is often ingested multiple times by different users of the data lake. It becomes very easy for the data to be uncontrollable. Data scientist often needs to integrate data from various silos. It is very often that data scientists unable to identify the data where it was stored. Even when the data is found, diverse data is very hard to integrate. These are all due to the lack of design methodology.

It is well known that 70% of the data lifecycle is spent on finding, interpreting, cleaning, integrating data, whereas only 30% is spent on the analysis itself. With the majority of time is spent on data preparation rather than data analysis, schema-on-read provides no help since the schema is unknown until the data is read; and even after the data is read, the schema is still very hard to interpret, and often the data needs further preprocessing, transformation, integration, and so on. Therefore, there must be a clear strategy, policy, or even design methodology for data lakes to be able to transform its features and promises into a reality.

6   Conclusion

Data management is now experiencing the latest wave created by data lakes. Data lakes, with all the features and promises, have created a new challenge on the way we store, manage, process, and analyze data. Combining with Big Data capabilities, in terms of high-velocity data ingestion, a variety of data formats, and handling of big data volume, data lakes address the current need for handling big data more efficiently and effectively.

Like any other waves in the past, such as object-oriented databases, then object-relational databases, and recently the NoSQL databases, data lake faces challenges whether it will just be another passing wave. There are some critical elements for the uptake of the new data technology. One of them is simplicity in the solution, as opposed to complexity. With simplicity, uptakes by practitioners will be more guaranteed, and as a result, widespread adoption will likely more to happen. Another key to the success of the new wave is the coexistence with the existing technologies, which in this case is the relational technology, such as the current operational/transactional databases, as well as data warehousing. Therefore, it is critical for data lakes to embrace existing technologies rather than separating themselves from the existing paradigm. Last but not least is that the challenges faced by data lakes must be addressed to some degree in order to make data lakes useful.

References

1. Panwar, A., Bhatnagar, V.: Data lake architecture: a new repository for data engineer. Int. J. Organ. Collect. Intell. **10**(1), 63–75 (2020)
2. Azevedo, L.G., de Souza Soares, E.F., Souza, R., Moreno, M.F.: Modern federated database systems: an overview. In: Filipe, J., Smialek, M., Brodsky, A., Hammoudi, S. (eds.) Proceedings of the 22nd International Conference on Enterprise Information Systems, ICEIS 2020, Prague, Czech Republic, 5–7 May 2020, vol. 1, pp. 276–283. SCITEPRESS (2020)
3. Endris, K.M.: Federated Query Processing over Heterogeneous Data Sources in a Semantic Data Lake. Ph.D. thesis, University of Bonn, Germany (2020)
4. Endris, K.M., Rohde, P.D., Vidal, M.E., Auer, S.: Ontario: federated query processing against a semantic data lake. In: Hartmann, S., Küng, J., Chakravarthy, S., Anderst-Kotsis, G., Tjoa, A., Khalil, I. (eds.) Database and Expert Systems Applications - 30th International Conference, DEXA 2019, Proceedings, Part I, Lecture Notes in Computer Science, Linz, Austria, 26–29 August 2019, vol. 11706, pp. 379–395. Springer (2019)
5. Berger, S., Schrefl, M.: From federated databases to a federated data warehouse system. In: 41st Hawaii International International Conference on Systems Science (HICSS-41 2008), Proceedings, Waikoloa, Big Island, III, USA, 7 10 January 2008, p. 394. IEEE Computer Society (2008)
6. Rusu, L.I., Rahayu, W., Taniar, D.: A methodology for building XML data warehouses. Int. J. Data Warehous. Min. **1**(2), 23–48 (2005)
7. Rusu, L.I., Rahayu, W., Taniar, D.: On building XML data warehouses. In: Yang, Z.R., Everson, R.M., Yin, H. (eds.) Intelligent Data Engineering and Automated Learning - IDEAL 2004, 5th International Conference, Exeter, UK, 25–27 August 2004, Proceedings, LNCS, vol. 3177, pp. 293–299. Springer (2004)
8. Chen, L., Rahayu, W., Taniar, D.: Towards near real-time data warehousing. In: 24th IEEE International Conference on Advanced Information Networking and Applications, AINA 2010, Perth, Australia, 20–13 April 2010, pp. 1150–1157. IEEE Computer Society (2010)
9. Le, D.X.T., Rahayu, W., Taniar, D.: A high performance integrated web data warehousing. Clust. Comput. **10**(1), 95–109 (2007)
10. Rusu, L.I., Rahayu, W., Taniar, D.: On data cleaning in building XML data warehouses. In: Bressan, S., Taniar, D., Kotsis, G., Ibrahim, I.K. (eds.) iiWAS 2004 - The sixth International Conference on Information Integration and Web-based Applications Services, 27–29 September 2004, Jakarta, Indonesia, vol. 183. books@ocg.at. Austrian Computer Society (2004)
11. Maurer, D., Rahayu, W., Rusu, L., Taniar, D.: A right-time refresh for XML data warehouses. In: Zhou, X., Yokota, H., Deng, K., Liu, Q. (eds.) Database Systems for Advanced Applications, 14th International Conference, DASFAA 2009, LNCS, Brisbane, Australia, 21–23 April 2009. Proceedings, vol. 5463, pp. 745–749. Springer (2009)

12. Rahayu, W., Pardede, E., Taniar, D.: The new era of web data warehousing: XML warehousing issues and challenges. In: Kotsis, G., Taniar, D., Pardede, E., Ibrahim, I.K. (eds.) iiWAS 2008 - The Tenth International Conference on Information Integration and Web-Based Applications Services, Linz, Austria, 24–26 November 2008, p. 4. ACM (2008)

13. Tjioe, H.C., Taniar, D.: Mining association rules in data warehouses. Int. J. Data Warehouse. Min. **1**(3), 28–62 (2005)

14. Taniar, D., Leung, C.H.C., Rahayu, W., Goel, S.: High Performance Parallel Database Processing and Grid Databases. Wiley, Hoboken (2008)

15. Taniar, D.: Big data is all about data that we don't have. In: 2017 International Conference on Advanced Computer Science and Information Systems (ICACSIS), pp. 1–8 (2017)

16. Camacho-Rodríguez, J., Chauhan, A., Gates, A., Koifman, E., O'Malley, O., Garg, V., Haindrich, Z., Shelukhin, S., Jayachandran, P., Seth, S., Jaiswal, D., Bouguerra, S., Bangarwa, N., Hariappan, S., Agarwal, A., Dere, J., Dai, D., Nair, T., Dembla, N., Vijayaraghavan, G., Hagleitner, G.: Apache hive: from mapreduce to enterprise-grade big data warehousing. In: Boncz, P.A., Manegold, S., Ailamaki, A., Deshpande, A., Kraska, T. (eds.) Proceedings of the 2019 International Conference on Management of Data, SIGMOD Conference 2019, Amsterdam, The Netherlands, 30 June–5 July 2019, pp. 1773–1786. ACM (2019)

17. Ono, K., Nonaka, J., Kawanabe, T., Fujita, M., Oku, K., Hatta, K.: HIVE: a cross-platform, modular visualization framework for large-scale data sets. Future Gener. Comput. Syst. **112**, 875–883 (2020)

18. Mami, M.N., Graux, D., Scerri, S., Jabeen, H., Auer, S.: Querying data lakes using spark and presto. In: Liu, L., White, R.W., Mantrach, A., Silvestri, F., McAuley, J.J., Baeza-Yates, R., Zia, L. (eds.) The World Wide Web Conference, WWW 2019, San Francisco, CA, USA, 13–17 May 2019, pp. 3574–3578. ACM (2019)

19. Bagui, S., Devulapalli, K.: Comparison of hive's query optimisation techniques. Int. J. Big Data Intell. **5**(4), 243–257 (2018)

20. Giebler, C., Gröger, C., Hoos, E., Schwarz, H., Mitschang, B.: Leveraging the data lake: current state and challenges. In: Ordonez, C., Song, I.-Y., Anderst-Kotsis, G., Tjoa, A.M., Khalil, I. (eds.) Big Data Analytics and Knowledge Discovery - 21st International Conference, DaWaK 2019, Linz, Austria, August 26–29, 2019, Proceedings, LNCS, vol. 11708, pp. 179–188. Springer (2019)

21. Ravat, F., Zhao, Y.: Data lakes: trends and perspectives. In: Hartmann, S., Küng, J., Chakravarthy, S., Anderst-Kotsis, G., Tjoa, A., Khalil, I. (eds.) Database and Expert Systems Applications - 30th International Conference, DEXA 2019, Proceedings, Part I, LNCS, Linz, Austria, 26–29 August 2019, vol. 11706, pp. 304–313. Springer (2019)

22. Sangat, P., Indrawan-Santiago, M., Taniar, D.: Sensor data management in the cloud: data storage, data ingestion, and data retrieval. Concurr. Comput. Pract. Exp. **30**(1) (2018)

23. Bogatu, A., Fernandes, A.A.A., Paton, N.W., Konstantinou, N.: Dataset discovery in data lakes. In: 36th IEEE International Conference on Data Engineering, ICDE 2020, Dallas, TX, USA, 20–24 April 2020, pp. 709–720. IEEE (2020)

24. Nargesian, F., Pu, K.Q., Zhu, E., Bashardoost, B.G., Miller, R.J.: Organizing data lakes for navigation. In: Maier, D., Pottinger, R., Doan, A., Tan, W.-C., Alawini, A., Ngo, H.Q. (eds.) Proceedings of the 2020 International Conference on Management of Data, SIGMOD Conference 2020, Online Conference [Portland, OR, USA], 14–19 June 2020, pp. 1939–1950. ACM (2020)

25. Zhang, Y., Ives, Z.G.: Finding related tables in data lakes for interactive data science. In: Maier, D., Pottinger, R., Doan, A., Tan, W.-C., Alawini, A., Ngo, H.Q. (eds.) Proceedings of the 2020 International Conference on Management of Data, SIGMOD Conference 2020, Online Conference [Portland, OR, USA], 14–19 June 2020, pp. 1951–1966. ACM (2020)
26. Eichler, R., Giebler, C., Gröger, C., Schwarz, H., Mitschang, B.: HANDLE - a generic metadata model for data lakes. In: Song, M., Song, I.-Y., Kotsis, G., Tjoa, A.M., Khalil, I. (eds.) Big Data Analytics and Knowledge Discovery - 22nd International Conference, DaWaK 2020, Proceedings, LNCS, Bratislava, Slovakia, 14–17 September 2020, vol. 12393, pp. 73–88. Springer (2020)
27. Schönhoff, M.: Version management in federated database systems. DISDBIS. Infix Akademische Verlagsgesellschaft, vol. 81 (2002)
28. Nargesian, F., Zhu, E., Miller, R.J., Pu, K.Q., Arocena, P.C.: Data lake management: challenges and opportunities. Proc. VLDB Endow. **12**(12), 1986–1989 (2019)

Using Object Detection Technology to Measure the Accuracy of the TFT-LCD Printing Process by Using Deep Learning

Ting-Wei Yeh(✉) and Fang-Yie Leu

Computer Science Department, Tunghai University, Taichung City, Taiwan
cyfox0305@gmail.com, leufy@thu.edu.tw

Abstract. Currently, manufacturing industries have begun to employ Artificial Intelligence (AI) technology to enhance production efficiency, and reduce management costs and other problems encountered in their manufacturing processes, e.g., TFT-LCD manufacturing companies are typical examples. During their manufacturing processes, if a dedicated measurement machine is used, it takes time, also lengthening the production time and increasing the production cost. This study would like to measure the printing accuracy of the PI (Polyimide film) process in the cell segment process of a TFT-LCD panel manufacturing factory in central Taiwan by proposing a object detection architecture called Object Detection Measurement (ODM) system which measures PI Coater accuracy by using a deep learning object detection technology. In the ODM, an AOI (automatic optical inspection) machine is utilized to take photos for PI Coater. After that, Yolo v3, an object Identify algorithm, is used to evaluate accuracy of the PI Coater. When PI Coater's printing accuracy is abnormal it notifies users to check the parameters of the proposed system. It also builds SPC (statistical process control) control graph for PI Coater accuracy. With the SPC, process engineers can query and monitor whether the accuracy has trended to abnormal.

Keywords: PI coaster · TFT-LCD panel manufacturing · Object detection measurement · Automatic optical inspection

1 Introduction

The goal of this research is to measure the printing accuracy during the PI Coater process of the cell segment process [1] performed by an TFT-LCD manufacturing company in central Taiwan. If the distance of the PI film printed in this PI Coater process (PI process for short) and the edge of the LCD panel is too short, the light will leak at the edge of the panel. Consequently, customers will complaint and/or cancel their product orders. Then, lots of manpower is required to comprehensively invent and inspect products produced in the time period the same as that of these defected products. In the end, a large number of products were often scrapped.

When the company would like to measure the PI printing accuracy, it consumes some time to deliver glass substrates to the measurement machine. Currently, we sample one

© The Author(s), under exclusive license to Springer Nature Switzerland AG 2021
L. Barolli et al. (Eds.): EIDWT 2021, LNDECT 65, pp. 358–369, 2021.
https://doi.org/10.1007/978-3-030-70639-5_33

piece from 5000 glass substrates that have so far produced for measurement: this prolongs the production time of products. This research aims to enhance the printing accuracy of the PI film edge after the PI process. To achieve this, an object detection scheme called Object Detection Measurement (ODM for short) is proposed for measuring PI Coater accuracy for products.

In the PI process, we send products to a station for AOI (automatic optical inspection) image inspection [2], which takes a photo for the surface of each piece of glass substrate and checks to see whether any unexpected matter remains. Wishing to use image detection technology to measure the accuracy of PI printing, so that the glass substrates can be inspected comprehensively and automatically.

Because photographing takes a very short time, the affection on the product's production cycle can be ignorable. Next, Yolo v3, an object detection algorithm of deep learning, is employ to identify the blocks to be measured from the photo, calculate the length and width of the block (called box), and then multiply them by the ratio of actual-size/measured-size to obtain the real length and width of this box.

The rest of this thesis is organized as follows. Section 2 explains the production process we study in this research, including the panel-related process and the technology used. Section 3 introduces our system architecture. The effectiveness by adopting the system and the comparison between this system and other existing systems are presented and discussed in Sect. 4. Section 5 concludes this study and address our future directions.

2 Literature Review and Background

2.1 PI Process

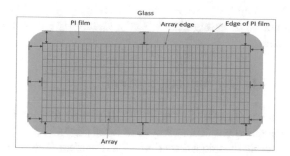

Fig. 1. PI accuracy measurement diagram

The TFT-LCD producing process can be mainly divided into three sub-processes (1) Array and CF process [3]. (2) Cell process [4]. (3) Module process [5].

The PI process as the first step of the Cell process contains Cleaner → PI Coater → Pre Baking → PI Inspection → Post Baking → Rubbing.

This research aims at confirming whether the PI film covers the components in the glass substrate and whether the area of the circuit array (Array for short) is deviating from Array edge after the completion of PI Coater. To achieve this, we measure the distance between the edge of the PI film (the blue periphery in Fig. 1) and the Array's edge in the TFT and CF glass substrate (the periphery of the middle Array area in Fig. 1). The measurement positions are marked with 12 red arrows.

2.2 Yolo

Yolo (You only look once) as a neural network algorithm for object detection published in 2015 [6, 7] is evolved from R-CNN [8], fast RCNN [9], and faster RCNN [10]. Its object detection is composed of three stages: Object Localization, Feature Extraction and Image Classification.

Currently, three versions of Yolo have been released, i.e., Yolo v1 [6], Yolo v2 [11] and Yolo v3 [12, 13] in which Yolo v3 is adopted in this research.

In this research, the Yolo v3 algorithm is used to identify edge images for the photo images on a glass substrate. There are three reasons for adopting this algorithm:

1. Backgrounds of different product's photo images are very different. Traditional image detection methods cannot effectively calculate the distance of the arrow shown in Fig. 1. In fact, using the Yolo v3 algorithm can reduce the influence coming from complex background of an image. Consequently, when a model is trained for the images in the same type of products, the class, e.g. Side and Angle, can be effectively identified.
2. A larger glass substrate contains many smaller panels, and we need to identify images located in multiple areas (detailed below) for each panel. In other words, the ODM system needs to identify a large number of images in the glass substrate within a limited period of time. After our observation, Yolo v3 does not seriously affect the production time of the concerned company's production line.
3. During model training [14], Yolo v3 algorithms are compared with other state-of-the-art algorithms on detecting classes. Only a relatively small amount of photos is required for training.

3   Architecture of the Proposed System

The ODM's system structure as shown in Fig. 2 can be divided into 7 subsystems, including: (1) AOI Equipment: This subsystem takes photos, e.g., P, for the objects on the glass substrate before its object detection; (2) AI System: The subsystem captures P, measures the accuracy of PI Coaters on P, and generates recognized images; (3) Transfer System: It grabs accuracy-data and images recognized by AI System and saves them to DB Server and Image Server, respectively; (4) DB Server: This database stores accuracy data of the recognized images; (5) Image Server: It stores those recognized images; (6) WEB Server: This Server not only captures product's accuracy data stored in DB Server, displays the data on the web in a graphical format, but also captures images stored in Image Server and displays them through web links (hyperlinks); (7) Alarm System: The subsystem retrieves product's accuracy data stored in the DB Server. When the accuracy of PI Coater exceeds its predefined spec, it will automatically send an email to notify users, i.e., process engineers.

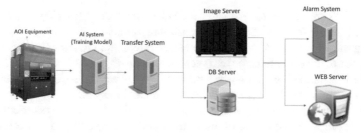

Fig. 2. System structure of ODM

In the following, we will explain (1) the location of the images involved in the training phase of the detection model; (2) the method used to mark boxes in an image; (3) the pre-processing of the image before training. At last, the architecture of this system will be introduced.

3.1 Training Model

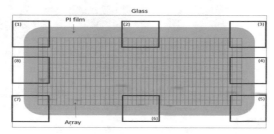

Fig. 3. The 8 locations of the image numbered from 1 to 8 are the focus of the following training model

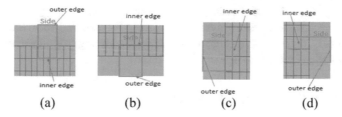

Fig. 4. The way to label Side classes ((a) The labeling box no. (2) in Fig. 3. (b) The labeling box no. (6)in Fig. 3. (c) The labeling box no. (8) in Fig. 3. (d) The labeling box no. (4) in Fig. 3.)

PI Inspection in the PI Coater process uses AOI technology to automatically photograph TFT and CF glass substrates for detecting defects on the glass-substrate surface. As mentioned above, a glass substrate will be cut into several panels depending on the size of a panel. The photo areas of a panel include 4 angles and 4 sides as shown in

Fig. 3. The files created for the 8 images are numbered from 1 to 8. These 8 locations of a panel are also the focus of the following training model.

During training stage, Yolo v3 model needs to label input images (such as Fig. 4) and assign classes to the 8 locations/boxes (such as Side or Angle). We adopt the LabelImage program [15] described previously to label the 8 boxes which are also the measurement positions shown in Fig. 1. There are two ways to label images.

(1) The method for labeling Side is shown in Fig. 4, in which the outer edge of a box is as close as possible to the edge of the PI film, and the inner edge of a box will contain an Array to improve the training accuracy. The reason is that, when the inner edge is only overlapped with the edge of the PI, the model may catch an incomplete range of a box due to the impurity of the PI film. If the impure part is a long strip of dust near and parallel to the edge of the glass substrate, the strip may be mis-identified as the line of an array, resulting in detection and measurement errors.

(2) The method for labeling Angle is the same as that for labeling Side classes. As shown in Fig. 5, an Angle class includes an angle and an Array, and the outer two edges like edges a and b in Fig. 6(a) are as close as possible to the two corner edges of the PI film.

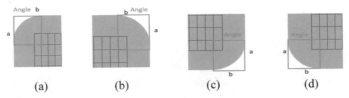

Fig. 5. The way to label Angle classes ((a) The labeling box numbered (1) in Fig. 3. (b) The labeling box numbered (3) in Fig. 3. (c) The labeling box numbered (5) in Fig. 3. (d) The labeling box numbered (7) in Fig. 3.)

Photographs shooting machines often make edge brightness of the PI film a little different from that of its central part due to the slight difference in lighting and/or focal length, causing Yolo v3 model's recognition failure. Our processing method is that we first converts the color images to grayscale of 0–255, and then transform these images by using Sigmoidal Contrast Enhancement Function [16], aiming to increase the brightness of the image and enhance the edge of the PI film.

3.2 Component-Based System Architecture

The component-based architecture of our proposed system as shown in Fig. 6 (please also refer to Fig. 2) consists of 8 steps, including Image Capture, Load Picture, Sigmoid Transfer, Yolo v3, Label Image/Transfer Value, Save Result, Web Control Chart and Alarm System which will be described in the following.

Step 1. Image Capture, adopts AOI inspection machine as the AOI Equipment to take photos for the four sides and four angles, on a panel, and save them in a file, the name of which consists of the glass substrate's serial number, product number and number of

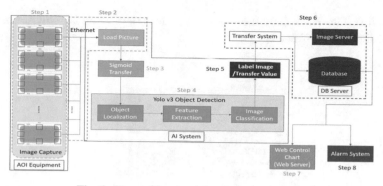

Fig. 6. The architecture of our proposed system

shooting position. In fact, this file is saved in AOI Equipments, i.e., all AOI inspection machines.

Step 2. Load Picture, as one of the functions of the AI System, is the first step of the main detection system. This detection system retrieves all the photos currently taken by the AOI Equipment in sequence periodically.

The Load Picture system containing a GPU is connected to the AOI Equipment via Ethernet. All AOI Equipments store their photos in image files in a folder. The AI System server retrieves these photos from the folder every 5 min. Each time, the AI system retrieves those photos saved in a unretrieved folder of a glass-substrate serial number from all AOI Equipments for step 3.

Step 3. Sigmoid Transfer, as the second function of the AI System, converts the photos/images loaded in the previous step from color to grayscale and applies Sigmoid function to clearly illustrate the edges of a PI film. In fact, during the training process, all photos/images are processed by the same procedure for the training model. The converted images are temporarily stored in the memory.

Step 4. Yolo v3 Object Detection as the third function of the AI System receives the images processed in the previous step from memory and applies the Yolo v3 algorithm to identify their Side/Angle classes and the Bounding Box size.

Object detection results contain the following information:

1. Class: The object is Side or Angle.
2. Confidence Score: The similarity between the object and its training data ranges between 0 and 1 as the detection reliability of this object.
3. Bounding Box: The position of an object in the concerned image includes the coordinates (x, y) as the starting point of this object, and its width (W) and height (H).

Because an image shown in Fig. 7 may identify multiple objects, i.e., Angles and/or Sides, the AI system filters images with the following two steps:

1. First, detecting category and location of this object in an image. The four sides (corners) of an image / TFT-LCD panel should be Side (Angle) class as shown in

Fig. 7(a) (Fig. 7(b)). Only those classes at their right locations will be measured, including class's W and H.

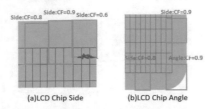

(a)LCD Chip Side   (b)LCD Chip Angle

Fig. 7. Multiple objects (Class Name: Confidence Score) are identified in a single image

2. Select the object with the highest Confidence Score (CF). As the LCD-Chip Sides shown in Fig. 8(a), when multiple objects are identified simultaneously at the position of the object or near the object, the Yolo v3 model chooses the one with the highest Confidence Score. In Fig. 8(a), only the object with Score = 0.9 is retained. The purpose is to reduce the burden of the subsequent calculation of the measured values.

Step 5. Label Image/Transfer Value, as the fourth step (also the last step) of the AI System, mainly generates an image file for saving all drawn boxes, and a CSV file for storing the actual measurement results. Both the two files are temporarily stored in different folders in the AI System.

This system uses the detected Angle/Side classes and Bounding Boxes yielded in the previous step to draw object classes and these Bounding Boxes on the original photos retrieved from AOI machine in the AI Load step. The images of these Bounding Boxes are temporarily stored in the Image folder also in the AI machine.

After that, the system retains the lengths and widths of the boxes at the concerned positions, for example, the width (w) and height (h) of a Bounding Box at one of the 4 corners or one of the 4 sides. However, the length of the box line is measured in pixels, rather than its real length, meaning the length must be multiplied by the length-pixel ratio defined as the actual length over the pixel length of a box edge. So, the Yolo v3 model needs to convert the length and width of an edge into their actual values.

Finally, the system saves the object detection results, including Angles/Sides and Bounding Boxes, of all images in the CSV folder named as the serial number of the concerned glass substrate in CSV format.

Step 6. Transfer to Image Server/Database, as shown in Fig. 6 builds a data conversion system, saves the images temporarily in AI system in the previous step to Image Server, and stores the CSV files of the detection results (the image's Angle/Side classes and the class's w/h) to the database (DB server). To achieve this, we set up an FTP server in the AI System Server.

The conversion system retrieves all images stored in the Image folder in the AI System Server via the FTP server, saves them in the Image Server and reads the content of the CSV file. After that, all the accuracy values (w/h value) of an image in the content, their angle position information and some related product information (e.g., the serial number of glass substrates, product number... Etc.) are stored together in the DB Server.

Step 7. The Web Control Chart, draws the data generated during PI Coater's production time in a control chart. After viewing this chart, user can remotely log in the Image Server to verify whether current products are normal or not via web. Users can also remotely access the DB Server to retrieve the accuracy information stored in the previous step, and then draw the SPC control chart for current products.

Step 8. The Alarm System, sends an Alarm to users when necessary. The system calculates the measured value of each Angle/Side position per hour. If more than 3 points exceed the upper limit or lower than the lower limit, of the SPC control chart within an hour, the system will send an email to notify the personnel/users to check and confirm product quality.

4 Simulation and Discussions

In this study, two experiments were performed. The first compared the performance before and after the invocation of the ODM technology. The second evaluated the measurement accuracies of PI Coater when the traditional AOI technology [17] and the ODM technology were individually employed.

4.1 After Using the ODM Technology

In the first experiment, we evaluate the measurement accuracies when the ODM technology is used. The comparison between manual inspection and utilizing ODM technology are shown in Table 1.

Table 1. Comparison between sampling measurement by users and measurement by using ODM technology

	Sampling measurement	ODM technology
Measuring frequency	measurement 1 pc pre 5000 pcs	Per piece
Measurement time	≒40 min	2 s
Accuracy	100% (actually measured)	85–95%
Number of training images	0	≒1200
The need for manpower	Yes	No

4.2　AOI Technology and ODM Technology

Our second experiment evaluated the performance of traditional AOI and the ODM technology. The AOI as mentioned above retains the PI film's edge lines and the Array edge lines for calculating the distance between the two lines.

This experiment uses three different products each with its own characteristics (based on business secrets, only denoted by products A, B, and C). As listed in Table 2, the last column shows the numbers of training images consumed.

Table 2. Description of three products and the amount of their training images

Product	Description	Number of training images
A	There are no lines on the edge of the product, and on device	1000
B	There are simple lines on the edge of the product, and on device	1500
C	There are complex lines and device on the edge of the product	1800

Among the three products, product A's background image is the simplest, while product C's is the most complex. When training our detection model, we expect reaching 85% of detection accuracy. It is one of the reasons why the amounts of training images for the three products vary.

Both the AOI technology and ODM technology mark the calculated area on the original picture/photo, generate an image file, and then provide users with the measurement accuracies. The verification method is the same as that described above, i.e., users use the box line function of the retouching software to verify whether the deviation of a box-line length is less than 10 pixels from the detected box-line length.

The number of images utilized and the success rates of detection is shown in Table 3.

Table 3. Number of training images on product x/success rate, x = A, B or C, when AOI technology and ODM technology are individually employed.

Item	AOI technology	ODM technology
A	24/100%	1000/95%
B	24/100%	1500/91%
C	24/0% (Cannot calculate)	1800/86%

The verification method is to select 100 images, which are not a part of training images, for each product and utilizes AOI technology and ODM technology to measure boxes in an image. The measurement results are then retouched. The software actually calculates the length of the identified boxes. Table 4 presents the verification results of products A, B and C.

Table 4. Comparison of the two concerned technologies (A/B/C represent the verification results of Products A, B and C, respectively)

Item	AOI (pcs)	ODM (pcs)
Number of images successfully measured	100/100/–	95/91/86
Number of images with measurement error which is large than 0.3cm	2/5/–	1/2/2

As shown, AOI technology is unable to detect the boundaries of product C. But it is fine for products A and B. While using ODM technology, some images cannot determine to which categories they belong. Therefore, it is hard for the system to judge whether the measured value is correct or not.

Here, we put the actual measurement results (errors) on products A and B together when the AOI technology is used. The advantages are as follows.

(1) Low error rate: the error rate of AOI technology is 3.5% ($= (2 + 5) / (100 + 100) \times 100$) on average, while that of the ODM technology is 1.6% ($= (1 + 2) / (95 + 91) \times 100$) which is lower.

(2) Solving the problem of material impurities by increasing number of training images: when we inspect those images with measurement error on the AOI technology, we found that there are 7 pieces of images (2 and 5 for products A and B, respectively), in which many connected bubbles exists at the locations between the edge of an PI film and the edge of the Array. The AOI only calculates the distance between the edge of the PI film and the bubbles. Although the algorithm has been modified, the error still exists. The reason why the ODM technology fails to detect some boxes is that no such kinds of images are involved in the training images. After adding this type of images to the training data set, the detection system correctly detects them, and accurately calculates the measured values.

Initially the error rate of ODM technology is 1.6% which is lower than that of AOI technology, i.e. 3.5%. But after increasing the number of training images, the detection rate of boxes on product A reaches 97% and product B is 96%; Also, AOI technology cannot detect images of complex background.

5   Conclusions and Future Studies

In this paper, we propose an ODM technology, which uses an object detection approach to measure the accuracy of the PI Coater for products; this system is developed based on the Yolo v3 deep learning algorithm and labeled images for model training. It then employs an AOI Equipment to automatically photograph each piece of glass substrate that has been processed by PI Coater. We use the ODM technology to measure the printing accuracy of the products and establish a complete query on SPC control chart and a monitoring mechanism to detect abnormal products.

This system has been verified that it is helpful in improving product quality and avoiding financial loss for the TFT-LCD company. It is confirmed that this system can effectively monitor the PI Coater's printing accuracy on the products along a production line, and comprehensively measure all products, consequently avoiding the case in which only one is sampled from every 5,000 glass substrates.

Our experiments also shows that although traditional AOI technology has a higher object-detection accuracy than ODM technology does, the measurement error rate of ODM technology is lower.

Furthermore, the detection rate of ODM technology is about 80 – 90% which is not as high as the detection rate in our test stage (86–95%). There are still some images that have not been successfully measured; and for the reduction of detection rate and accuracy, there is no processes that can automatically tune the process after the model fails. At present, this system can only manually sample products so as to check the measurement result. Also, when users receive an alarm, currently they manually make sure whether this is a true alarm or a false alarm with which to determine whether to retrain the model or to adjust the model's parameters; these will be a part of our future research to continue enhancing the system.

References

1. TFT LCD Manufacturing Process. https://insightsolutionsglobal.com/tft-lcd-manufacturing-process/
2. Huang, S.C., Chang, Y.C.: Examine the effect of the Image Preprocessing on Defect Detection of TFT-LCD Panels by Automatic Optical Inspection, Master Thesis, National Chiao Tung University (2017)
3. Sabnis, R.W.: Color filter technology for liquid crystal displays. Displays **20**, 119–129 (1999)
4. Lai, C.H., Cheng, S.: Improvement on Fat Edge Defects of Polyimide Coating Film in LCD Alignment Process, Master Thesis, National Chiao Tung University (2010)
5. Hayashi, N., et al.: Development of TFT-LCD TAB modules. In: Proceedings of Japan IEMT Symposium, Sixth IEEE/CHMT International Electronic Manufacturing Technology Symposium, Nara, Japan, pp. 79–82 (1989). https://doi.org/10.1109/IEMTS.1989.76114.
6. Redmon, J., Divvala, S., Girshick, R., Farhadi, A.: You only look once: unified, real-time object detection, arXiv preprint arXiv:1506.02640 (2015)
7. Yolo: Object detection based on deep learning (including YoloV3). https://mropengate.blogspot.com/2018/06/yolo-yolov3.html
8. Girshick, R., Donahue, J., Darrell, T., Malik, J.: Rich feature hierarchies for accurate object detection and semantic segmentation. In: IEEE Conference on Computer Vision and Pattern Recognition, pp. 580–587 (2014)
9. Girshick, R.: Fast R-CNN. In: The IEEE International Conference on Computer Vision, pp. 1440–1448 (2015)
10. Ren, S., He, K., Girshick, R., Sun, J.: Faster R-CNN: towards real-time object detection with region proposal networks. IEEE Trans. Pattern Anal. Mach. Intell. **39**(6), 1137–1149 (2017)
11. Redmon, J., Farhadi, A.: YOLO9000: better, faster, stronger. In: The IEEE Conference on Computer Vision and Pattern Recognition, pp. 7263–7271 (2017)
12. He, K., Zhang, X., Ren, S., Sun, J.: Deep residual learning for image recognition. In: The IEEE Conference on Computer Vision and Pattern Recognition, pp. 770–778 (2016)
13. Lin, T.Y., Dollar, P., Girshick, R., He, K., Hariharan, B., Belongie, S.: Feature pyramid networks for object detection. In: The IEEE Conference on Computer Vision and Pattern Recognition, pp. 2117–2125 (2017)
14. Redmon, J.: Darknet: open source neural networks in C. https://pjreddie.com/darknet/
15. Tzutalin. LabelImg. Git code (2015). https://github.com/tzutalin/labelImg

16. Braun, G.J., Fairchild, M.D.: Image lightness rescaling using sigmoidal contrast enhancement function. J. Electron. Imaging **8**(4), 380–393 (1999)
17. Chen, S.L., Jhou, J.W.: Automatic optical inspection on mura defect of TFT-LCD. In: Proceedings of the 35th International MATADOR Conference, pp. 233–236 (2007)

Web-Based 3D and 360° VR Materials for IoT Security Education Supporting Learning Analytics

Wei Shi[✉], Akira Haga, and Yoshihiro Okada

The Innovation Center for Educational Resources,
Kyushu University Library, Kyushu University, Fukuoka, Japan
{shi.wei.243,akira.haga.879,
okada.yoshihiro.520}@m.kyushu-u.ac.jp

Abstract. Internet of Things (IoT) technology is developing rapidly in recent years. With the maturity of this technology, many device companies have already start to provide various IoT devices to consumers, such as smart speaker and smart meter. Then, the security problems of IoT devices become more and more important. Except that device companies ensure the security of IoT devices in technology level, users and IoT professors should have enough security knowledge as well. In this paper, we introduce a novel framework for generating web-based 3D and 360-degree VR materials for IoT security education. In order to enhance the educational effort of these materials, we important visual analyzing tools, Time Tunnel and Cubic Gantt Chart, into this framework. By using these two tools, we can easily obtain the common week points of the education target, and such results can help us improve the development of IoT security education materials.

1 Introduction

Internet of Things (IoT) has attracted a lot of attentions in recent 10 years. In recent years, many companies start to provide their IoT devices, including smart speakers, smart lights, smart sensors, and so on. By using these devices to build smart homes or smart buildings, they bring the great convenience to our daily lives. However, the security problems of using IoT devices also become more and more serious. To solve the security problems, there are two common methods: (1) improving the security service of the device; (2) improving users' ability to deal with security problem. Except the IoT device users, IoT professors also need to update the related knowledge of different devices. For realizing the second method, E-learning are wildly used. Comparing to traditional paper materials, E-learning materials have the following main merits: (1) multi-media contents are wildly used to help learners to know the devices much better; (2) E-learning materials are very easily shared through the web; (3) E-learning materials are easy to update for ensuring the leaners can obtain the latest knowledge; (4) E-learning system can easily record learners' activities for further analyzation.

Our research group also focuses on the research of E-learning. We have already proposed a new E-learning materials development framework, which supports the automatic generation of these materials through a Linked Data set [1]. Then we extended

L. Barolli et al. (Eds.): EIDWT 2021, LNDECT 65, pp. 370–378, 2021.
https://doi.org/10.1007/978-3-030-70639-5_34

this framework for supporting the generation of quizzes for evaluating the educational effects. Our proposed framework has already been applied to create E-learning materials about IoT security [2]. This framework support to use many types of media contents, such as images, 2D/3D animations, and AR/VR contents, to generate E-learning materials. Especially, we extend our framework to support 360-degree images and videos in this paper [3]. Such kind of environment can help users to simulate the real world to realize better educational effects. In this paper, we introduce our framework and how to implement this framework to create web-based IoT E-learning materials. Then, we use these materials to collect learners' activities and perform visual analyzation to the collected data. In this paper, we use Time Tunnel and Cubic Gantt Chart as the visualization tools [4–7].

In the following sections, the paper is organized as following: Sect. 2 introduces the related works to our research, Sect. 3 introduces the details of our framework, and Sect. 4 introduces our visualization tools and how to use them to analyze learners' activities. Last, we conclude this paper and introduce our future work.

2 Related Work

In this section, we will introduce what is linked data, and show other research results in this field.

2.1 Linked Data

Our framework requests users to store the resources for generating E-learning materials as Linked Data, which refers to "a set of best practices for publishing structured data on the Web" [8]. A data item of Linked Data is defined as a Resource Description Framework (RDF) triple [9]. Each RDF triple is composed by a subject, a predicate, and an object. Every data element has a URI as its identifier. Figure 1 shows an example of the graph to represent three triples, *(S1, P1, O1)*, *(S1, P2, O2)* and *(O1, P3, O2)*. The triples with the same subject *S1* will be combined in the graph, and one triple may link to other triples. We can realize the knowledge discovery through such kind of linkages among triples. The features of Linked Data allow users to easily share, extend and reuse the data. Currently, many organizations publish their Linked Data sets. Different data sets can also connect with each other. One of the famous Linked Open Data (LOD) set is the DBpedia (https://wiki.dbpedia.org/).

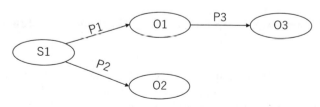

Fig. 1. An example of three triples

Users can write SPARQL [10] query to retrieve necessary data. Its grammar is different to the Structured Query Language (SQL) of querying the relational databases. A basic SPARQL query is shown as following:

```
SELECT ?v
WHERE
{
  S1 P1 ?v.
  ?v P3 O3.
}
```

In this query, "*?v*" is a user-defined variable indicating what we want to retrieve. Users can name the variable as they want. In the "Where" part, users define the conditions that the variable should satisfy. In the above example, the retrieved result of this query is one or a set of values, which satisfy both of the following two conditions: 1) the result should be the objects of the tuples whose subjects are "*S1*" and the predicts are "*P1*"; and 2) the result should be the subjects of the tuples whose predicts are "*P3*" and objects are "*O3*". In this example, the result should be the "*O1*".

2.2 E-learning Material Generation Systems

IntelligentBox is a powerful system proposed by Okada and Tanaka [11]. Comparing to the E-learning materials development system, it is a visual development system for 3D graphics applications. It can be also used to develop E-learning materials. Unfortunately, IntelligentBox do not support to create web-based materials. However, it is a very important feature because current learners are more familiar to use smart phones, tablet PCs to study at anywhere they want. Also, there is another similar system called Webble World [12], which can be used to create web-based E-learning materials through simple operations. It can be also used to render 3D graphics assets. However, it does not support 360VR images/videos.

To realize the functions mentioned above, users can use some game developement engines, such as Unity [13]. Unity is very powerful, and can be used to create applications which can work on iOS and Android smartphones and tablets. Of course, Unity can be used to create web-based E-learning materials. However, Unity is not easy for the users, such a medical professors, who may not have a professional programming skills. One of our research purpose is to propose a framework which can be used to easily to create E-learning materials without programming.

3 A New Framework for Generating Web-based E-learning Materials

In this section, we introduce how our framework is used to generate web based IoT security materials. The generated materials are organized as HTML pages. To define these pages, we need to first define page templates. Each part of a template will be related with an RDF triple's subject or object. In Sect. 3.2, we will show the details. We have built a database to store all resources as the RDF format. We will show the details of how to build our database in Sect. 3.2. In this database, we store the texts, images, videos. And furthermore, we add 3D animations and 360-degree images and videos into this database.

3.1 Linked Data Set for Storing Resources for Creating E-learning Materials

Before we start to create E-learning materials, we need to prepare necessary E-learning resources, including texts, images, videos, and other multi-media contents as RDF format. These resources are used to describe IoT devices' features. Figure 2 is a sample of our RDF data store. In this example, each circle represents a subject or object of an RDF triple, and each line represents a predict. In this figure, we can know that "IoTSecurity:0" is a Device and can be attracted by the virus. And we also can obtain the information about "Coutermeasure" and "Damage".

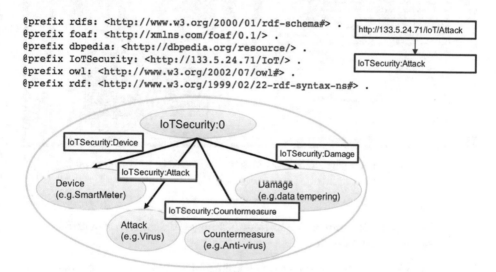

```
@prefix rdfs: <http://www.w3.org/2000/01/rdf-schema#> .
@prefix foaf: <http://xmlns.com/foaf/0.1/> .
@prefix dbpedia: <http://dbpedia.org/resource/> .
@prefix IoTSecurity: <http://133.5.24.71/IoT/> .
@prefix owl: <http://www.w3.org/2002/07/owl#> .
@prefix rdf: <http://www.w3.org/1999/02/22-rdf-syntax-ns#> .
```

Fig. 2. A sample of our RDF data store

Using this method, our framework supports learners to explorational access the E-learning materials. For example, we can explore the triples whose subject is "Attack" in our data store to obtain further information. This is one important merit of using RDF format to store our resources. In this data store, we also can store the 3D contents and 360° images/videos.

3.2 Authoring Tools for Defining E-learning Materials of IoT Security

In this framework, we have proposed an authoring tool for defining templates for generating E-learning materials for IoT Security. Figure 3 shows an example of a HTML page of our E-learning materials of IoT Security. In this page, users can input a keyword which is the subject of some triples. Then in this page, users can define how to display the objects of these triples (their sizes and positions). In this example, the text description, the image, and the videos of introducing the smart meters will show in this page.

When learners read our defined materials, our system will automatically record learners' activities as log data. We will visually analyze the data using Time Tunnel and Cubic Gantt Chart. The details of how to analyze the data in Sect. 4.

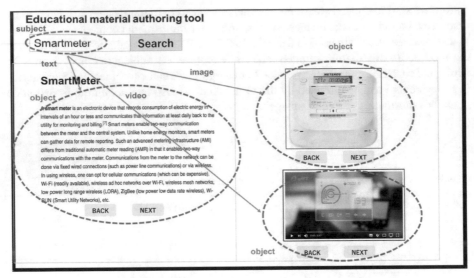

Fig. 3. A sample page of the E-learning materials created by our framework

3.3   Including 3D Contents and 360-Degree Images/Videos into E-learning Materials of IoT Security

Our framework also supports to create E-learning Materials using 3D contents. This method can simulate the real environment, so we believe we can obtain better educational effects. Comparing to the HTML page introduced in Sect. 3.2, we need to implement a 3D environment using Three.js. (https://threejs.org/). Figure 4 shows the components of how to realize the 3D environment and the interaction between leaners and E-learning contents. To create such kind of E-learning materials, users also need to use our authoring tool. In each page, learners can directly click a 3D object which represent an IoT device to replace the keyword input action in the system introduced in Sect. 3.2. Figure 5 is an example of the E-learning materials. In this 3D environment, if you create an IoT device, the related text and videos will be shown.

However, the 3D environment is still different with the real world. We extended our framework to support the 360-degree image and video contents to create E-learning materials. Replacing to the virtual 3D environment, users can take a 360-degree images and videos as the system's working environment. Then, In this paper, we use the Cascade Classifier provided by OpenCV to realize the object detection and recognition. We first searched the images of these IoT devices and train the classifier. Then, the system will find out all the IoT devices in the 360-degree images or videos automatically. Then, when learners watch these images and videos and focus on an IoT object (place this object in the center of the screen), our system will show the information related to this device.

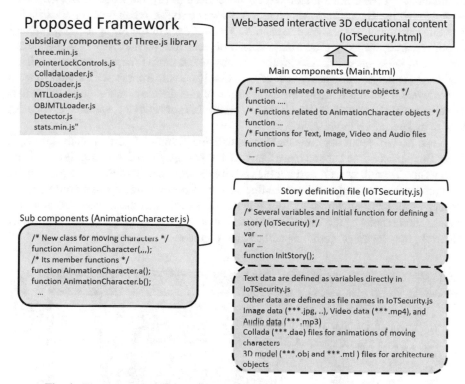

Fig. 4. The components of our framework for creating 3D E-learning materials

4   Visually Analyzing Learners' Log Data by Cubic Gantt Chart and Time-Tunnel

As we introduced in Sect. 3.2, our system can automatically collect learners' activities during they are using our 3D or 360-degree VR materials. We collected the following data from learners:

(1)   button clicking operation,
(2)   camera direction, and
(3)   time spent on a page.

Form the collected, we want to know when a learner uses our 3D or 360° VR environment, which IoT device he can find out by analyzing the camera direction. We also can know how long time spent on finding out each device. Then, we can know how long time they spent on learning the knowledge of a device, and know what kind of contents (text, image, or video) are more attractive to learners.

We will use the Cubic Gantt Chart and Time Tunnel to analyze all learners' activities for finding out their common features. As introduced in [6], the Cubic Gantt Chart is a 3d version of the Gantt Chart, which uses a 3D space so that it can treat three attributes. The three attributes are assigned to each X, Y and Z axes of the 3D space, respectively. In our research, we assign the time, focused objects, and learner ID to the three axes. We wish to know on which devices the learners spend more time, and to know if there any device is not found out by most of learners. If most learners spend a long time on some device, we think it means the explanation of this device is not enough so learners cannot understand it well. If most learners cannot find some device, we need to consider how to make it more impressive.

Next, we will input our collected data into Time Tunnel for further analyzation [5]. Time Tunnel has a time bar and more than one data-wings. Each data-wings is a 2D chart. Using Time Tunnel, we can find out the activity pattern change of different date. In our research, each data-wings will use the line chart to represent time spent on different IoT devices of all learners whose use our e-learning materials in the same date. From the visualization results of the Time Tunnel, we can know when we update our E-learning materials, how learners' activities will change together.

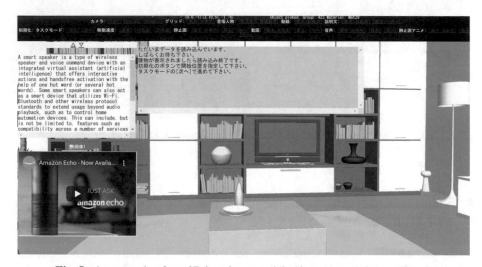

Fig. 5. An example of our 3D learning materials (the smart speaker is clicked)

5   Conclusion

In this paper, we proposed a new framework, which support the creation of E-learning materials for IoT security education. These created materials by our framework is web-based and automatically generated from the Linked Data. Web-based materials allows learners can easily access their desired contents. Linked Data can support learners to explorational study. In this framework, users can create a HTML page including texts, images, and videos. Furthermore, users can also create 3D and 360° VR materials. We believe these two kinds of materials can simulate our real environments for obtaining better educational effects. Our system also supports to collect the data of learners' activities for further analyzation. We will use Cubic Gantt Chart to analyze which IoT device is focused by most learners and which is difficult to be found out. This analyzation result can be used to improve our E-learning materials. Next, we will use Time Tunnel to analyze the learning activities of the learners who use our E-learning materials in different dates. From this result, we can find out the different activity patterns of these learners to evaluate if our materials are improved. Next, we will practically perform the experiment. We will ask the students from our university to help us. From the analyzation result of these students, we also hope to find out if there any rules to create good e-leaning materials for IoT security education.

References

1. Ma, C., Srishti, K., Shi, W., Okada, Y., Bose, R.: Interactive educational material development framework based on linked data for IoT security. In: Proceedings of the 10th Annual International Conference of Education, Research and Innovation (iCERi 2017), pp. 8048–8057, 12–14 November 2017
2. Shi, W., Ma, C., Kulshrestha, S., Bose, R., Okada, Y: A framework for automatically generating IoT security quizzes in a virtual 3D environment based on linked data. In: Barolli, L., Xhafa, F., Khan, Z., Odhabi, H. (eds.) Advances in Internet, Data and Web Technologies. EIDWT 2019. Lecture Notes on Data Engineering and Communications Technologies, vol. 29, pp. 103–113. Springer, Cham (2019)
3. Okada, Y., Haga, A., Wei, S., Ma, C., Kulshrestha, S., Bose, R: E-learning material development framework supporting 360VR images/videos based on linked data for IoT security education. In: Barolli, L., Xhafa, F., Khan, Z., Odhabi, H. (eds.) Advances in Internet, Data and Web Technologies. EIDWT 2019. Lecture Notes on Data Engineering and Communications Technologies, vol. 29, pp. 148–160. Springer, Cham (2019)
4. Nakamura, S., Kaneko, K., Okada, Y., Yin, C., Ogata, H.: Cubic gantt chart as visualization tool for learning activity data. In: The 1st workshop on e-Book-based Educational Big Data for Enhancing Teaching and Learning of ICCE 2015, 30 November–1 December 2015, pp. 649–658 (2015)
5. Nakamura, S., Okada, Y.: Learning analytics using BookLooper and time-tunnel. In: Proceedings of 11th annual International Technology, Education and Development Conference (INTED 2017), Valencia, Spain, 6–8 March 2017, pp. 8767–8776 (2017). ISBN 978-84-617-8491-2
6. Ma, C., Srishti, K., Shi, W., Okada, Y., Bose, R.: Learning analytics framework for IoT security education. In: Proceedings of 12th annual International Technology, Education and Development Conference (INTED 2018), Valencia, Spain, 5–7 March 2018, pp. 9181–919 (2018). ISBN 978-84-697-9480-7
7. Okada, Y.: Time-tunnel: 3D visualization tool and its aspects as 3D parallel coordinates. In: Proceedings of 22nd International Conference Information Visualisation (IV), pp. 50–55, July 2018. https://doi.org/10.1109/iV.2018.00019
8. W3C: LinkedData - W3CWiki (2016). https://www.w3.org/wiki/LinkedData
9. Berners-Lee, T: Linked Data – Design Issue (2009). https://www.w3.org/DesignIssues/LinkedData

10. W3C: SPARQL Query Language for RDF (2008). https://www.w3.org/TR/rdf-sparql-query/
11. Okada, Y., Tanaka, Y.: IntelligentBox: a constructive visual software development system for interactive 3D graphic applications. In: Proceedings of Computer Animation 1995, pp. 114–125. IEEE CS Press (1995)
12. Okada, Y.: Web Version of IntelligentBox (WebIB) and its integration with Webble World. In: Communications in Computer and Information Science, vol. 372, pp. 11–20. Springer (2013)
13. Unity. https://unity3d.com/jp

Author Index

© The Editor(s) (if applicable) and The Author(s), under exclusive license
to Springer Nature Switzerland AG 2021
L. Barolli et al. (Eds.): EIDWT 2021, LNDECT 65, pp. 379–380, 2021.
https://doi.org/10.1007/978-3-030-70639-5

Printed in the United States
By Bookmasters